KAPITZA
in Cambridge and Moscow

GW00724561

Peter Kapitza, Moscow 1937.

KAPITZA

in Cambridge and Moscow
Life and letters of a Russian Physicist

Compiled and Edited by

J.W. BOAG
Emeritus Professor of Physics Applied to Medicine
London University

P.E. RUBININ
formerly Personal Assistant to P.L. Kapitza
Institute for Physical Problems, Moscow

D. SHOENBERG
Emeritus Professor of Physics
Cambridge University

1990

NORTH-HOLLAND
AMSTERDAM · OXFORD · NEW YORK · TOKYO

© Elsevier Science Publishers B.V., 1990

All rights reserved. No part of this publication may be reproduced, stored in a retrieval system, or transmitted, in any form or any means, electronic, mechanical, photocopying, recording or otherwise, without the prior permission of the publisher, Elsevier Science Publishers B.V., P.O. Box 211, 1000 AE Amsterdam, The Netherlands.
Special regulations for readers in the USA: This publication has been registered with the Copyright Clearance Center Inc. (CCC), Salem, Massachusetts. Information can be obtained from the CCC about conditions under which photocopies of parts of this publication may be made in the USA. All other copyright questions, including photocopying outside of the USA, should be referred to the publisher.

ISBN: 0 444 98753 3   (hardbound)
       0 444 98749 5   (paperback)

North-Holland
Elsevier Science Publishers B.V.
P.O. Box 211
1000 AE Amsterdam
The Netherlands

Sole distributors for the USA and Canada:

Elsevier Science Publishing Company, Inc.
655 Avenue of the Americas
New York, NY 10010
USA

Library of Congress Cataloging-in-Publication Data

Kapitsa, P. L. (Petr Leonidovich), 1894–1984
   [Correspondence.   Selections]
   Kapitza in Cambridge and Moscow : life and letters of a Russian
physicist / compiled and edited by J.W. Boag, P.E. Rubinin, D.
Shoenberg.
      p.   cm.
   Includes bibliographical references and Index.
   ISBN 0-444-98753-3. -- ISBN 0-444-98749-5 (pbk.)
   1. Kapitsa, P. L. (Petr Leonidovich), 1894–1984.   2. Physicists –
Soviet Union – Biography.   3. Physicists – Soviet Union –
Correspondence.   I. Boag, J. W., 1911–   .   II. Rubinin, P.E.
III. Shoenberg, D. (David)   IV. Title.
QC16.K25A4 1990
530'.092 – dc20
   [B]   90-42875
                                                            CIP

Printed in The Netherlands

Preface

This is a book about the famous Soviet physicist Peter Kapitza and his unusual career, mainly divided between Cambridge and Moscow. In the Cavendish Laboratory of the 1920's and 30's he was a protégé of Rutherford, whom he irreverently nicknamed the Crocodile. During his 13 years in Cambridge he acquired a considerable reputation for his skill and originality and for his eccentricities. He opened up a new area of research in magnetism and low-temperature physics in the Cavendish and became part of the British scientific establishment, although remaining a Soviet citizen. However, in 1934, during a summer visit to the Soviet Union, he was prevented from returning to Cambridge and remained in Moscow for the rest of his long life. In spite of many ups and downs and considerable struggles with top political figures in the Kremlin, he continued to enhance his scientific reputation and late in life was awarded the Nobel Prize.

The subtitle of the book defines rather more precisely the scope of the book. The "life" is based on a published biographical memoir by one of us (D.S.), but with rather less emphasis on the details of his scientific work, and rather more on the human side. The "letters" are extracts from his voluminous and lively correspondence over many years, which vividly illustrates many matters touched on in the "life". Together they provide a rounded picture of a remarkable personality who contributed so much to the scientific and cultural life of both England and the Soviet Union. The letters should be of interest to the historian of science in illuminating Kapitza's relations with Rutherford and other important scientific figures, his courageous interventions with the Kremlin on behalf of unjustly repressed colleagues in the Soviet Union and his vigorous views on sociological and organizational matters boldly offered to Stalin and other top political figures. But above all they are of considerable human interest in revealing Kapitza's emotional reactions to the major blows he had to suffer on several occasions, and also for his penetrating and often amusing comments on English life and institutions as seen by a foreigner, and on Soviet life as seen from the inside.

It is these aspects we have had particularly in mind in selecting the extracts to go into this book from the immense amount of material available. In our rather free translations into English we have tried to retain something of the almost conversational style and the raciness and humour of the Russian original, but many of the staccato-like short sentences and paragraphs in the original have been run together to achieve a freer-flowing English style. Except in a few instances – usually at the start of each series to serve as illustrations – the opening and closing phrases of the letters have been omitted, as have most of the purely family matters and frequent repetitions of topics in successive letters. To keep the length of the book within reasonable bounds many of the selected letters have also been shortened from the full text of the original letters in other respects. Omissions from the full text are indicated by . . . Square brackets indicate editorial additions introduced to clarify the meaning; numbers in round brackets indicate references in the bibliography of Kapitza's publications. In the transliteration of Russian names we have usually followed a system which helps to appreciate the Russian pronunciation, but a few well-known names, such as Semenov and Joffé have been left in the forms usually given in bibliographies. It should be noted that styles of name and title in Russian convey subtle shades of difference in familiarity or respect. Thus Kolka, Kolya (diminutives), Nikolai, Nikolai Nikolaevich, Semenov, and the use of the titles Professor, Comrade (Tovarishch) and Citizen (Grazhdanin), represent increasing degrees of formality, while Mister (Gospodin), unless applied to a foreigner, implies that he is "not one of us".

The extraction of many of the letters from Kapitza's archive in Moscow, the deciphering of Kapitza's almost illegible handwriting and some of the selection and editing have been the work of Kapitza's former personal assistant (P.E.R.). The other two editors have also had a share in the selection and have been responsible for the translation into English of the Russian letters as well as the selection and editing of English language correspondence, mostly from the Rutherford Archive in the Cambridge University Library. Most, but not all, of the Russian material translated in this book has already appeared in Russian periodicals and books, and some of it has been translated into English before. A complete list of the Russian and English publications is appended and in the text, code symbols are used to indicate where the relevant letter or an extract from it has previously appeared, or its source if unpublished. To avoid a plethora of footnotes, essential information about people mentioned in the text is summarized in the name index on pp. 423 to 430. Detailed acknowledgements for help in preparing the biographical memoir on which chapter 1 is based and a more detailed bibliography of Kapitza's publications are given at the end on the orginal version (Biogr. Mem. Fellows Roy. Soc. 31 (1985) 327). A number of errors in the original version, noticed since it was published, have been corrected in chapter 1.

We should like to thank Mrs Anna Kapitza for her generous permission to use the letters in her late husband's archive and Elena [Alyona] Kapitza (granddaughter of Peter's brother Leonid, so often mentioned in the letters) for help in sorting out from the Moscow Archive many of the photographs which illustrate the book. We are also grateful to Professor Peter Fowler for permission to reproduce the letters from his grandfather, Lord Rutherford and to the Churchill Archive Centre, Cambridge for permission to reproduce a few items in the Cockcroft collection.

Sources and previous publications of Kapitza's letters

Russian (translated titles)

(A)   Hitherto unpublished material in the Kapitza Archive, Moscow.

(B)   D. Danin, *Rutherford,* Molodaya Gvardiya, Moscow, 1966 (pp. 504–517)

(C)   P.E. Rubinin (ed), Letters to his mother 1921–1923, *Priroda* 1985, No. 1 (pp. 56–68)

(D)   P.E. Rubinin (ed), Letters to his mother 1921–1926, *Novy Mir* 1986, No. 5 (pp. 192–216), No. 6 (pp. 194–218)

(E)   P.E. Rubinin (ed), Letters to his mother, in B.G. Volodin and V.M. Strigin, *Paths into the unknown,* Sovetskii Pisatel, Moscow (pp. 445–478)

(F)   P.E. Rubinin (ed), Letters as a student 1916–1919 in V.M. Tuchkevich (ed), *Readings in Memory of A.F. Joffé* (pp. 5–29) Nauka, Leningrad, 1988

(G)   P.E. Rubinin (ed), Letters to Molotov, Mezhlauk, Stalin and Andropov, *Sovetskaya Kultura,* 21 May 1988 (p. 6)

(H)   P.E. Rubinin (ed), Letters about Landau: "What happened, happened", *Ogonyok,* No. 3 January 1988 (pp. 13–15) (also in *Recollections of Landau,* ed I.M. Khalatnikov, Nauka, Moscow, 1988 (pp. 334–347)). "The scientist and the powers that be", *Ogonyok,* No. 25, June 1989 (pp. 18–22)

(I)   P.E. Rubinin (ed), *Letters about science,* Moskovskii Rabochii, Moscow, 1989

English

(a) Material in the Rutherford Archive at the Cambridge University Library

(a') Material in the Cockcroft collection at the Churchill Archive Centre, Cambridge

(b) A. Parry (ed), *Peter Kapitsa on Life and Science* (full translation of the extracts in (B)) Macmillan, New York, 1968 (pp. 122–139)

(c) L. Badash, *Kapitza, Rutherford and the Kremlin*, Yale University Press, New Haven and London 1985. This contains Mrs. Kapitza's translations of some of the letters to her in 1934–1935, all available in (a); the Russian originals are all in the Moscow Archive and some appear in (I). It also contains some of the Rutherford correspondence in (a)

(d) D.J. Lockwood (ed), *P.L. Kapitsa: Letters to Mother,* Nat. Res. Council, Canada: Ottawa, 1989 (full translations of (D))

(e) J.W. Boag and D. Shoenberg, Letters from Kapitza to his mother, *Notes Rec. R. Soc. Lond.* 1988, Vol. 42, pp. 205–228 (translations of extracts from miscellaneous letters to his mother)

(f) Peter Kapitsa: the scientist who talked back to Stalin, Bull. Atomic Scientists, April 1990 (adaptation and translation by T. Hoisington, of the June 1989 *Ogonyok* article in (H) above)

(g) P.E. Rubinin (ed), Peter Kapitza and Paul Dirac: letters 1935–37, *Science in the USSR,* No. 1, 1990, pp. 21–24

Note: The code letters (b), (c) and (d) will not ordinarily be used in the text since they correspond almost exactly to the original sources in (B), (a), and (D), respectively. Code (A) will be used only if some part of the extract has not previously been published.

Contents

Illustrations

Note: Some of the dates given above are only approximate.

Sources of illustrations

The illustrations come mostly from the Kapitza Archive in Moscow, except for 52, 56, 60, 71 and 76, which come from the Cavendish Photographic Archive. Illustrations 9, 10, 11, 14, 16, 23, 24, 25, 26, 31, 32, 33, 35, 39, 65, 66, 67, 73 and 74 were photographed by D. Shoenberg and 54 by J.W. Boag; 77 is by the late Lotte Meitner-Graf and is reproduced by kind permission of her executors, 12, 15, 22, 38, 40 and 79 are from D. Shoenberg's personal collection, 7 is reproduced from the reprint held in the the Cockcroft Collection at the Churchill Archive Centre, Cambridge, 53 is reproduced from "Paul Ehrenfest" by M.J. Klein by kind permission of North-Holland, 69 is reproduced from a photograph kindly lent by Professor R.H. Dalitz and 68 is copied from a Soviet newspaper illustration in the Rutherford Archive at the Cambridge University Library.

1. Portrait by B.M. Kustodiev, Leningrad 1926.

1

Biographical introduction

Peter Kapitza* was a legendary figure both in Rutherford's Cambridge of the 1920's and 30's and subsequently in Moscow to the end of his long life. In his scientific work he showed great versatility and brought the skills of an engineer and mathematician to bear on important problems in physics and technology in an entirely original way. It could be said that he was a brilliant exponent of the concept of lateral thinking long before the term became fashionable. He also had broad cultural and social interests and his original ideas on scientific education and organization have had a profound influence, particularly in the development of Soviet physics. Although some of the legends about him may be more than somewhat apocryphal, many contain a sufficient basis of truth and serve to illustrate Kapitza's colourful personality and his scientific style and will be retold in this memoir.

1.1. Russia (1894 – 1921)

Kapitza was born on 8 July** 1894 (26 June 1894, old style) at Kronstadt, the island fortress in the Gulf of Finland near St. Petersburg, into a family with a strong tradition both in the sciences – particularly engineering – and the arts. His father, Colonel (later

---

* In England his first and last names were usually rendered in this way but his first name was given as Pierre when he was elected to the Royal Society and it is often transliterated from the Russian as Pyotr; the tz in his surname is often replaced by ts, both being equally valid equivalents of the Russian letter ц.

** There is some confusion about the correct date. After 1918, 13 days were added to old style dates to change to the Gregorian calendar and this made the birth date 9 July. In fact, however, for dates in the 19th century only 12 days should have been added making it 8 July. The mistake was realized only some time later and within the family his birthday continued to be celebrated on 9 July.

General) Leonid Petrovich Kapitza, was a military engineer involved in modernizing the fortifications of Kronstadt. The Kapitzas had been landed gentry with Polish antecedents and the family was well represented in the professions, particularly law and engineering. His mother, Olga Ieronimovna, née Stebnitskaya, to whom he was very close until her death in 1937, was a specialist in children's literature and folklore and was an important figure in the literary world of St. Petersburg in the early years of the century and especially during the difficult times of the Revolution and its aftermath. Together with S.I. Marshak, the well-known children's author, she organized a kind of club for young writers where they read their works and exchanged ideas. Many of them subsequently became famous, particularly in the field of children's literature. Her father, Ieronim Ivanovich Stebnitski, also of Polish origin, was an army general, also a military engineer, specializing in cartography. He was a geographer of international repute, a corresponding member of the Imperial Academy of Sciences and an ardent traveller all over the world; it is perhaps from him that Peter Kapitza inherited his own love of travel. A great deal of Stebnitski's military service was in the Caucasus, where he personally mapped out all the important mountain peaks and made gravitational measurements. Later he was head of the Military Topographical Department of the Imperial General Staff in St. Petersburg and lived opposite the Winter Palace. Rather unusually for his time, he decided that his daughters Olga and Alexandra should receive a higher education and they attended the Bestuzhev Courses (the only form of university level higher education open to women),

2. Kapitza family, Kronstadt 1905.
(l. to r. Father, Mother, Leonid, Evgenia Kalishevskaya (cousin), Peter).

Olga in the humanities and Alexandra ("Aunt Sasha") in mathematics and science.

Aunt Sasha played an important part in Peter's upbringing for it was she who discovered, rather to the surprise of the family, that Peter had an unusually quick grasp of arithmetic; although a lively boy, he had seemed a bit backward in some other respects. He never overcame a certain indistinctness and sloppiness in speech and he never learned to spell properly in any language, which was a considerable handicap in his early career. He and his older brother Leonid were admitted to the "classical gymnasium", which specialized in the humanities and classics, and was the essential preliminary for entry to the University. However, Peter had to leave this school in 1907 at the end of his

3. In school uniform, 1911.

second year because of inadequate progress and transferred to the more scientifically oriented "Realschule". He enjoyed a kind of revenge 70 years later when, to mark his second award of the title Hero of Socialist Labour, his bust was erected in a public square of Kronstadt close to the ill-starred gymnasium from which he had been excluded! Although this exclusion evidently rankled, the Realschule was much more appropriate for developing his talents and in due course (1912) he graduated with honours and entered the electrotechnical faculty of the St. Petersburg Polytechnical Institute – for lack of Latin and

4. Bust by A.M Portyanko, Kronstadt 1979.

Greek he was not eligible to enter the St. Petersburg University, which was then regarded as the more prestigious institution.*

The outbreak of war in the summer of 1914 caught Kapitza in Scotland where he was staying with the Millar family in Glasgow to learn English (pp. 88 to 94). His return to St. Petersburg was delayed until the end of November 1914 and early in 1915 he volunteered as an ambulance driver at the Polish front (pp. 94 to 97). However, after a few months he returned to civilian life and resumed his studies. In May 1916 he took leave from the Institute and made a romantic trip to the Far East (pp. 97 to 104). He collected his fiancée. Nadezhda Kirillovna (fig. 47), daughter of General Chernosvitov, in Shanghai and after a visit to Japan they returned to Petrograd** where they were married at the end of July. Shortly before Kapitza's graduation in

5. Joffé seminar, Petrograd 1916.
(l. to r.: Ya.I. Frenkel, N.N. Semenov, P.L. Kapitza, A.P. Yushchenko, P.I. Lukirski, A.F. Joffé, M.V. Kirpicheva, J.R. Schmidt, Ya.G. Dorfman, I.K. Bobr, K.F. Nesturch, N.I. Dobronravov).

---

* It is a pity to spoil a good story, but since this was written, Anna Kapitza (his widow) has informed me (D.S.) that the story was by way of a joke, invented by Kapitza himself. In fact, he was transferred to the Realschule by his mother, who realized that he was better suited to a scientifically oriented education.
** The Germanic sounding St. Petersburg was renamed Petrograd at the start of the 1914 war and after Lenin's death in 1924 became Leningrad.

1919 he was appointed to a teaching post in the physico-mechanical faculty of the Institute, which he held until he moved to Cambridge in 1921.

During these years the moving spirit in the physics world of Petrograd and indeed of Russia, was A.F. Joffé, who was actively developing a physics school with an outlook more modern than had been traditional in Russia up to then and with an emphasis on research and on regular seminars to keep abreast of developments in the Western world. After 1918, the research effort was concentrated in the newly established Petrograd Physico-Technical Institute (at first known as the Röntgenological Institute) where Joffé succeeded in collecting a group of young and enthusiastic scientists around him who gradually got things going and created the nucleus of the enormous growth of Soviet physics in later years. Besides Kapitza, this group included many future leaders of Soviet research, such as his lifelong close friend N.N. Semenov, later a Nobel Prize winner for his work on reaction kinetics, the quantum theoretician Ya.I. Frenkel, the magnetician Ya.G. Dorfman and others.

6. With N.N. Semenov, portrait by Kustodiev, Petrograd 1921.

The creation of this new institute for basic physics research must have seemed wildly visionary and impractical at that particular time. Following the chaos of the war years and the upheaval of the Revolution, the economy of the country was in a disastrous condition. There were great shortages of food and fuel and practically no scientific equipment, so that research could be done only on a "do it yourself" basis. Moreover, many of the country's academics had emigrated and contacts with developments in the West were very limited for political reasons. In spite of all these difficulties, however, quite a lot got started.

Kapitza became involved in several projects of topical interest. One was the measurement of the angular momentum associated with magnetization and his first scientific paper* was a critical review (2) of the experiments by Einstein and de Haas, who had demonstrated the effect in 1915 and by Barnett who almost simultaneously demonstrated the inverse effect (i.e., the magnetization produced by rotation). At that time there appeared to be a contradiction between the two sets of experiments and Kapitza concluded that there must be some error in Barnett's experiment since it gave only half the classically expected ratio of angular momentum to magnetization. In fact, Einstein and de Haas later discovered an error in their experiment and confirmed Barnett's "anomalous" result. It was only later that this gyromagnetic anomaly was understood in terms of electron spin. A second paper (3) describes an ingenious improvement in the technique of preparing Wollaston fibres**. Kapitza also became adept at using Boys' bow and arrow method of producing very fine quartz fibres; this made a somewhat spectacular demonstration which he was fond of showing off to students. Kapitza's interest in fine fibres was partly in connection with the Einstein–de Haas effect project and partly for delicate electrometer suspensions. However, although Joffé mentioned in a letter to Paul Ehrenfest (18 June 1920) that he and Kapitza had successfully observed the angular momentum produced by thermally demagnetizing a nickel sample, no results were ever published.

---

* This was, in fact, not his first publication. He had already published a popular article (1) about the production of cod liver oil, based on his travels in the far North of Russia, and illustrated by his own photographs.
** These are very fine platinum wires made by drawing down a platinum wire covered with a thick layer of silver and finally removing the silver.

Another line of work was the study of atomic and molecular beams and together with Semenov, Kapitza proposed a technique for measuring atomic magnetic moments by observing the spread of an atomic beam after passing through a strongly inhomogeneous magnetic field (5). Although their paper mentions that "experiments in this direction have already begun" nothing further was published. In fact, between the time of submission of the paper in December 1920 and its publication in 1922, a whole series of papers by Stern and by Gerlach and Stern had appeared which not only proposed essentially the same idea, but carried it out in practice and so provided the demonstration of spatial quantization for which many years later Stern was awarded a Nobel Prize. Kapitza and Semenov do not seem to have been aware of the prediction of spatial quantization and in their discussion they took it for granted that there would be a continuous spread of deflected atoms. A third field to which Kapitza contributed was that of X-rays and his proposal (4) to focus a broad beam of X-rays by a bent crystal seems to have anticipated what much later became an important practical technique.

A son, Ieronim, was born to the Kapitzas in July 1917 and a second child was expected towards the end of 1919, but at this point disaster struck the family. In the terrible conditions following the Revolution and the Civil War, epidemics were rife and Kapitza's father died of Spanish influenza on 9 December. Then Ieronim caught scarlet fever and died on 13 December. Nadezhda was quite struck down by this blow; three days after she had given birth on 6 January 1920 to a daughter (also called Nadezhda), and weakened by 'flu, she too succumbed and her daughter survived for only a few more hours. Kapitza, who was himself very ill at the time with nephritis, was quite overwhelmed by these tragic losses and although he gradually recovered, he was for a time unable to work. His friends wondered what could be done to help him forget his unhappy situation and by an odd coincidence something turned up opportunely which not only distracted Kapitza from his grief, but, as it turned out, completely changed the course of his life.

This *deus ex machina* was the setting up, at Joffé's prompting, of a "Commission of the Russian Academy of Sciences for renewing scientific relations with other countries". The Soviet government had a very liberal policy toward science and in spite of its economic difficulties was able to provide the commission with a generous allowance of foreign currency to buy scientific and technical equipment abroad. The whole idea of a mission abroad was sanctioned by Lunacharsky, the Commissar for Education, and had Lenin's personal blessing. Besides Joffé, an important member of the commission was Admiral A.N. Krylov, the well-known naval engineer and applied mathematician, who was later to become Kapitza's second father-in-law.* Both not only had a high opinion of Kapitza's scientific gifts but as friends wanted to help in his difficult personal situation, and appointing him to travel abroad with the commission seemed an ideal solution.

Travel abroad at that time was very complicated since most of the outside world had no diplomatic relations with Soviet Russia. Joffé set out in February 1921, but had to wait a month in Estonia before getting visas for Germany and England. Once in Berlin, Joffé began trying to get a visa for Kapitza while he and other members of the commission were occupied with buying the equipment needed to bring Soviet research up to Western standards. Germany, France and Holland did not want to risk admitting someone who might prove to be a young communist agitator, but England was more accommodating and eventually in May, Kapitza obtained an English visa. Once the visa was assured, Kapitza began to wonder whether it might not be possible to spend a short time under Rutherford at the Cavendish. He had already heard exciting things about Rutherford and his school from Jadviga Schmidt, a Polish lady who had worked with Rutherford in Manchester and had since married a Russian physicist, Chernyshev, and settled in Petrograd, where both were colleagues of Kapitza. On the off chance, Kapitza, who had been cooling his heels in Reval since April (pp. 112 and 113) wrote to ask her to send a letter of recommendation to Rutherford. Joffé had had the idea of sending Kapitza to spend some time in Leiden, where his friend, the theoretician Paul Ehrenfest, had settled after several years in St. Petersburg before the war, but this idea fell through because the Dutch would not give a visa.

At last, late in May, Kapitza arrived in England by ship and early in June he and Joffé started a round of scientific visits, which culminated in a visit to Rutherford in Cambridge on 12 July (pp. 116

---

* A fascinating account of Krylov's part in the work of the commission is given in his memoirs (Krylov 1984).

and 257). They were received by Rutherford in a very friendly way, but when Kapitza asked if he might come back to work in the Cavendish for a few months, Rutherford was rather discouraging and said the laboratory was already too crowded and it would be difficult to accommodate one more. Rutherford was rather taken aback when Kapitza replied by asking what accuracy he aimed at in his experiments. The answer to this seemingly irrelevant question was 2 or 3% and this enabled Kapitza to point out that since there were about 30 researchers in the Cavendish, one more would hardly be noticed since he would come within the experimental error! This ingenious approach persuaded Rutherford to admit Kapitza after all.* As a postscript it may be mentioned that a year later Kapitza asked Rutherford why he had agreed to take him on and Rutherford laughed and said "I can't think why, but I'm very glad that I did".

1.2. Cambridge (1921 – 1934)

Kapitza started work in the Cavendish in July 1921 and although the original plan was for him to stay only over the winter, he remained for 13 years. His introduction to Cambridge life and his impressions of Rutherford and the Cavendish are vividly described in the letters he wrote at frequent intervals to his mother in Petrograd (see chapter 3). The usual initiation of new research students was a month or two of practical work in the Cavendish attic on various relevant techniques under Chadwick's supervision and a few days after his arrival Kapitza wrote (24 July):

> For the time being I am getting acquainted with radioactive measurements and I am working only at my practical course; as to the future, I don't know. I plan nothing and don't try to guess what will happen. Time will tell . . .

---

* Kapitza assured me (D.S.) a few years ago that although this story may have been somewhat embellished with age, something like it did actually happen. Rutherford may also have been swayed by Chadwick, who showed Kapitza round the laboratory and was much impressed by the originality of Kapitza's comments.

In fact, his skill and assiduity were such that Chadwick was satisfied after only two weeks or so and on 6 August Kapitza wrote:

> Now comes the riskiest moment – choosing a topic for my research. Not an easy matter. At such a moment I don't like to say a lot and so it is difficult to write anything definite . . .

The project chosen at Rutherford's suggestion was a study of how the energy of an α-particle passing through matter falls off towards the end of its range. Previously, this had been measured by observing the deflection of the α-particle in a magnetic field, but the limitation of this method is that particles of low energy cannot be detected so that it had not been possible to observe particles of much less than a quarter of the initial energy. Kapitza's method was to measure the energy in a collimated beam of particles by the heating it produced in a plate attached to a Boys radiomicrometer – essentially an extremely sensitive thermometer. This project was brought to a successful conclusion in an amazingly short time. Within 9 months from the conception of the idea he was already drafting a paper for publication and by mid-June the paper was sent off. He described his success in an euphoric letter to his mother (19 June 1922, see p. 150):

> Today the Crocodile summoned me twice to discuss my paper . . . It will be published in the Proceedings of the Royal Society, which is the greatest honour a piece of research can achieve here . . . It is only now that I have felt my strength. Success gives me wings and I am carried along by my work.

"Crocodile" was a nickname Kapitza invented for Rutherford early on – it first appears in a letter of 25 October 1921 (see p. 131). Many years later he told Ritchie Calder (1951, p. 66):

> In Russia the crocodile is the symbol for the father of the family and is also regarded with awe and admiration because it has a stiff neck and cannot turn back. It just goes straight forward with gaping jaws – like science, like Rutherford.

A more fanciful, but probably entirely apocryphal version popular in the Cavendish, was that the nickname was after the crocodile in

Peter Pan which swallowed an alarm clock to give warning of its somewhat frightening appearance – like Rutherford, it could be heard well before it could be seen. This version finds some slight support in an early letter (12 October 1921, p. 129):

> Rutherford greets me increasingly pleasantly when he sees me and asks how my work is getting on . . . But I am a little afraid of him. I work almost next door to his office. This is bad since I have to be very careful with my smoking: if he should catch you with a pipe in your mouth you're in trouble. But thank God he has a heavy tread and I can recognize his footsteps a long way off . . .

Yet another suggestion (Boag and Shoenberg 1988) is that the inspiration for the nickname may have come from a well-known Russian poem for children entitled "The Crocodile" by Kornei Chukovsky. However, Anna Kapitza insists that although Kapitza may have had some quiet fun encouraging various speculations, none of them is true. The real explanation is simply that when Kapitza came to the Cavendish he was at first rather terrified of Rutherford and chose the most frightening nickname he could think of. Later he came to use it (see, for instance, pp. 187 and 196) for any important personage.

Kapitza's enthusiasm about the success of his first Cambridge research was typical of how he reacted when things went his way, in contrast to his gloom and despair when things were against him. The published paper (6) shows that the enthusiasm was indeed justified. Boys said of his radiomicrometer that "its extreme delicacy of construction requires more than ordinary skill on the part of the user" and Kapitza not only demonstrated this skill and exploited his experience of fine quartz fibres, but incorporated many ingenious features to avoid various stray effects which in his preliminary experiments completely swamped the genuine small heating effect of the α-particles. He wrote to his mother (4 December 1921) that his device is "so sensitive that it can detect the flame of a candle 2 versts (roughly 2 km) away and respond to a temperature rise of a millionth of a degree! . . ." * His results, which showed how the energy fell off

---

* Kapitza was always very fond of homely illustrations to make his point, but this particular one is probably based on an actual experiment described by Boys, in which, however, the candle light was collected by a telescope of 40 cm diameter.

steeply to a negligible value towards the end of the α-particle range, were made particularly convincing by his thorough discussion of the theory of the instrument, of the possible residual systematic errors and of the comparison with earlier results.

This publication illustrates several characteristic features of much of Kapitza's later work: his ability to work out in practice rather complicated schemes to give extreme sensitivity, his thorough grasp of the potential pitfalls of the method, his speed of execution once he had started and his thoroughness in interpretation, particularly as regards slight discrepancies and subtleties which are so often ignored. One other interesting feature which should also be mentioned is Kapitza's fondness for regarding each project as part of a larger group of investigations, but then getting excited by a new idea and forgetting about continuing the original plan. Thus the sub-title of the paper is "Part I: Passage through air and CO_2" and the abstract mentions that passage of α-rays through solid bodies will be described in another paper. However, no Part II ever appeared, probably because Kapitza had become completely engrossed by a new idea which came up at that time.

This idea was a possible new method of studying the velocities of α-particles – Rutherford's favourite particles – by measuring the curvature of their tracks in a magnetic field. Existing magnets were not capable of producing large enough steady fields to curve the tracks sufficiently and Kapitza's novel idea was to use fields which were much larger, but lasted for only a very short time; even a "very short" time is long compared to the time of travel of an α-particle. This was the beginning of a turning point in Kapitza's career. Once the large impulsive magnetic fields had been achieved, he saw whole new vistas opening up for exploiting the technique and this soon led him into pioneering work in solid-state and low-temperature physics. It was also the beginning of the transition of the Cavendish from the string and sealing wax tradition to the age of large-machine physics.

One of the basic problems of producing an impulsive field is how to provide a large store of energy which can be converted into electrical power and rapidly discharged through a coil. An obvious candidate for such an energy store was an electrical condenser of large capacity charged to a high voltage, but such a condenser would have been expensive and Kapitza chose instead a chemical store, essentially a large accumulator of special design, which he could build himself.

He was encouraged by Rutherford's reaction to his idea; as he wrote to his mother (15 June 1922, see p. 150):
(15 June 1922, see p. 150)

> The Crocodile is keen on my idea and thinks it will succeed . . . He
> has the devil of a nose for experiment and if he thinks that something
> will come of it, that is a very good omen. He treats me better and better
> which makes me very happy . . .

He also mentions that a "young physicist" (P.M.S. Blackett, who
was 3 years his junior) is going to work with him. Blackett (see fig. 60)
introduced him to the subtleties of the Wilson cloud chamber, which
was used for revealing the α-particle tracks, but the collaboration did
not continue beyond the initial stages of the project. Indeed, it is
evident from remarks in a letter of 19 June (see p. 150) that there was
some friction in the collaboration between Kapitza and Blackett – not
perhaps surprising since both had extremely independent characters.

Another significant sign of Rutherford's rising esteem for Kapitza
was that he gave permission for Kapitza to bring over an Estonian,
Emil Yanovich Laurmann*, who had collaborated with him in
Petrograd. Laurmann was remarkably skilled in all sorts of
techniques, such as electrical engineering, photography and, most
importantly, the handling of delicate equipment. He was a little deaf
and visitors to the laboratory were often startled by Kapitza's loud
cries of "Baron" (pronounced Baarón) when he needed Laurmann's
assistance. Baron was a nickname coined rather perversely by
Kapitza, because of the hatred of most Estonians (including
Laurmann) for the German barons who were for long the ruling class
in the Baltic provinces. Laurmann continued as Kapitza's personal
assistant throughout the Cambridge period and later spent some time
in Moscow helping Kapitza to get started in his new Institute.

Although the new project was technically much more complicated
and on a larger scale than anything Kapitza had tackled before,
Kapitza's drive, backed by Rutherford's support and Laurmann's
assistance, produced results very quickly. After some early failures he
was able to write to his mother, only two months after he had started,
that the preliminary experiments had been completely successful and
that he had been given additional money and working space to
develop the idea on a larger scale (p. 155). And then at last on 29
November, after the construction of the Wilson chamber as well as the

---

* An obituary note (Shoenberg 1954) gives some biographical details of Laurmann,
and a photograph appears in fig. 25.

impulsive field equipment, he was able to report complete success in the most euphoric terms (p. 160).

By March 1923 – again only nine months after starting – a paper had been submitted, this time to the Proceedings of the Cambridge Philosophical Society, and was duly published soon after (8). This paper outlined the main features of the method and discussed some of the results, but another year's work was needed to complete the research properly and this resulted in two papers (9, 10), the first giving details of the impulsive field method and the second describing the special design of the Wilson chamber, the elaborate switching and timing devices involved and, finally, a detailed account of how the curvature of the α-ray tracks varied with position along the track. In the preliminary paper the curvature at any particular point of a track seemed to vary considerably from track to track – by as much as a factor of 5 very near the end of the track. Kapitza ingeniously speculated that this might be explained by an asymmetric shape of the α-nucleus, for which Rutherford had earlier found some evidence. Such an asymmetry might then make the velocity dependent on how the shape was orientated with respect to the direction of motion. The later work, however, showed that the variation between tracks occurred for a much simpler reason. This was that even in the absence of a magnetic field the tracks were not perfectly straight, especially near the end of the range. This random "natural curvature" was probably due to slight scattering of the α-particle and could roughly account for the random differences between the curvatures in a field. With this interpretation, the appropriate procedure was to average the curvature over many individual tracks and this produced a reasonably smooth curve of curvature against position on the track. Effectively, the curvature measured the ratio of velocity to charge and the variation of the average charge with position could be deduced from other experiments and some appeal to theory. Thus, finally, a graph was obtained of velocity against range, just as in the heating experiment described earlier and much to Kapitza's satisfaction the two agreed quite reasonably.

While all this work was progressing, Kapitza's position in Cambridge was rapidly being consolidated. Towards the end of January 1923 he was officially admitted as a research student for the Ph.D. degree, with back-dating to October 1921, and he obtained a year's remission in view of the work he had done in Russia. This meant he could complete his Ph.D. in the summer of 1923. He was also admitted as a member of Trinity College. He continued to write ecstatically to his mother about the almost fatherly care and consideration he got from Rutherford – more than he had ever had from any other boss. He also writes of the stressful nature of his work and of the effort involved in setting up his private seminar which became known as the Kapitza Club and which will be described more fully in a section of its own (p. 40). By mid-June he had been awarded the Ph.D. degree and gives his mother an amusing account of the ceremony and of how he pulled Rutherford's leg about it (pp. 168 to 170).

Another example of the kind of joke Kapitza liked to perpetrate on Rutherford was recounted many years later in a lecture he gave about Rutherford (34). Kapitza had been rather shocked when, at the start of his Cambridge career, Rutherford warned him he would not tolerate him making Communist propaganda in the laboratory. When a year later Kapitza's first major paper was published he tried a mild leg-pull. In his own words:

> I presented Rutherford with a reprint and I made an inscription on it that this work was proof that I had come to his laboratory to do scientific work and not Communist propaganda. He got extremely angry with this inscription, swore and gave me the reprint back. I had foreseen this and I had another reprint in reserve with an extremely appropriate inscription with which I immediately presented him. Obviously, Rutherford had a characteristically hot temper but cooled down just as quickly.

The original inscribed reprint can be seen in the Cockcroft collection at the Churchill Archive Centre, Cambridge and the inscription is reproduced in fig. 7.

The paper on the curvature of α-tracks in strong magnetic fields was Kapitza's last research in Rutherford's personal area of interest. From then on he was concerned with applying the impulsive field technique to other problems and with extending the technique to produce still higher magnetic fields. The first application was a study of the Zeeman effect in fields up to 130 000 gauss (11) in collaboration with H.W.B. Skinner (see figs. 21 and 56). The hope was that something new might turn up at such high fields – typically three or four times

[*From the* PROCEEDINGS OF THE ROYAL SOCIETY, A, VOL. 102, 1922.]

The Loss of Energy of an α-Ray Beam in its Passage through Matter. Part I.—Passage through Air and CO_2.

By P. L. KAPITZA.

The author presenting this paper with his most kind regards, would be very happy if this work will convince Prof. E. Rutherford in two things. 1) That the α particle have no energy after the end of his range. 2) That the author came to the Cavendish Laboratory for scientifical work and not for communistical propaganda.

7. Reprint rejected by Rutherford, 1922.

larger than had ever been used before. However, in spite of the ingenuity of the technical arrangements (particularly as regards producing an intense light source at exactly the right moment) nothing very new did turn up. It was of course a great feat of virtuosity to make everything work properly at the same time, but apart from a few slight discrepancies, the results at the high fields proved to be no more than reasonable extrapolations of what had been observed before at lower fields. Patents were taken out in respect of the special type of battery capable of rapid discharge and of the method of producing an intense light source lasting a very short time.

About this time, Kapitza started an even more ambitious project for the production of strong impulsive fields. The accumulator method was in practice limited to fields of little more than 100 000 gauss in a reasonable experimental volume and moreover the accumulator required difficult repairs after comparatively few discharges because of the development of local "hot spots". The idea of a new method was conceived by Kapitza in 1924 in the course of discussions in London and Cambridge with a Russian electrical engineer, M.P. Kostenko (later an Academician), who calculated that it should be practicable to design a large dynamo which on short-circuit through a suitable coil could generate very high power for a very short time and so produce a field of the order of a million gauss. The stored energy in this scheme would be the kinetic energy of rotation of the dynamo rather than the chemical energy of the accumulator.

Once again Kapitza was enthusiastically supported by Rutherford. But the expense involved was of a bigger order of magnitude than could be provided from Cavendish funds, and Rutherford had to turn to the Department of Scientific and Industrial Research (DSIR) to enable his first step into the "machine age" to become reality. A grant of £8000 was made to be spread over four years and, yet again in an extremely short time, the centre piece of the scheme – the big dynamo – had been designed, built and tested at Metropolitan Vickers in Manchester (see pp. 184 and 185). On 7 July 1925, Kapitza was able to write to Kostenko:

Dear Misha,

I have been meaning to write to you for a long time, but have been very busy and am seizing the opportunity to write from Göttingen where I am giving a lecture during the vacation. The machine has been quite a success: it is finished and tested and gave a maximum 168 000 kW at full excitation at 3500 r.p.m. If it were short-circuited at the right instant it should give 220 000 kW. For the experiment we could use 1/4 of this power, i.e., 55 000 kW, which is a little more than we had calculated. This is of course very good and even better than we expected. The machine has been brought to Cambridge and is already installed. All this has taken a lot of work and energy, but as you see in the one year and four months since we first conceived the idea of this machine it has been built and installed. That's not bad . . .

The dynamo was patented in the names of Kapitza and Kostenko, but some difficult problems had to be tackled before it could be used to generate large fields. These were solved in the course of another two years and a detailed account of the whole scheme was published in 1927 (12).* Of these problems the most challenging were the switch and the coil. The switch had to close and open exactly synchronously with the dynamo cycle and was operated by a roller engaging on a cam attached to the rotor. Between its closing (at zero voltage) and its opening (at zero current) the switch had to carry a current of as much as 30 000 A. The coil had to be made strong enough to withstand the enormous stresses produced by the field and its detailed design was worked out by Cockcroft, who had joined Kapitza soon after starting research in the Cavendish in 1924. The problem of the coil design is vividly illustrated by a letter Kapitza (see p. 260) wrote to Rutherford on 17 December 1925, when Rutherford was in Egypt on his way home from a trip to the Antipodes. In typical Kapitza style, he emphasized the instructive value of the spectacular bursting of the coil at a field of nearly 300 000 gauss, although in fact the strengthening of the coil proved a major problem and was still worrying him nearly a year later (see p. 199).

The correspondence of Kapitza with DSIR over the financing of the scheme throws an amusing side-light both on the rather niggling accounting methods of the DSIR and on Cockcroft's assistance in the project. When Kapitza wrote in his own handwriting asking for an allowance of something like £2 a week to pay for an assistant who also helps with correspondence and accounts, one of the DSIR administrators at first suggested that £1 would be sufficient, but eventually wrote: "I am not sure whether Mr Cockcroft can type. If he can, perhaps an alternative arrangement would be to pay him rather more, say, £1.10s.0d. a week and arrange for him to take over the whole of the clerical work". At this stage Kapitza appealed to Rutherford, who pointed out that the assistant concerned had first-class degrees in engineering and mathematics and eventually a more satisfactory arrangement was arrived at.

Although the high-field machine worked successfully and enabled Kapitza to open up several new and fruitful fields of research over the next few years, it is puzzling that he should have chosen the dynamo

---

* In this paper, Kapitza expresses his indebtedness for help and advice in the construction of the dynamo to Kostenko, Miles Walker and Kuyser. Kostenko did not consider this an adequate recognition of his role in the conception of the whole project and this led to an acrimonious correspondence between him and Kapitza.

method rather than what with hindsight would seem to have been a simpler and cheaper approach – namely storing the energy in a large capacitor. In his earlier paper on the accumulator method (9) he mentions this as a possibility but dismisses it without any very convincing argument. It is true that large capacitors at that time may not have been as reliable or as cheap as they have become since, but a capacitor of, say, 10 mF which could be charged to 2.4 kV would almost certainly have been cheaper and simpler than the special dynamo. Discharge of such a capacitor through a coil comparable to that used by Kapitza should have produced fields quite comparable to those achieved with the dynamo. Indeed, T.F. Wall, an engineer at Sheffield University, did propose such a method in 1924 and later (Wall 1926) described experiments in which fields as high as 200 000 gauss were achieved (but for rather shorter times than Kapitza's fields) using a capacitor of about 1.4 mF charged to 2 kV. However, Wall's coils burst in more extreme conditions and he does not appear to have overcome the difficulty, nor continued the experiments.

Wall's first announcement of his project was accompanied by a big splash in the Press with somewhat extravagant speculations about what the high fields might do. A rather sarcastic comment in the Kapitza Club minute book: ''Will Dr. Wall disintigrate the Univerce. P.K.?'' (Kapitza's spelling) on 7 October 1924 suggests that Kapitza was well aware of Wall's plans and perhaps his reason for not trying the capacitor method was his dislike of following a trodden path. In a report to the Solvay Congress in 1930, Kapitza referred to Wall's limited success as evidence for the unsuitability of the condenser method. In fact, with the development of more reliable capacitors and of electronic methods of switching and recording, the condenser method came into its own soon after the War (for a review see Cotti 1960). The dynamo has been twice reproduced: once (about 1936) by the Soviet factory *Elektrosila* in a more powerful version, designed by Kostenko for the use of Ya.G. Dorfman in Sverdlovsk and again (in 1939) by Y. Tanabe at Tohoku University in Sendai. The Japanese installation was successfully tested but was destroyed during an air raid on 9 July 1945; I have not been able to trace any record of what was done with the Sverdlovsk machine.

Kapitza loved showing off the special features of his installation. Even though the machinery was mounted very elaborately to avoid transmitting vibrations, the considerable mechanical shock of the discharge, with a loss of 20% of the kinetic energy of the dynamo, might easily have upset the delicate recording instruments. To avoid any disturbance, the coil and the recording instruments were placed about 20 metres away from the machine, so that the seismic wave travelling through the ground arrived more than 1/100 s after the start of the discharge, i.e., after the experiment was over. It was, therefore, necessary to install the equipment in a room at least 20 m long and the necessary space was found in an outbuilding of the Chemistry Department, which Rutherford persuaded the head of department, Sir William Pope, to make available. Later, when the work was transferred to the Royal Society Mond Laboratory (see fig. 9), the long "magnet hall" became the central feature of the building with research rooms on either side of it. This turned out to be a very successful design even after the high-field machinery had gone, since the long hall provided convenient opportunities for researchers to meet and exchange ideas.

To bring home the novelty of an experiment lasting such a short time Kapitza was particularly fond of telling two little jokes. One was that 1/100 s was a very long time if you know what to do with it, and the other that he must certainly be the highest paid scientist in the world since he received a full academic salary for work lasting in total only a few seconds in a year. Yu.B. Khariton, a Russian physicist who spent a year in the Cavendish, recalls one occasion – a rare one – when Kapitza was upstaged in showing off his high-field equipment. The occasion was a visit from R.W. Wood, the American physicist, himself famous for his ingenuity in experiment and his scepticism of what others could show him. Kapitza prepared a striking experiment to demonstrate the great strength of his fields by shooting a glass rod immersed in liquid oxygen (strongly paramagnetic) up to the ceiling, when the impulsive field was applied. Wood was accompanied by Rutherford, C.T.R. Wilson, Aston and Blackett, as well as Khariton and other spectators, and they looked expectantly to see what impression had been made on Wood. He, however, did not seem particularly impressed. Instead, he walked up to the coil, took out the small vessel containing the liquid oxygen, raised it to his lips and calmly drank Kapitza's health! The spectators, not familiar with this rather difficult trick, which depends on keeping the liquid in the spheroidal state, were rather relieved to see Wood spit out the liquid again after a few seconds.

Kapitza's first major research with the new equipment was an extensive study of how the electrical resistance of metals increases with magnetic field (13, 14). This was chosen partly because it was the simplest kind of measurement to make in a pulsed field – although even so, considerable care and inventiveness were needed to avoid a number of potential pitfalls – and partly because it had been relatively little explored before. A general feature of the results was that the resistance, after starting off quadratically in field, eventually changed asymptotically to a linear variation – this is often known as Kapitza's law of magnetoresistance. Kapitza suggested an ingenious hypothesis to account for this behaviour but although this bold attempt to give a phenomenological theory of magnetoresistance at a time when the electronic theory of metals was still in its infancy was useful in providing a guide to further experiments, it did not stand the test of time. Later experiments, particularly at lower temperatures and with single crystals, coupled with a better understanding of the theory showed that Kapitza's linear law was not really basic but a consequence partly of the relatively high temperature of his measurements (only down to liquid-nitrogen temperature) and partly of his use of polycrystalline samples in most of his experiments (for a recent discussion see Pippard (1979)). Kapitza himself realized that experiments at lower temperatures would probably be essential for a better understanding of metals and this was one of the main motivations for his work on hydrogen and helium liquefaction, a year or two later (see p. 26). He also emphasized the importance of chemical purity and of working with single crystals, although his study of crystalline anisotropy was limited to bismuth (13). The tradition he established of great attention to purity and to the use of single crystals was a valuable legacy to those who followed in his footsteps.

Following the magnetoresistance work, Kapitza carried out two more researches at high fields, both, once again of an exploratory nature. The first was a study of the magnetization of a variety of substances: ferromagnetics in the saturation regime, paramagnetics and single-crystal bismuth (diamagnetic) (15). The second was a study of magnetostriction, mainly of single-crystal bismuth (16). At the heart of both studies was an ingenious instrument which could be used either as a balance to measure the force on a sample in an inhomogeneous magnetic field and hence the magnetization, or as an extensometer to measure the change of length of a rod fixed at its other end. The idea was to apply the force or the change of length to a diaphragm sealing the bottom of a small vessel containing oil and itself completely immersed in a bath of the same oil. Displacement of the diaphragm squirted oil through a small hole between the inner vessel and the outer bath and the oil motion tilted a light mirror mounted on a pivot near the hole. By suitable design the deflection of a light beam reflected by the mirror could be made proportional to the force or to the extension. The magnetization measurements did not reveal anything very novel, for it turned out that the laws established at lower fields could safely be extrapolated into the new range of fields, a factor of 10 or so higher than previously used. The observation of magnetostriction in bismuth was in fact the first such observation on a diamagnetic and Kapitza also developed the theory for specifying how the crystal anisotropy of the effect was determined by the components of a fourth-rank tensor (as is appropriate for a quadratic dependence on field, such as was observed). Kapitza's interest in the somewhat anomalous properties of bismuth started yet another tradition which continued long after he had left Cambridge and indirectly led to studies which helped clarify the electronic structure of metals (see, for instance, Shoenberg 1978).

In the course of all this work, there were several important developments in Kapitza's life and career. In January 1925, he was appointed to an official university position, that of Assistant Director of Magnetic Research, although this was a non-stipendiary post. In October 1925, he was elected to a Research Fellowship at Trinity College (see pp. 191 and 258) and he became a very popular member of the High Table, mixing easily with young and old of all specialities. He greatly valued his association with the College where he lived until he was married and where he continued to dine frequently during the rest of his time in Cambridge. He was particularly pleased when many years later (in 1966), he was elected to an Honorary Fellowship.

In the spring of 1927 he went to Paris and from there, on 27 April, announced his marriage in the letter to Rutherford, reproduced in fig. 61 (see also p. 260), which incidentally illustrates Kapitza's very idiosyncratic spelling and English style. Rutherford's congratulatory reply (see p. 261), carries a characteristic reminder that work should not be forgotten. The lady referred to in the letter was Anna Alekseyevna Krylova, daughter of the Krylov mentioned earlier, the eminent applied mathematician and naval architect who had been the

8. Photograph taken after election to Royal Society, 1929.

senior member of the Soviet commission which brought Kapitza to England. Although Anna's father had stayed in Russia after the Revolution, her mother emigrated to Paris and it was there that Anna completed her education as an archaeologist; she was also an accomplished artist. The marriage was a very happy one and Anna was not only a charming hostess to their many friends at their attractive home in Cambridge (fig. 42), but a great support to Peter in the difficult times he had to go through on several occasions later. Two sons were born in Cambridge: Sergei in 1928, who is a distinguished physicist and a very successful presenter of popular science on Soviet television, and Andrei in 1931, a well-known Antarctic explorer and geographer. In 1929, Kapitza was elected to Fellowship of the Royal Society at the first election after he had been proposed. This in itself

9. Royal Society Mond Laboratory, 1933.

was a rare distinction but even rarer was the election of a foreigner –
for Kapitza had always retained his Soviet citizenship. In fact, the
Statutes at that time did not exclude the election of a foreigner provided he was an "inhabitant of His Majesty's dominions", though such
elections were indeed rare, the previous one having been in 1914! But
to cap everything, his scientific distinction had also been recognized
in his own country by his election a few months earlier to Corresponding Membership of the Soviet Academy of Sciences.

Early in 1930 Kapitza discussed with Rutherford the possibility of
setting up a special new laboratory which could house not only the
high-field equipment, but also provide cryogenic facilities enabling his
researches to be extended to much lower temperatures. Rutherford
backed this idea enthusiastically and persuaded the Royal Society to
put up the money required, £15 000, for building such a laboratory.
The money was found from a bequest to the Royal Society by Ludwig
Mond (the founder of ICI) and the new laboratory was to be called
the Royal Society Mond Laboratory. At the same time Kapitza was
appointed to a Royal Society Messel Professorship. All the complicated negotiations between the Royal Society, the University and

10. New helium liquefier, 1934.

DSIR were completed remarkably quickly and smoothly – partly because of the dominant position of Rutherford in all three bodies and partly through the drive of Kapitza and Cockcroft – and by 1931 a handsome modern building designed by the architect H.C. Hughes was going up in the courtyard of the Cavendish (fig. 9).

The first step in the advance towards lower temperatures was a hydrogen liquefier of novel design and this was completed by Kapitza in collaboration with Cockcroft a year or so before the Mond Laboratory was ready (17). A special feature of the room where this liquefier was set up in the Mond was a very light roof designed to blow off and relieve the pressure in the event of an explosion. Kapitza was fond of warning nervous visitors of this possibility, although actually it never occurred. In constructing a helium liquefier, Kapitza as usual chose a completely original method (19) rather than following existing designs. Instead of relying on cooling with liquid hydrogen to get the

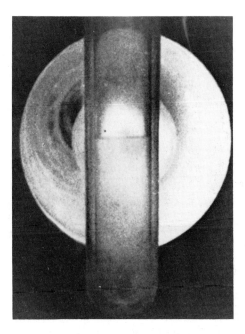

11. First visible liquid helium made in England, 21 April 1934.

helium below its Joule – Thompson inversion temperature, the main cooling was achieved by adiabatic expansion in a piston and cylinder engine. This of course was a well-known principle, but up to then the problem of how to lubricate the piston had never been solved. Kapitza achieved this by the ingenious expedient of using the helium gas itself as the lubricant. The machine (see fig. 10) was completed in 1934, a year after the opening of the Mond Laboratory, and proved important not only in providing liquid helium for Cambridge research over 15 years or so, but in providing the basic idea for a factory-built helium liquefier designed by S.C. Collins at MIT in 1947. The commercial availability of these Collins machines revolutionized low-temperature physics by making liquid helium easily accessible all over the world, rather than at only a few specialized centres, such as Leiden.

Kapitza had hoped that Cambridge would have the first liquid helium in England, but he was upstaged (much to his annoyance) by F.A. Lindemann at Oxford, who brought Kurt Mendelssohn over from Germany with one of Simon's miniature liquefiers (also based on adiabatic expansion, but single shot rather than continuous) and

12. *Daily Mail* picture of crowd at the official opening of Royal Society Mond Laboratory, 3 February 1933.

13. Gilded crocodile key used for opening.

triumphantly announced the first liquid helium in England a year ahead of Kapitza. Ironically, the announcement appeared in the same issue of *Nature* (11 February 1933) as an article describing the new Mond Laboratory, following its formal opening. This opening, by Stanley Baldwin, then Chancellor of the University, was a great occasion (fig. 12) with a large luncheon party at Corpus Christi College hosted by Sir William Spens, the Vice Chancellor, followed by a procession (everyone in full academic dress) to the Arts School, where a number of speeches were made, and finally the actual opening with a gilded key (fig. 13) in the shape of a crocodile and a visitation of the laboratory. Baldwin's speech accepting the Royal Society's gift to the University caused some amused puzzlement, for much of it was almost identical with what Rutherford had said earlier in the proceedings. It seems that Rutherford had quite forgotten that he had already used his remarks in briefing Baldwin as to what he should say! During the visit round the laboratory Kapitza provided a nice example of his style of irreverent teasing. He was overheard replying to a query from Baldwin: ''Yes indeed. You can believe me. I'm not a politician''.

In the course of the opening a lifesize carving of a crocodile (fig. 14) on the wall just outside the main entrance and a bas-relief of Rutherford himself, inside the foyer, both by Kapitza's friend, the sculptor Eric Gill, were revealed for the first time. The bas-relief gave rise to a considerable public controversy on the question of the shape of Rutherford's nose, which was eventually settled in favour of Gill by an approving letter from Bohr (for details, see Oliphant 1972). The opening received a great deal of attention in the Press with reports ranging from factual (although even *The Times* described the crocodile as ''the dragon of science guarding the entrance'') to the sensational and entirely inaccurate (for instance, implying that the extremely high fields and low temperatures would help split the atom). The building received favourable comment in the architectural press for its functional and yet attractive design and it is still there, over 50 years later, although no longer a physics laboratory. After the Cavendish moved to new quarters in Madingley Road in 1972, the building was taken over by the Department of Aerial Photography. Kapitza himself worked in the new Laboratory for only just over a year, but was unable to continue exploiting the excellent facilities he had created there. He and his wife left for what was intended to be a short visit to the Soviet

14. Crocodile carved by Eric Gill on Mond Laboratory, 1933.

Union in August 1934, but he was refused permission to return to England and eventually built a new Institute in Moscow. The story of this affair, which became a *cause célèbre* at the time will be told in a later section (p. 45; see also chs. 4, 5 and 6).

Something should now be said about other aspects of his work and leisure in Cambridge. The Kapitza Club, which will be dealt with in the next section, represents his major contribution – albeit informal – to teaching activities. His other teaching took the form of supervision of research students and some lecturing. The latter activity consisted of eight lectures on ''Recent researches in magnetism'' given once a week in the Lent Term. I (D.S.) have vivid memories of my own attendance at this course in 1932. I found it fascinating not only for its content, but also for the intriguing way he presented it. His strong Russian accent and rather high-pitched voice, his peculiar English constructions and his habit of occasionally writing on the blackboard

15. With Stanley Baldwin after opening of Mond Laboratory, 3 February 1933.

something quite different from what he had said, had considerable entertainment value, but sometimes made it difficult to follow. I remember one occasion when I came up to him after the lecture and asked him to clarify a contradiction in the notes I had taken. His reply was "If I make everything so clear that there are no contradictions, there is nothing left for you to think about and so you won't learn anything". He also gave many semi-popular lectures about his work to University scientific societies and the like and once described his technique*: "I try to pitch it so that 95% understand 5% and that 5%

---

* This was over dinner at the home of Professor S.R. Milner, the head of the physics department at Sheffield University, where Kapitza had lectured to the Physical Society.

16. W.L. Webster outside the Kapitza cottage, Moscow, 1937.

fully understand 95%. I always tell a joke in the first 5 minutes and if they laugh I know they are understanding my English well enough''.

Kapitza supervised only a few research students. Mention has already been made of J.D. Cockcroft, who subsequently became famous for his pioneer work with E.T.S. Walton on nuclear disintegration and another was W.L. Webster, who did some pioneer work on ferromagnetism but then abandoned physics for economics. The only others prior to the building of the Mond were D.S. Kothari, who in 1932 turned to theory halfway through his Ph.D. course and later made distinguished contributions to astrophysics in India, and A.G. Hill who joined Kothari for a short while. At that time the usual practice in the Cavendish was for the Cavendish Professor to be the official supervisor of all the research students, although the detailed supervision was often done by Rutherford's lieutenants. However, after the move to the Mond, Kapitza was named as the official supervisor of two more research students. The first one was myself (D.S), in 1932, and the second one, in 1933, Professor S.R. Milner's son, C.J. Milner who later became Professor of Applied Physics in Sydney.

Both Milner and I recall Kapitza's stimulating influence as a supervisor and his original way of helping us to overcome our experimental difficulties. I was put on to the measurement of the transverse magnetostriction of bismuth crystals in the field of an ordinary electromagnet in order to complement Kapitza's own results on the longitudinal effect in his much higher impulsive fields. The expected change of length was very small – only of order 10^{-7} cm – and when I was feeling a bit desperate because the sensitivity of my apparatus was still 100 times too small, Kapitza would point out half a dozen simple improvements, each of which should give a factor 2 of improvement and then throw in a couple of rather unorthodox suggestions – typical lateral thinking – each of which should give a factor 3. So the problem seemed to be solved, with even a bit in hand, since $2^6 \times 3^2$ is nearly 600! But when he had gone and I started trying his ideas, the factors of 2 proved to be only perhaps 1.2 and the factors of 3 perhaps 1.5 so that the gain was only perhaps 6 rather than 600. However, I was that much nearer to my goal and was ready for the next batch of suggestions. Milner was also much impressed by the fertility of Kapitza's inventiveness, although he gradually learnt that many of the ingenious suggestions Kapitza would make, carried some fatal flaw. But even if only perhaps one in five proved sound, this one might well hit the jackpot and overcome the difficulty.

Kapitza had many interests and skills outside his scientific work. One of his early enthusiasms was for mechanical aids to travel. Soon after he had arrived in Cambridge he bought a motorcycle and even earlier he had advocated the advantages of owning a bicycle (see p. 90). The motorcycle features several times in his letters home. On 16 August 1921 (see p. 120), he gave a graphic account of a bad spill he and his passenger, Alex Müller, had had a few days earlier on an intended outing to the seaside. Even though he was not seriously injured his various bruises and scratches made him look like a Chinese scarecrow, as he put it, and since the machine needed extensive repairs the outing had to be abandoned. This letter ends with a good resolution "now I shall be much more careful", but in spite of this he had a very similar accident only 4 months later (see p. 141). This time he blames his passenger: ". . . I was stupid enough to let Chadwick drive and he managed to upset the machine and send us both flying". Once again he was lucky that neither of them was seriously injured.

Before long he graduated to a Lagonda car (fig. 17) and later a

17. In his Lagonda car, 1924.

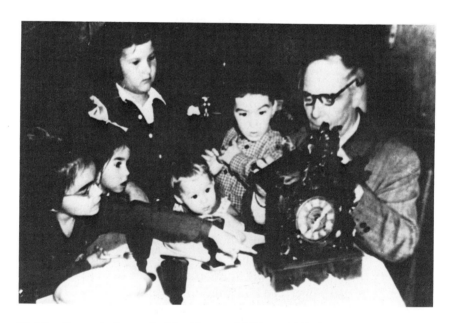

18. Mending a clock, watched by his grandchildren, 1959 (l. to r.: Anna, Maria, Elena, Varvara, Fyodor).

Triumph and a Vauxhall, the latter chosen because full working drawings were available, but although he had a reputation for rather reckless driving he was not involved in any more accidents. When he was going particularly fast he would reassure a nervous passenger by telling him that the high speedometer reading was in km per hour rather than m.p.h. On one occasion he overcame the apparently insoluble problem of overtaking a slow car in Richmond Park (i.e., one keeping to the legal speed limit) by going off the road on to the grass and passing rapidly on the wrong side. Perhaps the best story is of his Trinity friend, the Reverend F.A. Simpson, sitting behind him on a country drive. As they approached a dangerous corner Kapitza pressed the accelerator and turned round to say "Pray God Simpson, Pray God". It seems the prayers were effective. He had, indeed rather a weakness for teasing clergymen; another example was when a clergyman, a guest dining in Hall, asked Kapitza who Eddington was and was told "He knows far more about the Heavens than you do".

As has already been mentioned, he was very skilled at constructing and handling delicate equipment and he loved any activity involving

19. Playing chess with P.A.M. Dirac, 1928.

fine mechanisms. One of his hobbies was repairing watches and clocks (fig. 18) and he enjoyed making small replacement parts on a jeweller's lathe. Another hobby involving manual skill was conjuring. Sergei Kapitza recalls how fascinated he was as a small boy seeing his father appear to swallow table knives and bite bits off china plates and hearing stories about Houdini and other famous illusionists. E.T.S. Walton remembers a train journey when Kapitza showed off the three-card trick ("find the lady") as convincingly as any professional trickster. The famous, but somewhat staid, applied mathematician Timoshenko was evidently rather shocked to see his young compatriot Kapitza showing off tricks one evening during a British Association meeting in Edinburgh. In his memoirs (1968) he comments on Kapitza's glee in making various eminent scientists look a little silly by involving them as stooges in his demonstrations. A rather different type of skill at which Kapitza excelled was chess (fig. 19). He played regularly at the University Chess Club, but never took the game seriously enough to interfere with his work. Even so, his chess was of a high class and Smyslov, the Russian one-time world champion, said that he could not take a game with Kapitza lightly.

Finally, some comments on Kapitza's personality and style as remembered by those who knew him in his Cambridge days. At times he could be almost a text-book example of the absent-minded professor. If a question was addressed to him at such a time his typical reaction would be "What you say?" and only after several repetitions of the question and his "What you say?" would he deal with the question – although sometimes he might simply walk off leaving the question unanswered. Occasionally, this may have just been a ploy to give himself time to think, but usually it was genuine absent-mindedness, or rather an indication that his mind was fully occupied by other problems. Indeed when he was trying out a new idea in the laboratory he would become so completely absorbed as to lose all sense of time and sometimes Anna would have to come in to remind him that they were expecting guests to dinner. He paid little attention to his personal appearance and one Cambridge legend – an authentic one – is that he was once refused admission to a formal gathering at the Senate House because he was wearing a blazer and carpet slippers rather than the obligatory dark suit and black shoes. Sometimes, particularly when he could not see his way through a difficulty in his work, he would look sad and it was perhaps such an occasion that prompted the Trinity historian G. Kitson Clark to remember him looking like "a tragic Russian prince".

But more often, he was cheerful and outgoing, a great charmer and very good company. He found something interesting to say on almost any subject and was an excellent raconteur with an enormous repertoire of stories and anecdotes, some traditional and some made up on the spur of the moment. Almost everyone I consulted in writing this memoir remembered some such story, often rather of the shaggy dog variety. One typical example, although dating from much later, when J.H. Fremlin visited Kapitza in Moscow, illustrates Kapitza's scepticism about professional theoreticians, although in fact he had a great admiration for Landau as will appear later:

> Two theoreticians were stirring their tea when one said to the other "I wonder what makes the tea taste sweet. Is it the sugar or the stirring?" They argued about it for a long time but couldn't find a convincing answer, so they decided to consult Landau. He thought a little and said he thought he could see the answer, but there was one difficulty that needed further thought and he asked them to come back the next day. When they saw him again he said "Now everything is clear. It is obviously the stirring that makes the tea sweet. What held me up at first was that I couldn't see the reason for putting the sugar in, but now I have realized that if you didn't put the sugar in, the tea wouldn't need stirring".

Occasionally, the point of the story would be obscure to someone not familiar with Russian traditions or because of Kapitza's peculiar English, but his laughter over his own joke was so infectious that those around him found themselves joining in, even if they had not altogether understood the joke.

It was perhaps the combination of his outgoing personality, his genuine interest in people and his curiosity about what "made them tick" that enabled him to overcome shyness and reserve and make friends of people like the rather reserved Chadwick, the taciturn Cockcroft, the usually silent Dirac and even the austere A.E. Housman of Trinity (it is said that what he and Housman talked about was the Church of England). These qualities also enabled him to gain the friendship and support of Rutherford, although as discussed in Wilson's (1983) biography of Rutherford, other factors were also relevant here. Rutherford seems to have been captivated by Kapitza's boldness – both his boldness of scientific vision and his boldness in treating Rutherford with much less reverence and respect than the great man was accustomed to from his juniors, while at the same time openly showing his admiration of Rutherford's genius. Rutherford seems to have enjoyed the kind of lighthearted teasing exemplified by Kapitza's congratulations on his 60th birthday. Kapitza presented him with a silver propelling pencil and wrote (p. 263) that Rutherford's fondness for blunt stub ends of pencil – a long standing joke in the Cavendish – was no longer appropriate now that he had become a lord. The gift did not, however, wean Rutherford from his old habits. After Kapitza became a Fellow of Trinity, there were new opportunities for informal contacts with Rutherford. After Sunday dinner in Hall they would often sit together discussing all sorts of topics ranging from politics to drama and continue their talk as Kapitza walked Rutherford home along the Backs. A few years later when Kapitza was back in Moscow, Rutherford wrote to say how much he missed their evening walks and talks – indeed Wilson in his biography suggests that Kapitza was something like a surrogate son to Rutherford.

Two other aspects of Kapitza's personality will be evident from much of what has already been said: his fondness of boasting and exaggeration and his enormous self-confidence.* However, these were usually displayed in a somewhat tongue-in-cheek manner which endeared rather than repelled. For instance, N. Kurti recalls that when Kapitza visited Berlin in 1930 and F.E. Simon (who liked precision) asked for his exact English address, he said with a chuckle "Kapitza, England" will be enough. Another example was when the famous Soviet politician Nikolai Bukharin visited him in Cambridge (fig. 20) and Kapitza took him to dinner in Trinity College and introduced him to the rather conservative Fellowship as "Comrade Bukharin". He loved above all to achieve something which was unique or at least exceptional. Thus, he was proud of his unique distinction of being both an FRS and a member of the Soviet Academy of

---

* However, some of his early letters to his mother reveal that behind this outward self-confidence were hidden considerable doubts about his ability to cope with the problems that faced him.

20. With N. Bukharin, Cambridge, 1931

Sciences and of his extremely rare status of remaining a Soviet citizen yet holding an important post in England. C.P. Snow wrote of him:

> He once asked a friend of mine if a foreigner could become an English peer; we strongly suspected that his ideal career would see him established simultaneously in the Soviet Academy of Sciences and as Rutherford's successor in the House of Lords.

More examples will appear in the account of his later life in Moscow and in his letters. His self-confidence was based on realistic assessment of the problem in hand, whether scientific or administrative and usually his self confidence proved justified by the successful outcome of whatever he was trying to achieve. Rutherford was certainly impressed to see that Kapitza's confidence in the ingenious methods he adopted to deal with his α-ray problems was justified by the successful results he obtained so quickly. This made him all the more ready to accept Kapitza's persuasive arguments about the potential importance of high magnetic fields and low temperatures and give his full backing in developing a new branch of physics outside his own special area. However, Kapitza's self-confidence did not always carry him through: the most notable occasion was his inability to continue indefinitely his remarkable feat of heading an English laboratory while remaining a Soviet citizen and travelling freely between the two countries.

1.3. The Kapitza Club (1922 – 1966)

When Kapitza came to Cambridge in 1921 he missed the lively discussions of the Joffé seminars in Petrograd and soon after he had established himself in Cambridge he decided to start an informal discussion group of his own, providing an injection of Russian temperament into his more phlegmatic English colleagues. The first meeting was on 17 October 1922 with Kapitza giving a talk on magnetism. Although the emphasis was always on informality, Kapitza did insist on certain rules. No one could remain a member if he missed more than a small number of consecutive meetings, and members were regarded as permanent only after they had given a talk of their own.

Although this weekly seminar soon came to be called the "Kapitza Club" (fig. 21), Kapitza himself used to refer to it modestly as "The Club", although he was well aware of the name others used, and of the significance of the club in stimulating Cambridge physics. In a letter to Anna from Moscow in 1935 (see p. 225) while his future was still uncertain he wrote:

> . . . Even in Cambridge I left a mark. Take the club which is usually connected with my name, and of which, like old Pickwick, I was a permanent president, I think it will stay a long time . . .

Soon after its inception, a minute book was started and the tradition

21. 46th Meeting of Kapitza Club, 18 March 1924.
(l. to r. K.G. Emeleus, P.L. Kapitza, E.C. Stoner, D.R. Hartree, H.W.B. Skinner
(standing) and E.G. Dymond; the figure in front has not been identified).

was established that the speaker should add his signature to the title
of his talk. The original first volume of the minute book is in Moscow,
but Kapitza had a photocopy made which, together with the second
(final) volume, is now in the Cockcroft Collection at the Churchill Archive Centre, Cambridge. The record provided by these minutes is an
impressive one. Among the speakers were not only the members, but
also nearly every important physicist from England and abroad and
many of the key discoveries of the 1920's and 30's were reported and
eagerly discussed at the Kapitza Club, sometimes ahead of publication. To mention only a few, Franck talked in 1924 on ''New experiments on resonance fluorescence'' – the experiments which provided such convincing evidence for discrete energy levels, Heisenberg,
and later Dirac, spoke about the principles of quantum mechanics in
1925, Landau presented his theory of diamagnetism of metallic electrons in 1930, Chadwick, Lea and Feather reported the discovery of
the neutron in February 1932, although the title was cautiously

entered as "Neutrons?" and in June 1932 Cockcroft reported on the artificial disintegration of lithium.

Kapitza greatly encouraged interruptions and to break the ice he would himself often ask seemingly naive questions. Sometimes such questions would provide a useful hint to the speaker that he was talking over the heads of his audience and nearly always they would start an argument and so clarify an obscure issue, which might otherwise be forgotten by the time the talk had finished. R.W. Ditchburn, who was a member in the late 1920's, recalls how on one occasion a distinguished American visitor described a theory of the current distribution in a discharge tube, based on the theorem that the currents in a network adjust themselves to make the energy production a maximum. Kapitza interrupted to ask "Don't you mean a minimum?" and others supported him. To decide the issue Kapitza took down Maxwell's famous treatise from his bookshelves (the meeting was in his Trinity rooms) which he regarded almost as a Bible and said "Ah yes, it says a minimum in Book II, Chapter 7" and someone chipped in "Verse 12", causing a big laugh (for once Kapitza could not see the joke). This judgement caused some difficulty to the speaker although he claimed that it did not matter what kind of extremum it was.

Some of the spirit of the discussion and argument comes over in brief remarks in the minute book. For instance, in 1923 G.N. Lewis talked on "What is the origin of radioactive substances?" and Kapitza recorded a characteristic remark:

> It is not advisable to look on the phenomena on the stars if you do not understand the phenomena on the earth.

In the same year, Skinner spoke about the newly discovered Compton effect and the comment "Compton is wrong" appears over the initials of Kapitza, Skinner, Hartree and Lennard-Jones. But a few months later "Compton right we hope" is initialled by Stoner and Blackett, while Kapitza and others still backed the wrong horse and wrote "We hope wrong". Everyone I have asked who remembers the old days of the Kapitza Club speaks warmly of its stimulating influence and, as already mentioned, it was probably Kapitza's most valuable contribution to physics education in Cambridge. I experienced its value myself when in 1933, as a young research student, I was asked by Kapitza to give a paper on superconductivity to the club. Not only did this introduce me to a new and little understood subject, but the lively discussion at the meeting suggested new experiments which I was able to realize a year or so later.

The 377th meeting of the club on 21 August 1934 was the last with Kapitza in charge (the paper was by Otto Stern on the magnetic moment of what was then called the deuton or diplon and soon became the deuteron). When Kapitza was unable to return from his visit to the Soviet Union, Cockcroft took over and meetings continued until the 523rd on 27 February 1940 – although at irregular intervals after the outbreak of war in 1939. Meetings were resumed about a year after the war was over with the 524th meeting on 11 June 1946 and with a membership eager to get back to physics after the long interval of doing other things. The post-war organization was for a time shared by J.F. Allen and myself, but when Allen left for St. Andrews in 1947, I continued on my own, except for a year when D.H. Wilkinson took over while I was away. For a while, I think the club did maintain its old Kapitza tradition of lively discussion between physicists from many different specialities, but as time went on physics became rather more compartmentalized and specialized and it was only rarely that physicists from one specialty could be persuaded to come and learn what was going on in a different field. And, indeed, with increasing specialization, even the language used by, say, a high-energy particle physicist was not readily intelligible to someone specializing in, say, solid state. At any rate, by the late 1950's, I felt that the club had had its day and the last regular meeting, the 675th, was held on 4 March 1958 with a paper by R.A. Fisher on "Probability and scientific inference".

This, however, was not quite the last meeting. In May 1966, Kapitza was at last given permission to visit Cambridge again and a special meeting was arranged on 10 May with dessert wines served rather than coffee and with quite a few present who had been members when Kapitza was in charge more than 30 years earlier (fig. 22). The scientific highlight of the meeting was a review by Kapitza and Dirac of a project (18) they had mooted in 1933 of scattering electrons by standing light waves. In 1933, light sources were too weak and electronic detectors too insensitive to produce any detectable Bragg reflection but with the technical advances since then, particularly the advent of lasers, the project no longer looked as "way out" as it had seemed then.

22. Final (676th) meeting of Kapitza Club, Caius College, Cambridge, 10 May 1966
(l. to r.: D. Shoenberg, J.D. Cockcroft, P.A.M. Dirac, P.L. Kapitza).

It is convenient to conclude this section with a brief account of the
seminars which Kapitza organized in his Moscow Institute after 1936.
These, like his Cambridge club, acquired a considerable reputation.
As in Cambridge, they were at first intimate affairs with a dozen or
so regulars from the Institute itself and occasionally a few special
guests from outside. The audience sat in comfortable easy chairs in
Kapitza's huge office and instead of the Cambridge coffee and
biscuits, Russian tea and caviare sandwiches were provided. Kapitza
presided in his characteristic manner and there was the same tradition
of irreverent interruptions from him and the audience. Many years
later, one of the senior members of the Institute, A.I. Shalnikov,
wrote a skit called "At the seminar", published in *Literaturnaya Gazeta*
(1975) and vividly recalled the style of the early days. In this parody,
Shalnikov makes fun of the various participants and describes how a
seminar speaker ran out of material rather early. Kapitza is then made
to say:

"We still have 2½ hours left. I mean to say 3". Landau looks at his watch and says "There are 47 minutes left". "Yes, exactly. There are 57 minutes left. Perhaps er . . er . . Deryagin can tell us about his work", says Kapitza looking fixedly at Kusakov. Kusakov says gloomily "My name is Kusakov". "Yes, of course, just as I said, Deryagin". Kusakov looks even more gloomy "My name is Kusakov". "What's that you say?" "My name is Kusakov". "What's that you say?" The audience begins to get restive, but Kapitza gets out of the awkward situation by a skilful manoeuvre: "Well, if there's nothing new for today I think we can close the meeting for today . . ."

After the war and after the eight year interval during which Kapitza was away from the Institute, the seminars resumed, but gradually became much larger, with audiences of a hundred or so in a large lecture room, rather than the small group in Kapitza's office. Although little of the informality and intimacy survived with the greatly increased numbers of participants, the seminars still served a useful purpose in bringing together physicists from all over Moscow, and it was regarded as a great privilege to be invited to give a seminar. Kapitza continued to preside in his own idiosyncratic way, but eventually, although regular dates were set aside, only a few seminars were actually given and during the last ten years or so of his life they practically ceased.

1.4. Return to the Soviet Union (1934)

As already mentioned, Kapitza was rather proud of his unusual status of directing a prestigious laboratory in Cambridge while remaining a Soviet citizen and being able to go in and out of the Soviet Union at will. Starting in 1926, he visited the Soviet Union nearly every summer, and always had a return visa underwritten by high-up figures in the Soviet political and scientific establishment. During these visits he was able to spend time with his mother, gave lectures and consulted, and usually managed to have a good holiday in the Caucasus or Crimea.

As early as 1929, he had been approached by L.B. Kamenev (at that time an important political figure with special responsibility for science, but later executed in the great purges) inviting him to act as a consultant to the big Ukranian Physico-Technical Institute being set up in Kharkov and enquiring when he would return permanently to the Soviet Union. Kapitza was able to convince Kamenev that for some years to come he could more usefully develop his work in Cambridge, while acting as a consultant to the new Institute during short visits in the summer (see p. 313). Although his friends wondered whether his exclusive status could continue indefinitely, Kapitza laughingly shrugged off any warnings and in August 1934 made his usual summer trip, accompanied by his wife and on this occasion travelling in their Vauxhall across Scandinavia. After attending a congress in honour of Mendeleev and lecturing in Kharkov, he and Anna were preparing to return to England from Leningrad early in October when the blow fell. Kapitza was told that his ''guaranteed'' return visa was no longer valid and that he would have to stay in the Soviet Union. After a few days, Anna was allowed to return to Cambridge to look after the children and on 10 October she was able to tell Rutherford what had happened.

The story of the subsequent developments is one of almost Byzantine complexity and will be told only in outline here; a good deal more can be inferred from Kapitza's letters to his wife in 1934 and 1935 (ch. 4), to Rutherford (ch. 5) and to important Soviet Government figures (ch. 6); a more detailed account is given by Wilson (1983, ch. 16). The reason for Kapitza's retention has been the subject of much speculation but the basic reason was probably that suggested by Rutherford in a letter to Sir Frank Smith (then Secretary of DSIR):

> I think I told you that Kapitza in one of his expansive moods in Russia told the Soviet engineers that he himself would be able to alter the whole face of electrical engineering in his lifetime. I believe the Soviets now recognize that these chances are very remote. This seems to be a very probable explanation of their action and is due to our friend's love of the limelight.

Two other circumstances were also probably relevant. One was that George Gamow, one of the top young Soviet theoretical physicists, had recently failed to return from a visit to the West and Kapitza's retention may have been to some extent a tit-for-tat (see pp. 214 and 315). The other was the great expansion of the Soviet economy in the early five-year plans, of which a concomitant was the need for a rapid growth of the physical and technical sciences. In particular, Krzhizhanovski, an electrical engineer who had been a protégé of Lenin in promoting the rapid electrification of the Soviet Union and held an influential position in the political establishment, wanted Kapitza to build an Institute in Moscow on the lines of the Mond, which would, as Krzhizhanovski thought, revolutionize the production of electricity. Kapitza was absolutely devastated by his inability to return to Cambridge and tried to disabuse the authorities of the exaggerated ideas they had formed about the immediate technical relevance of his work. For a time he sulked and thought seriously of taking up biophysics with Pavlov in Leningrad if he could not continue his Cambridge work. But it was dangerous to sulk too long in those days and gradually Kapitza began to go along with the idea of working in Moscow, although he still hoped that a return to England might be possible.

Indeed, considerable efforts were being made by Rutherford on his behalf. He appealed to I.M. Maisky, the Soviet Ambassador, in very diplomatic terms, he got eminent scientists abroad, such as Langevin and Bohr, to make discreet representations to Soviet establishment figures, and representations were made to Stanley Baldwin – at that time Prime Minister – to raise the question at a high diplomatic level. Nothing came of all these attempts, even though the whole affair had been kept private, so that no loss of face would have been involved if the Soviet authorities had made a concession. Eventually, however, and inevitably, the affair leaked into the Press with a big splash in the *News Chronicle* on 24 April 1935 followed by sensational reports in many other papers. On 1 May *The Times* published letters from both Rutherford and Gowland Hopkins (then President of the Royal Society), setting out the facts in a very restrained fashion and appealing for Kapitza to be allowed to return to Cambridge for at least a while to complete the work he had started. The effect of these letters was somewhat undermined by a jingoistic letter to *The Times* on 7 May from the chemist H.E. Armstrong (then the senior Fellow of the Royal Society), who claimed that England had no need of foreign scientists in general and of Kapitza in particular:

> Instead of leading a lotus life at Cambridge, he, too, may well be doing national work [in Russia] of a far higher importance than even that involved in magnetizing atoms to destruction.

Finally, Maisky publicly defended the Soviet action in retaining Kapitza and concluded:

> Cambridge would no doubt like to have all the world's greatest scientists in its laboratories, in much the same way as the Soviet Union would like to have Lord Rutherford and others of your great physicists in her laboratories.

However, by the end of 1934, although still very unhappy with his situation and still often thinking of giving up physics, Kapitza began to cooperate in planning the new Institute of which, as he learnt from the newspapers, he had been appointed director. Together with A.I. Shalnikov, a very skilled young experimenter from Leningrad, whom he had intended to have working with him for a while in Cambridge, he toured Moscow looking for suitable sites. They settled on an attractive place on the Lenin Hills (originally called the Sparrow Hills) and the new Institute for Physical Problems (IFP) began to go up in May 1935. According to Khrushchev's memoirs – not the most reliable of sources – this choice site had originally been reserved for the American Embassy, but Stalin had taken against Bullitt, the then American Ambassador, and decreed that Kapitza should have it.

Kapitza continued to feel very frustrated and miserable all through the early months of 1935. He greatly missed his family and his work, and he was treated with some reserve by scientific colleagues who were either too afraid or else felt too important to see him, so that he felt particularly lonely living alone in the Metropole Hotel. In his negotiations with V.I. Mezhlauk (who was then an important political figure and Chairman of the State Planning Commission, but was executed only three years later) he had great difficulty in getting across the idea that his work lay primarily in pure rather than applied physics and that he could not do anything useful unless he had the same facilities as he had had in Cambridge. At times he had fits of depression and as he wrote to Anna on 13 April (see p. 235):

> Sometimes I rage and I want to tear my hair and scream. With my ideas, with my apparatus in my laboratory, others live and work. And here I sit all alone. What for? I do not understand. I want to scream and break furniture and sometimes I think I am beginning to go mad.

23. A.I. Shalnikov, 1937.

However, gradually Kapitza's views began to prevail and even before the newspaper publicity, the Soviet authorities had started negotiations with Rutherford for transferring Kapitza's special equipment from the Mond to the new Institute. The newspaper publicity seems to have acted as a spur to settle things more quickly, although Mezhlauk was unpleasant to Kapitza about it. As Kapitza wrote to his wife on 4 May (see p. 236):

> Altogether he [Mezhlauk] says I must obey them if the people gave them the right to rule the country. I said I obeyed all the orders I have been given. But some of the orders sound as if Beethoven was told to write the Fourth Symphony to order. Of course, Beethoven could conduct an orchestra to order, but he would scarcely have been willing to write a symphony to order, in any case not a good one . . .

Things began to get easier in other respects. The "guardian angels went back to heaven" (i.e., he was no longer under surveillance), he was allowed to travel freely within the Soviet Union, he was assigned a good flat, frequent theatre tickets and a good car which meant a lot to him (see pp. 224 and 237).

The negotiations went on with many ups and downs all through the summer of 1935; several times Kapitza threatened to resign and take up biophysics. At one stage there was an intervention by Molotov and Kapitza wrote to Rutherford saying that he was a loyal Soviet citizen and that he was prepared to work in the laboratory being built for him in Moscow. This letter (in mid-May 1935, see p. 266), evidently intended as a reply to the English Press reports, was considered unsuitably frank and was not approved by Molotov. It finishes rather pathetically:

> Personally, I am very miserable indeed that all this has happened. I miss you, my laboratory and especially my work and it is not to be expected that I soon will be able to resume it, and all this makes me very unhappy. The stupidity of the created position is that it is based on complete misunderstanding as everyone concerned really acts with the best intentions.

The most decisive stage of the negotiations came in August with visits to Moscow by Adrian, who was attending a physiological conference, and Dirac, who was travelling round the world, when both had the opportunity of frank discussions with Kapitza. Soon afterwards they were able to report to Rutherford what Kapitza hoped he might get from Cambridge to enable him to resume his work effectively (see p. 267). Eventually, by November a definite agreement had been reached between Cambridge University and the Soviet authorities, giving Kapitza most of what he wanted although not quite everything. On 5 November, he wrote to the University formally resigning his Directorship of the Mond Laboratory. His letter finishes:

> I must assure you that it is a terrific pain for me. Especially that it happened in this abrupt and unexpected way.

The agreement provided that in return for a payment of £30 000 the University would transfer to Moscow nearly all the Mond equipment

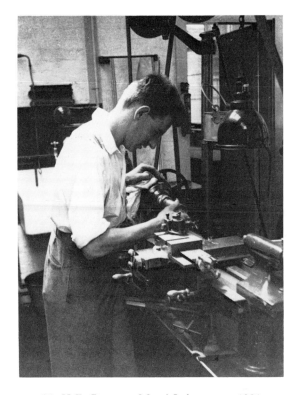

24. H.E. Pearson, Mond Laboratory, 1934.

or duplicates of those items (such as the liquefiers) which were needed in Cambridge to continue the Mond's activities. The high-field equipment was not duplicated, since it was not intended to continue its use in Cambridge without Kapitza. In addition, the University was to give a year's leave of absence to Kapitza's key assistants, Harry Pearson, the head mechanic who had built all the liquefying equipment, and Emil Laurmann, who was intimately acquainted with the high-field machinery, so that they could help Kapitza set up all the equipment in the new Institute as efficiently as possible.

During all this time Anna in Cambridge had been helping to keep Rutherford informed of the true state of affairs in Moscow as well as providing moral support to Peter by her frequent letters (ch. 4). At the end of September 1935 she visited him in Moscow for two months and then had the difficult task of winding up everything in Cambridge.

25. E.Ya. Laurmann, Mond Laboratory, 1934.

26. The big generator leaving for Moscow, December 1935.

Finally, in early January 1936 she and the children rejoined Kapitza in Moscow, where before long they moved into a comfortable and central apartment in Piatnitskaya Street. Even though things now began to hum, with Cockcroft in Cambridge organizing the immense task of sending the equipment to Moscow – the big dynamo (fig. 26) had already left by the end of December 1935 – and Kapitza struggling to get everything ready to instal the equipment as soon as it arrived, he still felt frustrated and unhappy that he could not return to scientific work immediately.

His mood may be judged from one of his letters to Rutherford (26 February 1936, see p. 279) in the course of which he tells about some of his difficulties in getting things done and says:

> I feel myself very miserable here, better than last year, but not so happy as I was in Cambridge. Anna's return brought me much comfort and happiness. In any case my family life is resumed and this is very important as I was here very lonely, quite alone, so the family is very much to me. Your letter reminded me of my happy years in Cambridge and then I felt you as you are, rough in the words and manners and good in your heart such as I like you, and this makes me feel rather happier. The lost Paradise!

This letter was in reply to one in which Rutherford, while sympathizing with Kapitza's difficulties had expressed himself rather bluntly about a previous letter from Kapitza to Cockcroft in which he had impatiently complained about various delays in the transfer arrangements and had made further requests which Rutherford thought excessive (see pp. 275 to 278).

By the summer of 1936 the building of the Institute was completed, most of the vital items of equipment had arrived and so had Laurmann and Pearson. They immediately got busy installing the machinery and familiarizing the Russian technicians with the subtleties of its operation and by the end of 1936 the Institute for Physical Problems was a going concern. Kapitza carried on his scientific work there to the end of his life apart from two interruptions: the first from 1941 to 1943, when the Institute was evacuated to Kazan during the war and the second, from 1946 to 1954, when he was dismissed from his Directorship.

1.5. Moscow and Kazan (1934–1946)

The name of the new Institute for Physical Problems (Institut Fizicheskikh Problem in Russian, or IFP for short) was intended to reflect that its programme of research would be determined by the particular interests of the scientists working there and that the emphasis should be on "pure" physics rather than technology. In fact, its main themes for its first 20 years or so were essentially the same as those of the Mond Laboratory – magnetism and low-temperature physics – although later the scope was extended to include plasma physics which became Kapitza's chief personal interest.

The Institute itself (fig. 27) was in part almost a Chinese copy of the Mond Laboratory, with a long magnet hall (fig. 28), and adjacent research rooms, a liquefier room with a light roof which could blow off in the event of an explosion and, of course, much of the original Mond equipment or duplicates, as, for instance, the special central switchboard with its distinctive look and the distribution boards in each room. However, in other respects the accommodation was on a rather grander scale than the Mond. The Director's office was immense and looked out on a large terrace, there was a large meeting hall

27. Main entrance of IFP, Moscow 1938.

28. Magnet hall at IFP with big generator at back, 1938.

and considerable space for offices to house the administration. Inevitably, the administration was much larger than it had been in the Mond because the Institute had its own living quarters, buffet, transport and library. However, Kapitza succeeded in keeping administration to a minimum by eliminating much of the elaborate paper work involved in planning and finance that was usual in most Soviet institutes. He was fond of saying that over elaborate planning in research was rather like a doctor having to prescribe medicine for an illness his patient would have in a year's time.

In a report to the Presidium of the Academy of Sciences (40) he says:

. . . We are given more freedom to dispose of our annual budget than ordinary state institutions. Some perseverance was required to convince the People's Commissariat of Finance of the necessity for such an arrangement. At that time the Finance people wanted to introduce the so-called thematic accounting whereby expenditure was reported in detail for each line of research. In my correspondence with the

29. The Kapitza's "cottage", their home from 1937 to 1946.

Commissariat I asked them: "When looking at a Rembrandt picture are you really interested in the great artist's expenditure for brushes and canvas? Then why are you so interested in the cost of equipment and materials when you consider research?"

A fruitful research is worth incomparably more than the expenditure involved in it. The cost of research in money terms is incommensurable with its cultural value. I asked: "How much funds in the Commissariat's opinion should Isaac Newton have been given for his work which culminated in the discovery of universal gravitation?"

Yet the Commissariat remained adamant. The arguments lasted over six months, and I think they would have gained the upper hand were it not for the intervention of the U.S.S.R. Council of People's Commissars. Finally, our Institute was given a simplified financial system which saved the Director from everyday troubles and endless "combinations" of work. Under the new system, for example, the Institute employs a single book-keeper, and when we are shorthanded he finds time to help us in testing equipment, taking records and making measurements. All this helps a great deal towards an efficient and healthy operation.

30. S.I. Filimonov in high-field control room, IFP, 1938.

31. A.V. Petushkov and his wife at dancing class, IFP, 1937.

As already mentioned, the Institute was attractively sited with gardens and parks around it. The staff living quarters looked out on a tennis court which became a skating rink in winter and there was a magnificent house close by for the Kapitzas, which the Russians referred to as the "Cottage" (under the impression that this was the English word for any detached house) even though the house looked more like a small palace (fig. 29). Not only was the Institute very well equipped, but it also provided a very high standard of technical assistance. Indeed, Kapitza had an extraordinary flair for picking good people to assist him. The glassblower, A.V. Petushkov (fig. 31), one of six glassblower brothers, was a virtuoso who loved the challenge of a difficult problem, which he would solve behind locked doors, producing the finished article with an enigmatic smile but no explanation

32. O.N. Pisarzhevski teaching modern ballroom dancing, IFP, 1937.

of how he had done it. Similarly, Minakov, the head of the workshop, Yakovlev, who had learnt all about liquefiers from Pearson and was the indispensable provider of liquid helium, and S.I. Filimonov (fig. 30), whom Laurmann trained to be Kapitza's personal technical assistant and who later got the equivalent of a Ph.D. and became an important member of the scientific staff, were all masters of their craft. Another key figure was Kapitza's personal assistant, Oleg Pisarzhevski (fig. 32), who was able to shield Kapitza from many administrative chores and also organized a class of modern ballroom dancing in the Institute. Later, he left to become a scientific writer and was replaced by Pavel Rubinin, a man of many parts, who again was well able to deal with many matters on his own initiative; he is now in charge of the Kapitza archive and is co-editor of this book. In 1937 there were only about seven research workers (fig. 33), but the number grew considerably later.

Kapitza's first project (20) in his new laboratory was to extend his earlier Zeeman effect experiments in collaboration with Strelkov (fig. 34)

33. Research group at IFP, 1938.
(l. to r. A.I. Shalnikov, N.A. Brilliantov, P.G. Strelkov, P.L. Kapitza, D. Shoenberg, N.E. Alekseyevski, M.A. Veksler).

34. P.G. Strelkov, 1938.

and Laurmann to the three times higher magnetic fields of the dynamo method. Once again the experiments confirmed the validity of the theory up to the highest fields and cleared up some features, such as the development of the Paschen – Back effect, which had not been fully resolved in the earlier experiments for lack of sufficient field. Although nothing very exciting turned up, this work was somewhat of a morale booster in demonstrating that the Cambridge equipment was able to work just as effectively in its new environment. It was also by way of being a swan song, since it was the last occasion on which the high-field equipment was used. It remained, however, enshrined in its magnet hall and became essentially a museum piece to show off to visitors. In parallel with this high-field work, rapid progress was being made in completing the cryogenic facilities and Kapitza took up two new lines of research which became his main preoccupation for the next ten years or so. One of these, the study of the transport properties of helium-II (i.e., liquid helium below its λ-point) led to a series of remarkable discoveries and the award of two Nobel Prizes – one to Landau for his theory of quantum liquids, based largely on Kapitza's experiments and the other, very belatedly, to Kapitza himself. The second line of research was more technically orientated – the development of a new and more efficient method of liquefying air and hence, by fractional distillation, of manufacturing oxygen on an industrial scale.

Kapitza's interest in liquid helium was stimulated by Keesom's discovery in Leiden (1936) of the extraordinarily high thermal conductivity of helium-II and by the Cambridge experiments of Allen, Peierls and Uddin (1937) which showed that the dependence of heat flow on temperature gradient was far from linear. This led him to the idea that the heat transport might be associated with convection, brought about by an extremely low viscosity rather than conduction. To observe such a low viscosity, however, the liquid flow must be through very fine channels, since otherwise the flow would be governed entirely by turbulence, and to achieve this Kapitza studied the radial flow under gravity of the liquid between two optical flats separated by a gap of order 1 μm or less. The result was that the viscosity was indeed many orders of magnitude less than had been inferred from earlier experiments in which the flow must have been turbulent rather than laminar. Indeed, Kapitza was able to set only an upper limit (of order 10^{-9} poise) to any possible viscosity and suggested, by analogy with superconductivity, that below the λ-point the liquid became a "superfluid".

By a rather remarkable coincidence, very similar results were being obtained in Cambridge by Allen and Misener at the same time, although entirely independently. Their flow was through fine capillaries rather than through a narrow gap and they too found an upper limit (comparable to Kapitza's result) to any possible viscosity. The Cambridge group first learnt of the Moscow experiments early in December 1937 from W.L. Webster (fig. 16) who had been visiting Kapitza and brought back a copy of the letter describing the results which Kapitza had sent for publication in *Nature*. Cockcroft, then in charge of the Mond Laboratory, wrote to Kapitza about the Cambridge results on 18th December and a few days later submitted a letter to *Nature* from Allen and Misener. In this letter, they pointed out that Kapitza's interpretation of the high heat conductivity as due to convection was too simple and that the real mechanism was not yet clear. The two notes appeared together ((21) and Allen and Misener 1938), although the appended dates of submission (3 and 22 December) make clear Kapitza's priority.

During the short remaining time before the interruption of the War, an intense effort both in Moscow and in the West went into trying to

35. L.D. Landau, 1937.

understand the behaviour of liquid helium-II more fundamentally. Important contributions were made by Allen and Jones, who discovered the thermomechanical ("fountain") effect, by H. London, who predicted the inverse mechanocaloric effect, by Daunt and Mendelssohn, who demonstrated this inverse effect, and by Tisza, who produced a phenomenological two-fluid theory – although without any sound fundamental basis. The decisive results were, however, obtained in a series of ingenious experiments by Kapitza (24, 25). It was this work which in 1945 enabled Landau, who was in charge of the theoretical work at the Institute, to develop a fundamental quantum theory of liquid helium which provided a sound basis for the two-fluid idea and also led to many important new predictions.

The major result of the experiments was to show that heat flow was accompanied by mechanical flow. Thus, when heat flowed from a closed bulb containing helium-II through a capillary into a bath of helium-II, a light vane suspended opposite the mouth of the capillary could be seen to deflect. A rather spectacular demonstration of the effect – involving a masterpiece of Petushkov's glassblowing art – was provided by a miniature reaction turbine consisting of six capillaries radiating out of a closed bulb and bent near their ends to point round the circumference of the circle. This whole "spider", as Kapitza called it (in spite of its entomological inadequacy), was mounted on a needle pivot and when light was shone on the bulb (blackened to absorb radiation), the spider began to spin rapidly – up to 120 r.p.m. – in reaction to the liquid flowing out of its legs.

The outflow of liquid accompanying the heat flow immediately raised the question of how the amount of liquid in the closed bulb was conserved. At first, Kapitza proposed that there was a compensating flow inwards restricted to a thin layer close to the capillary wall and having a different heat content from the bulk liquid because of its proximity to the wall. However, a more convincing explanation was provided by Landau's theory, according to which the inflowing liquid was the superfluid component of the two-fluid ensemble. This component exerted no force on the capillary or the suspended vane and carried no entropy, while all the entropy of the liquid was concentrated in the outflowing normal component which deflected the vane or made the spider rotate. In his second study, (25), Kapitza used a somewhat different experimental arrangement in which liquid could be made to flow into a bulb containing a heater and a thermometer through a very narrow channel (a gap between flats, as in his superfluidity experiments). He found that for sufficiently low power in the heater there was no temperature difference between inside and outside and in this regime the power in the heater was directly proportional to the rate of mass flow into the bulb. The fact that the constant of proportionality was independent of the channel width showed that his earlier hypothesis of anomalous heat content close to the wall was untenable. Indeed, this constant of proportionality proved to be just the product of the entropy per unit mass and the absolute temperature. This was a quantitative confirmation of the two-fluid theory rigorously established by Landau, since for the very narrow channels of the experiment it would be entirely superfluid, without any entropy, which flowed in and the entropy deficit was supplied by the heater to keep the temperature unchanged.

In the course of these studies, Kapitza also discovered that if heat

flows from a solid surface into helium-II there was a temperature
discontinuity at the surface. This has come to be known as the "Kapitza jump" and has been the subject of much study in post-war years.
It was only much later that Khalatnikov was able to explain it, at least
in principle, as being due to the acoustic mismatch of the solid and the
liquid helium. All this work was broken off in 1941 when the Soviet
Union was invaded by Germany, and Kapitza himself did not continue it after the war. However, he did actively encourage further
work and two of the most striking new predictions of Landau's theory
were confirmed in his Institute. One was that there should be a new
kind of wave propagation in helium-II, which E.M. Lifshitz pointed
out should be essentially a temperature rather than a pressure wave,
and this "second sound" was experimentally observed by Peshkov
very soon after the Institute had returned to Moscow in 1943. The
other was an elegant and direct demonstration of the two-fluid theory
by Andronikashvili, who showed that the moment of inertia of a pile
of closely packed discs undergoing torsional oscillations in helium-II
was enhanced by only the normal component of the liquid between the
discs (the superfluid was not carried around by the discs). In this way
he was able to determine what fraction of the whole density was normal and to show that this fraction fell off toward zero as the
temperature was lowered. In his very entertaining memoirs, Andronikashvili (1980) recalls Kapitza's comments at a seminar after
Andronikashvili had pointed out some slight discrepancies between his
results and Landau's theory:

> . . . [the discrepancy] is a good thing: exact agreement between experi-
> ment and theory represents a state of bourgeois satisfaction in science.
> If there is some disagreement then there is something to think about,
> which should be welcome. So here it seems Dau [Landau's nickname]
> will have to rethink some details of his theory. But, of course, after to-
> day's seminar we must first of all congratulate Dau on yet another in-
> disputable confirmation of his brilliant theory and we must con-
> gratulate Elevter [Andronikashvili] too. The experiment is a subtle one
> and he had to work very hard. . .

Kapitza's other major research effort at this time was the development of a new method of liquefying air, with a view to simplifying and
cheapening the large scale production of oxygen for industrial use.

Existing methods, in which the final cooling is achieved by the Joule – Thomson effect, involve the use of high pressure (typically 200 atm) and inevitably the machinery is bulky and expensive. Kapitza (22) proposed the use of an expansion turbine, requiring only about 5 atm pressure, to cool the air all the way to its liquefaction temperature and showed that with a novel radial – axial design of turbine a higher efficiency could be achieved than was possible with the conventional impulse type of turbine. The working out of the whole scheme involved a profound understanding of both physical and engineering principles and the successful construction of a laboratory scale liquefier in less than two years was a considerable achievement. This machine delivered about 30 kg of liquid air per hour after a very short start-up time of about 20 min and with an expenditure of 1.2 kWh/kg (rather less than that of a conventional machine of the same output and only a little more than has been achieved up to now with all the technical advances since 1939). An important part of the design was how to overcome the instabilities associated with the very fast rotation of the turbine motor. The analysis involved was presented in a separate paper (23) and the methods were patented. P.B. Moon, in his 1978 Rutherford lecture, which was mainly about other applications of high-speed rotors, commented enthusiastically on the elegance of Kapitza's analysis, which he found directly applicable to his own work.

Shortly after the Soviet Union entered the war, the Institute for Physical Problems, together with many other Academy institutions, was evacuated to Kazan, where the equipment was set up in the botany and zoology department of the University cheek by jowl with display cabinets of animal skeletons and the like. The Institute's building in Moscow was taken over by the staff of the 5th Moscow Army Division. The Kapitza family, together with Krylov who had been evacuated from Leningrad, occupied a small house which had once been the home of the famous mathematician Lobachevsky, the Rector of the University in the 19th century. One of the main efforts at Kazan was the rapid development of an industrial scale plant for oxygen production based on the expansion turbine method. It was a remarkable achievement that in spite of the very difficult conditions (lack of funds, extreme shortages of food, etc.) a large scale pilot plant was in fact successfully completed during the two years that the evacuation lasted. Kapitza was appointed head of a new government department dealing with oxygen production and after the return from evacuation in 1943 he was also in charge of a factory being built at Balashikha near Moscow. His design of an expansion turbine has proved to be the basis of much of the world's industrial production of oxygen and his pioneer work is still greatly respected by industrial engineers.

Kapitza received many official recognitions of his achievements – election as a full Academician in 1939, Stalin Prizes in 1941 and 1943, Orders of Lenin in 1943 and 1944 and for his oxygen work the title Hero of Socialist Labour together with a third Order of Lenin and the Hammer and Sickle gold medal in May 1945. But soon afterwards there were ominous signs that all was not well. To quote Andronikashvili (1980):

> All sort of odd things began to happen in the Institute in the spring of 1946. Commissions kept arriving to investigate the activities of the Institute and then the status of the commissions began to increase with the arrival of Ministers. They began to investigate Kapitza's oxygen project. They spoke badly of the factory at Balashikha – not productive enough, uneconomical . . . late in meeting its targets . . .

All this culminated in the summer of 1946 in a curt announcement on the bulletin board that:

> P.L. Kapitza, having shown a cavalier attitude to both Soviet and foreign achievement in the technology of oxygen production and having failed to meet the scheduled dates for introducing new installations into the metallurgical industry, is relieved of his duties as Director of the Institute for Physical Problems.

It is likely, however, that this sudden turnabout in Kapitza's fortunes occurred for quite different reasons than those officially stated – although no doubt the fact that Kapitza had trodden on many toes in putting through his oxygen project also played its part, as can be inferred from some of the letters in ch. 6. The real reason was probably to do with Kapitza's refusal to continue working in the organization set up under Beria (then head of the KGB, later executed soon after Stalin's death)* to develop the Soviet atomic bomb

---

* The familiar name KGB is used in this chapter, although in fact it was called the NKVD in the 1930's and MGB in the early 1950's.

immediately after the American bombs had been exploded (see, for instance, York 1976). Kapitza wrote to Stalin (see p. 376) that Beria was like the conductor of an orchestra with the baton in his hand but without any understanding of the score. Beria wanted to arrest Kapitza for such insubordination, but Stalin in his unpredictable way – perhaps because he admired Kapitza's boldness – vetoed this proposal and said dismissal would be sufficient.

This was not Kapitza's first brush with the KGB, and perhaps the memory of his victory on an earlier occasion may have rankled and also played a part in his dismissal. During the great pre-war purges Landau was arrested in 1938 and Kapitza showed great courage by intervening on his behalf – this was rather like jumping naked into a den of lions. He wrote to Stalin (see p. 348) asking for an explanation. and later sent a reminder to Molotov saying that his work on superfluidity could hardly go on without Landau (see p. 348). Eventually, he was summoned by high-ranking KGB officers (Merkulov and Kobulov, Beria's deputies) who kept trying to offer Kapitza a fat dossier about Landau. However, Kapitza refused to look at the dossier and succeeded in outfacing them, saying he understood nothing about legal technicalities, and insisted that although Landau may have had a sharp tongue he was certainly no spy, and that if he was not released it would be a disaster for science. Moreover, Kapitza kept asking what possible motive Landau could have had to commit the espionage he was accused of, but the KGB could find no answer. Landau was indeed released not long afterwards and many years later, in a tribute for Kapitza's 70th birthday he wrote:

> These years [the late 1930's] are memorable to me for another, but very sad reason. Because of a stupid denunciation I was arrested and accused of being a German spy. Now I can sometimes even find this funny, but then it was no joke. I spent a year in prison and it was clear that I couldn't last another 6 months – I was simply dying. Kapitza went to the Kremlin and announced that he would have to leave his Institute if I wasn't released. I was released. It is hardly necessary to say that such an action in those years required no little courage, great humanity and crystal-clear honesty.

36. Izba for Physical Problems, Nikolina Gora, 1954.

1.6. Nikolina Gora (1946 – 1954)

Although Kapitza was dismissed from his Institute he retained his position (and salary) as an Academician and chose to live at his dacha (country house) at Nikolina Gora (Nicholas Hill), where he managed to carry on scientific work until he was partially reinstated eight years later, and more fully soon after. Eventually (in 1958), it was officially acknowledged that the unfavourable report on his oxygen work was mistaken. He still occasionally visited Moscow to give lectures and to act as a consultant to the Institute of Crystallography, but most of his effort went into theoretical work and into building up a laboratory in the lodge of the dacha and various adjoining sheds and outhouses, where, aided by his sons (particularly Sergei) and Filimonov, quite a good deal of experimental work was done. This laboratory (fig. 36) was jokingly called the Izba for Physical Problems (izba means a peasant's hut) with the same initials as the IFP.

A whole series of theoretical papers appeared on diverse topics, such as heat transfer in two-dimensional turbulent flow (26), wave flow in a thin layer of viscous fluid (27), formation of sea waves by wind (29), and the hydrodynamic theory of lubrication in rolling (31). All this work was characterized by considerable mathematical skill and ingenuity in simplifying the analysis without losing the essence of the problem; the lubrication work is described as "classic" in a recent review (Dowson 1979). The theory of wave flow along liquid flowing over the outside surface of a cylindrical tube was verified experimentally in collaboration with Sergei (28), using carefully designed apparatus entirely built with the relatively primitive resources of the Izba; the style of this experiment is rather reminiscent of Kapitza's early work in the Cavendish.

However, a good deal of other work was going on at Nikolina Gora, which because of its potential applications, was at first classified and was published only in 1962 and later (33). An early hint of this work was a paper on ball lightning (32) in which it was suggested that the ball of hot gas was essentially a resonator fed by high-frequency radio waves of wavelength typically of order 50 cm (about four times the diameter of the ball), which were somehow generated by the storm. Later Kapitza tried to verify this idea experimentally (35), but no conclusive evidence was obtained and his hypothesis has not found general acceptance. Another and less direct hint was a paper (30) demonstrating both theoretically and experimentally the stability of an inverted pendulum with the point of support vibrating much faster than the period of the pendulum.

As was later made clear, both these studies were spin-offs from Kapitza's main thrust, the development of powerful high-frequency radio generators of the magnetron type. In the series of papers (33), he had worked out the detailed theory, in an explicit form, of a "planatron" (essentially a linear magnetron, sometimes called a "dicotron") and constructed and tested a working model. The merit of his approach was that it provided an objective method of design, in contrast to the trial and error methods based on numerical analysis which had mostly been used up to then. It was the formal analogy of the differential equations involved with those of the vibrating pendulum support problem that led him to tackle the latter by the same technique. The planatron gave several kilowatts output of 10 cm waves and a striking observation was that if the waves fell on a hollow quartz sphere of 10 cm diameter containing low-pressure helium, a vivid discharge took place, which after a few seconds raised the temperature enough to melt the quartz. It was this observation that led to the ball lightning hypothesis, but more significantly, it led to the idea that with a still more powerful microwave source a plasma might be heated sufficiently to produce a thermonuclear reaction. The next step was the development of a more elaborate and more powerful type of magnetron which was called a "nigotron" (after Nikolina Gora), but by the time this was done Kapitza was back in his old Institute. What had been achieved at Nikolina Gora was impressive considering the limited resources available and again illustrates Kapitza's drive and his intellectual powers in formulating and solving difficult problems. However, during the ten years before publication, much of this achievement was overtaken by developments elsewhere and its impact has not been great.

1.7. Moscow (1954–1984)

Once back in Moscow the scale of the high-power electronics work was greatly increased and a special wing was set aside for the "Physical Laboratory" where the construction of the nigotrons could be effectively carried on and moreover kept secret. The secrecy was partially lifted in 1962 and more completely in 1969, when a paper (36) appeared describing striking observations made from as early as 1958 of very hot glowing plasma. The glowing region had a sharply defined shape which could be varied from spherical to stringlike and it could be kept well away from the walls of the vessel into which the microwaves were fed. Spectral studies suggested that electron temperatures of order 10^6 K were reached, although the ion temperatures were much lower.

Kapitza was convinced that it should be possible to exploit such hot plasma strings to reach the conditions necessary for a sustained thermonuclear reaction and that his microwave heating approach offered a better chance of success than the Tokamak and laser heating methods which were being pursued so actively in the Soviet Union and in the West. He proposed and patented a schematic design for a realistic reactor (37) and continued hot-plasma experiments on an ever larger scale to the end of his life. Although he claimed (39) to have achieved electron temperatures as high as 5×10^7 K, it is likely that these were based on faulty diagnostic methods and some of the basic assumptions in his analysis of the problem have been questioned. The whole research must be regarded as a brave effort in Kapitza's style of "going it alone" and taking little notice of what others were doing around him, but he could hardly compete with the "big battalions". Although he produced much interesting plasma physics, his efforts were largely eclipsed by the work of the powerful teams at the Kurchatov Institute and in the West using the Tokamak system. It should be noted, however, that there is an echo of Kapitza's ideas in the latest Tokamaks that make use of microwave heating of plasma confined in quasi-D.C. magnetic fields.

In his style of running his Institute, Kapitza modelled himself closely on Rutherford and often explicitly said so. When I (D.S.) arrived to spend a year with him in 1937, I suggested several possible projects I might take up and Kapitza told me (as he often told others) how, many years ago, Moseley had suggested several projects to Rutherford and Rutherford with unerring intuition had picked the winner; in my case, too, Kapitza chose what proved to be a fruitful research. Like Rutherford, he insisted that laboratory work should stop at 6 pm (later 7 pm) except by special permission, so that research workers should have time to reflect. Just as Rutherford had once insisted that Kapitza should take a holiday following a period of hectic work, so Kapitza too insisted on occasion that his subordinates must take a break whether it suited them or not. Oliphant (1972) recalls an occasion when Rutherford was so impatient to see a photographic oscillograph record that he ruined it, and stained his clothes by awkward handling before the film had been properly processed; so too was Kapitza famous among his students for spoiling films and wetting his clothes by undue impatience. Appalling handwriting was another common feature, although this was perhaps coincidental. Again like Rutherford, he made a point of keeping in touch with what everyone was doing and making shrewd comments even if the topic was remote from his personal research at the time. Often he would achieve this contact through a rule he made that any new apparatus had to be approved by him before it was made by the workshop and this gave him the opportunity of reviewing progress from time to time. Also it was usual for any completed work to be presented at his seminar before it could be sent for publication.

Kapitza's fondness for nicknames such as Crocodile and Baron has already been mentioned – another was "Rat" for Anna because of her burrowing in archaeological archives (his calling out "Rat" to her at the theatre caused some alarm to neighbouring ladies on one occasion!). It was perhaps poetic justice that he and his Institute should also be widely known by nicknames. The Institute (and also Kapitza's seminar) was usually called the Kapichnik, which sounds a bit comic in Russian, although otherwise harmless, but his own nickname was less flattering. It came about when a visitor asked someone in the Institute "What sort of a chap is your boss? Is he a man or a beast?". The answer was a bit hesitant: "It's difficult to say – a bit of each perhaps" and the visitor immediately said "Oh, I see, a centaur" and "centaur" stuck for many years, although later it was often replaced by the more affectionate "grandfather".

There was indeed some truth behind the "centaur" characterization. He could on occasion fly into a rage and be very abusive (once again like Rutherford), he could be autocratic in dealing with his subordinates, and if, unreasonably, he took against someone, it was very hard and indeed sometimes impossible to make him change his mind. On the other side of the coin, however, he could be extremely generous in helping where help was needed – indeed not only generous, but courageous. His intervention to save Landau is only one example – several other interventions were made in letters to political leaders (see ch. 6). Other examples were his public denunciation of anti-semitism, speaking out for the dissident geneticist, Zhores Medvedev, who had been put in a psychiatric institution, his sponsoring talks on genetics at his seminar by opponents of Lysenko, at a time when Lysenko was still in favour, and opposing a move to expel Sakharov from the Academy by the reminder that even the Nazis had never expelled Einstein from the Prussian Academy*. Moreover, everyone in the Institute greatly appreciated the remarkable working conditions he had created, probably unique in the Soviet Union, and

---

* *Note added in proof.* Another version of what happened was that at a meeting of the Academy, the President (A.P. Aleksandrov) asked if there was any precedent for expelling a member of the Academy, and everyone understood that he had Sakharov in mind. Someone replied that there was no precedent (actually not quite true) and then Kapitza said "Yes, there was a precedent – Hitler expelled Einstein from the Prussian Academy". Although, again, this was not exactly true (in fact, Einstein resigned of his own accord, to spare embarrassing his colleagues), Kapitza's remark was effective and the matter was dropped.

the protection they enjoyed there in various ways. And, of course, above all, when he was in the mood, it was a delight to experience his sense of fun and warm cordiality – which he would extend to everyone irrespective of rank.

During the last half of his life, although Kapitza was very actively pursuing his laboratory work, he managed to find time for many other activities. He thought a great deal about the organization of science and its relation to technology and in various speeches and articles he was often sharply critical of the establishment. An example is his report (40) on the organization of his own Institute, where he emphasises the point that pure scientists should not be made to develop the technical applications of their work:

> . . . if Semenov [his close friend] were to attempt to build an internal combustion engine . . . the result would only be a waste of precious time and energy best spent in pure science, where he is nothing less than a virtuoso . . . If the singer cannot possibly accompany his own songs, why force him to do it? . . .

However, Kapitza points out that he himself is an exception to the rule being both scientist and engineer! A recurrent theme in many of his letters to Government figures (ch. 6) is the importance of proper respect for both science and the scientist – essentially the idea that "the scientist should be on top rather than on tap".

Closely related was Kapitza's active interest in education (40). He took a leading part in organizing a new and somewhat elitist kind of University institution – the Moscow Physico-Technical Institute (MPTI) where students came into active contact with research during their last undergraduate years. This enabled the research institute to get to know the students well and so to be able to pick the best of them for further research in a rational way. The MPTI was set up in 1946 just before Kapitza's demotion and he lectured there for a year or two on general physics while at Nikolina Gora; after his return to Moscow, he continued to be much involved in the organization of MPTI, but did not give many lectures himself. He took an active part in examining and invented (partly during his time at Nikolina Gora) many problems of a kind which required intuition and understanding for their solution rather than mere book learning. He was concerned with maintaining a high standard of scientific publication and from 1956 was Editor-in-chief of the Russian language *Journal of Experimental and Theoretical Physics*. He also took an active interest in global problems such as ecology and energy (40) and participated in the Pugwash movement.

He was much in demand for key-note addresses at important conferences and could on occasion make somewhat sharp remarks, such as in commenting on what he thought was rather an inadequate level of presentation at a low-temperature meeting: ''It is important not only that the dish should be nutritious, but that it should be well served''. He could always rise to the occasion and say something original when he had to respond after a ceremonial award or give an address commemorating some great figure in science. In such addresses and articles (40) he showed a particular talent for picking out what it was that characterized the life and work of his subjects and made them great. Particularly interesting are his recollections of his revered teacher Rutherford (34). In 1971 he organized a Rutherford Centenary Colloquium (fig. 37) in Moscow, to which he invited a few British physicists still alive who had personal memories of the great

37. Rutherford Centenary Colloquium, Moscow University 1971 (l. to r.: T.E. Allibone, D. Shoenberg, Yu.M. Tsipenyuk (colloquium secretary), P.L. Kapitza, N. Feather, S. Devons).

man. One of these, T.E. Allibone, in a letter to *The Times* (3 November 1971) complained that he had been unable to persuade the British Post Office to issue a Rutherford commemorative stamp, and asked Kapitza how he had managed to persuade the Soviet postal authorities to issue one (fig. 38). Kapitza pointed to a second telephone on his desk and said it was connected directly to the Kremlin. He had picked it up and said "Mr Brezhnev, I want a stamp to commemorate the centenary of the birth of Lord Rutherford" and Brezhnev had said "O.K.". Allibone regretted that he had no second telephone on his own desk!

For many years Kapitza was denied the possibility of travelling beyond the Soviet Union and the Eastern Bloc countries, even though he was frequently invited. According to Khrushchev, this was simply because Kapitza was thought to be too talkative and might give away some secret – probably an entirely fanciful explanation! However, in 1965 the ice was broken when he was allowed to travel to Copenhagen to receive the Niels Bohr Gold Medal of the Danish Engineering Society. A year later he was awarded the Rutherford medal of the Institute of Physics and Physical Society and came to England again after a lapse of 32 years. He made Cambridge his headquarters and was delighted to find so many of his old friends still there, most of them now establishment figures. He particularly appreciated the hospitality of Churchill College of which his old friend Cockcroft was then Master and while in Cambridge he arranged for his former house at 173 Huntingdon Road (fig. 42), which he had made over to the

38. Rutherford stamp, 1971.

Academy of Sciences, to be used by the College as a hostel – now known as Kapitza House – with priority for Soviet academic visitors. In subsequent years, he was able to indulge his love of travel to the full and visited Canada, the USA, India, Switzerland and many other countries to receive honorary degrees, honorary membership of academies, medals, and so on (for a full list see Kedrov 1984); posthumously, a street in Moscow was named after him (fig. 41). He came back to England on two more occasions: in 1973 to receive the Simon Memorial Prize of the Institute of Physics and in 1976 to give the Bernal lecture (38) at the Royal Society. On the second occasion, although he was saddened to find that so many of his old friends were no longer alive, he enjoyed the opportunity of making new friends

39. Back in Cambridge, 1966.

40. At Daresbury Nuclear Physics Laboratory, 1973.

41. Academician Kapitza Street, Moscow 1986.

42. Kapitza House, 173 Huntingdon Road, Cambridge.

during his stay at Churchill College, which had made him a Honorary Fellow in 1974.

The culminating event came in 1978 with the award of the Nobel Prize (fig. 43) for his work in low-temperature physics. In his Nobel lecture in Stockholm (39) Kapitza rather wittily made it clear that he thought the award was a bit belated.

> . . . I left this field some 30 years ago, although at the Institute under my direction low-temperature research is still being done. Personally, I am now studying plasma phenomena at those very high temperatures that are necessary for a thermonuclear reactor . . . I think that as a subject for the lecture it is of more interest than my past low-temperature work. For it is said "les extrêmes se touchent".

He then went on to review his work on hot plasma. The visit to Stockholm has been vividly described by Rubinin (1979).

The Kapitzas also led an active social life in Moscow, enjoying the company not only of their family (five grandchildren and seven great-grandchildren as well as the two sons, the family of his brother and

43. Nobel Prize award by King Carl XVI of Sweden, 1978.

44. 80th birthday party, Nikolina Gora, 1974.
(Kapitza seated at middle left).

45. Niels, Margrethe and Hans Bohr with the Kapitzas, Moscow 1937.

others) but of a wide circle of friends among the "intelligentsia" – artists, sculptors, writers, musicians, actors, film directors, and so on. On the occasion of his 80th birthday, almost everyone who was anyone in the Soviet cultural world was at the huge party given at Nikolina Gora (fig. 44). Kapitza supported many non-conformist artists by arranging exhibitions of their work at the Institute, particularly during the 1960's. These exhibitions were the talk of Moscow at the time and enabled the artists to sell some of their work. Kapitza himself bought a few to add to his collection of Kustodiev and other contemporary artists, some of whose works had been frowned on in the Stalin era. Kustodiev's fine portrait of Kapitza as a young man (fig. 1) was donated by Kapitza to the Fitzwilliam Museum in Cambridge and is at present on loan to Darwin College.

The Kapitzas' social life extended also to visitors from abroad, particularly those who were old friends, such as the Bohrs (fig. 45) and the Cockcrofts (fig. 46). Many people recall the warm hospitality they received in Moscow or Nikolina Gora and Kapitza's lively conversation. It is really astonishing that he managed to combine such a busy social and intellectual life outside the laboratory with his intensive

46. With John and Elizabeth Cockcroft and Anna outside the Kapitza home, Moscow 1965.

scientific work. This intense activity continued almost to the end of his life although towards the end he was becoming rather frail. At the end of March 1984 he had a severe stroke and died on 8 April after a few days in hospital. The announcement of his death in *Pravda* was signed by the whole of the Politburo (in alphabetical order after Chernenko) as well as by the top scientific establishment and many tributes have appeared in Soviet journals (particularly in *Priroda* of June 1984) as well as abroad. Great preparations had been on foot for his 90th birthday, but sadly it was instead a memorial meeting at the Institute, a moving occasion with many personal recollections by his close colleagues and friends.

Bibliography

The bibliography contains only references quoted in the text. Only the Sov. Phys. translations into English are quoted if available, but the year given is that of the Russian publication. The symbol R means the paper is in Russian. A complete list of Kapitza's publications has been given by Shoenberg (1985) and his collected papers are available in English in book form (ter Haar 1964, 1965, 1967, 1986).

Publications by Kapitza

(1) 1913   Cod-liver oil. *Argus No.* **10**, 76 (R).

(2) 1916   Electron inertia in molecular Ampère currents. *J. Russ. Phys. & Chem. Soc. Phys. Sect.* **48**, 297 (R).

(3) 1916   The preparation of Wollaston fibres. *J. Russ. Phys. & Chem. Soc. Phys. Sect.* **48**, 324 (R).

(4) 1919   A method of reflection from crystals. *Bull. Roentgenol. Radiol.* **1**, 33 (R).

(5) 1922   (with N.N. Semenov) On the possibility of an experimental determination of the magnetic moment of an atom. *J. Russ. Phys. & Chem. Soc. Phys. Sect.* **54**, 159.

(6) 1922   The loss of energy of an α-ray beam in its passage through matter. I. Passage through air and CO_2. *Proc. R. Soc. Lond.* A **102**, 48.

(7) 1923   On the theory of δ-radiations. *Philos. Mag.* **45**, 989.

(8) 1923   Some observations on α-particle tracks in a magnetic field. *Proc. Camb. Philos. Soc.* **21**, 51.

(9) 1924   A method of producing strong magnetic fields. *Proc. R. Soc. Lond.* A **105**, 691.

(10) 1924 α-ray tracks in a strong magnetic field. *Proc. R. Soc. Lond.* A **106**, 602.

(11) 1925 (with H.W.B. Skinner) The Zeeman effect in strong magnetic fields. *Proc. R. Soc. Lond.* A **109**, 224.

(12) 1927 Further developments of the method of obtaining strong magnetic fields. *Proc. R. Soc. Lond.* A **115**, 658.

(13) 1928 The study of the specific resistance of bismuth and its change in strong magnetic fields and some allied problems I. II & III. *Proc. R. Soc. Lond.* A **119**, 358, 387 and 401.

(14) 1929 The change of electrical conductivity in strong magnetic fields I & II. *Proc. R. Soc. Lond.* A **123**, 292 and 342.

(15) 1931 The study of the magnetic properties of matter in strong magnetic fields I & II. *Proc. R. Soc. Lond.* A **131**, 224 and 243.

(16) 1932 The study of the magnetic properties of matter in strong magnetic fields III, IV & V. *Proc. R. Soc. Lond.* A **135**, 537, 556 and 568.

(17) 1932 (with J.D. Cockcroft) Hydrogen liquefaction plant at the Royal Society Mond Laboratory. *Nature*, **129**, 224.

(18) 1933 (with P.A.M. Dirac) The reflection of electrons from standing light waves. *Proc. Camb. Philos. Soc.* **29**, 297.

(19) 1934 The liquefaction of helium by an adiabatic method. *Proc. R. Soc. Lond.* A **147**, 189.

(20) 1938 (with P.G. Strelkov and E. Laurmann) The Zeeman and Paschen–Back effects in strong magnetic fields. *Proc. R. Soc. Lond.* A **167**, 1.

(21) 1938 Viscosity of liquid helium below the λ-point. *Nature, Lond.* **141**, 74.

(22) 1939 Expansion turbine producing low temperatures applied to air liquefaction. *J. Phys. USSR*, **1**, 7.

(23) 1939 Influence of friction forces on the stability of high speed rotors. *J. Phys. USSR*, **1**, 29.

(24) 1941 The study of heat transfer in helium II. *J. Phys. USSR*, **4**, 181.

(25) 1941 Heat transfer and superfluidity of helium II. *J. Phys. USSR*, **5**, 59.

(26) 1947 Theoretical and empirical expressions for the heat transport in a two-dimensional turbulent flow. *C.R. Acad. Sci. URSS*, **55**, 591.

(27) 1948 Wave flow of thin layers of a viscous fluid I & II. *Zh. Eksp. & Teor. Fiz.* **18**, 3 and 19 (R).

(28) 1949 (with S.P. Kapitza) Wave flow of thin layers of a viscous fluid III. *Zh. Eksp. & Teor. Fiz.* **19**, 105 (R).

(29) 1949 On the problem of the formation of sea waves by the wind. *Dokl. Akad. Nauk* **64**, 513 (R).

(30) 1951 Dynamical stability of a pendulum with a vibrating point of support. *Zh. Eksp. & Teor. Fiz.* **21**, 588 (R).

(31) 1955 The hydrodynamic theory of lubrication in the presence of rolling. *Zh. Tekh. Fiz.* **25**, 747 (R).

(32) 1955 On the nature of ball lightning. *Dokl. Akad. Nauk*, **101**, 245 (R)

(33) 1962 Articles by Kapitza and colleagues in the series *High Power Electronics*,
     –1969 Vols. 1 to 6, Moscow, Academy of Sciences USSR (R).

(34) 1966 Recollections of Lord Rutherford. *Proc. R. Soc. Lond.* A **294**, 123.

(35) 1968 Ball lightning and radio emission from linear lightning. *Sov. Phys. Tech. Phys.* **13**, 1475 (1969).

(36)   1969   Free plasma-filament in high frequency field at high pressure. *Sov. Phys. JETP* **30,** 973 (1970).
(37)   1970   A thermonuclear reactor with a plasma filament freely floating in a high frequency field. *Sov. Phys. JETP* **31,** 199.
(38)   1977   Scientific and social approaches for the solution of global problems (Bernal Lecture). *Proc. R. Soc. Lond.* A **357,** 1.
(39)   1979   Plasma and the controlled thermonuclear reactions. *Rev. Mod. Phys.* **51,** 417.
(40)   1980   *Experiment, theory, practice,* ed. R.S. Cohen (D. Reidel Publ. Co., Dordrecht).

References to other authors

Allen, J.F., R. Peierls, and M.Z. Uddin, 1937, *Nature* **140,** 62.
Allen, J.F., and A.D. Misener, 1938, *Nature* **141,** 75.
Andronikashvili, E.L., 1980, *Recollections of liquid helium* (Ganat Leba, Tbilisi) (R).
Boag, J.W., and D. Shoenberg, 1988, *Notes Rec. R. Soc. Lond.* **42,** 205.
Born, M., 1978, *My life* (Taylor & Francis, London).
Calder, R., 1951, *Profiles of science* (Allen & Unwin, London).
Cotti, P., 1960. *Z. angew. Math. Phys.* **11,** 17.
Dirac, P.A.M., 1980, Dirac recalls Kapitza, *Physics Today,* May 1980, p. 17.
Dowson, D., 1979, *History of tribology.* (Longman, London).
Gamow, G., 1970, *My world line.* (Viking Press, New York).
Kedrov, F.B., 1984, *Kapitza: Life and Discoveries.* (Mir, Moscow).
Keesom, W.H. and A.P. Keesom, 1936, *Physica* **3,** 359.
Krylov, A.N., 1984, *My recollections.* (8th edition) (Sudostrogenie, Leningrad).
Landau, L.D., 1941, *J. Phys. USSR* **5,** 71.
Moon, P.B., 1978, *Proc. R. Soc. Lond.* A **360,** 303.
Oliphant, M.L., 1972, *Rutherford – Recollections of the Cambridge days.* (Elsevier, Amsterdam).
Pippard, A.B., 1979, *Philos. Trans. R. Soc. Lond.* A **291,** 569.
Rubinin, P.E., 1979, Kapitza's Nobel week. *Priroda* no. **6,** p. 122 (R).
Shalnikov, A., 1975, At the seminar, *Literaturnaya Gazeta,* 15 October (R).
Shoenberg, D., 1954, Mr. E. Laurmann (obituary), *Nature* **174,** 1129.
Shoenberg, D., 1978, Forty odd years in the cold, *Phys. Bull.* **29,** 16.
Shalnikov, A.I., 1975, At the seminar, *Literaturmaya Gazeta,* 15 October (R).
ter Haar, D., (ed.) 1964, 1965, 1967, 1985. *Collected papers of P.L. Kapitza,* (Pergamon Press, Oxford).
Timoshenko, S.P., 1968, *As I remember* (van Nostrand, Princeton, New Jersey).
Wall, T.F., 1926, *J.I.E.E.* **64,** 745.
Wilson, D., 1983, *Rutherford, simple genius* (Hodder & Stoughton, London).

47. Kapitza's fiancée, Nadezhda Chernosvitova, 1915.

2

Some early letters (1913 – 1920)

In the summer of 1913 Kapitza and his brother Leonid [Lyonya] went on a trip to the Kola Peninsula in the far north to make some anthropometric studies of the Pomors (Russians who colonised the far north from medieval times) and the Lopars (a native Finnish speaking tribe, similar to the Eskimos).

(A)   Gavrilovo, 17 June 1913

Dear Mother,

I am writing this by the light of the midnight sun – which I must confess I am heartily sick of. I wouldn't mind if it were to set for an hour or two so that I could sleep better. It doesn't give any warmth so that the stove has to be kept going in my room even though it's mid-June. Our work seemed to be going well but there was an unexpected difficulty – most of the Pomors go out to sea. Those that remain are busy with repairing their sails and preparing their nets for fishing so that in spite of all their good will their work can't be easily interrupted. So, only four people can be measured in a day.

The Pomors appreciate that we are students and scholars, so they trust us fully and treat us good naturedly. They are most hospitable and invite us to come in for tea. Lyonya and I let them think we are doctors and give them medical advice and sometimes we have to percuss those that are ill . . . We are also measuring some of those who work in the little factories for extracting cod-liver oil. We offer vodka to those willing to be measured as well as drinking tea and chatting with them. So far no one has refused to be measured and we have measured 35 in all.

Depending on the time and money available, we plan to return by way of Tromsö and Trondheim [in Norway]. I am taking a lot of photos. The Pomors love to be photographed and I can hardly cope

48. Leonid and Peter Kapitza by the yawl they used in the Kola Peninsula, 1913.

with their requests to be snapped – it's almost as if they preferred to be fed with photos rather than bread! They offer up to three rubles to be taken – pathetic! But I do it for nothing. We have also been to visit the Lopars and are shortly going again.

Please write and don't forget us.

Your loving son,
Peter

Kapitza went to stay with the Pearson family in Glasgow in June 1914 in order to learn English. In the following letters he usually doesn't distinguish between England and Scotland, and in discussing possible travel plans, the details sometimes seem inconsistent from one letter to another – a characteristic that continued throughout his life.

(A)   Glasgow, 24 June 1914

Dear Mother and Father,

I had three letters from you and am glad you haven't forgotten me. For three days I have been almost deaf and dumb, since I don't dare to speak in Russian, French or German and I can only stumble in English. I have been received in the most friendly fashion but life here is completely different . . . We meet in the morning at breakfast which consists of porridge and a meat or fish dish. You have to observe thousands of rules at table, but I find it very difficult to remember them. Each plate and knife has a strictly defined position and each dish has to be eaten in its own special way. For instance, I've already been ticked off a few times for not putting my knife and fork parallel when I finished eating, for holding my fork too near to the business end and so ad infinitum . .

It's very hard not to speak a word of Russian and even harder to learn table manners when you are twenty years old. However, I manage to get along. The Pearson sons, Charlie, George and a third whose name I've forgotten, are pleasant young people – in appearance at least (George borrowed five shillings from me which wasn't very nice). Pearson and his sons are all rather tall and thin, but jolly . . .

(A)   Heathcote, St. Fillans, Perthshire, 29 June 1914

I am at present in the hospitable home of Mrs Millar, because Mr Pearson is very ill. He has been on oxygen for the last few days and his pleurisy has developed into pneunomia, which is very bad. I couldn't stay any longer in their home so they sent me here. I agreed because it shouldn't be for more than a week and I had already met Mrs Millar and been invited to visit her. They have indeed received me most cordially, given me a splendid room and treat me *most kindly*. It seems she is very fond of Maria Nikolayevna and Volodya [friends of the Kapitzas with whom she stayed in Russia] and feels that by looking after me she is repaying the hospitality she received in Russia. I believe she even plans to invite me to stay for a longer period . . . time will tell! . . .

(A)   St. Fillans, Perthshire, 15 July 1914

Mr Pearson died the day before yesterday and I'm glad I wasn't with the family at the time as it would have been embarrassing both for them and for me. Mrs Millar is very kind to me and I feel much more at home here than at the Pearsons. On the very first day she invited me to be a guest rather than a boarder. Naturally I told her several times, as far as delicacy permitted, that it would be better if she took me as a boarder (i.e., for money) but she firmly refused to accept my offer. I should very much like you to write to her, Mother (in French or German), expressing your gratitude for her kindness.

Mrs Millar is very lively and much more intelligent than Mrs Pearson. The day I arrived she told me she would invite an English teacher from the village (for half a crown – about 1.25 rubles – per lesson) and he has indeed proved a very good teacher. I have a lesson every day and have had two so far. Mrs Millar herself sometimes reads with me. It's just a month since I last saw you.

(A)   St. Fillans, 17 July 1914

. . . The Millars have agreed to let me stay with them for the whole time I am in England and I am very happy with them. The only thing that worries me is that they won't accept any payment. But I suppose that when I get home I could send them some Russian craft things, which are very expensive here.

My life is quiet and even and they treat me very well, but I long terribly for home and I'm not sure if I'll be able to hold out till the end of August. I have rented a bicycle and I am thinking of buying it. It would cost only 80 rubles and if I use it enough, perhaps there would be no duty to pay. I ride a bike so much in St Petersburg that it would be well worth while to have my own, and indeed essential, since I live at Lesnoye [the location of the Polytechnical Institute on the outskirts of the city] . . .

(A)   St. Fillans, 5 August 1914

My soul aches not to be with you. Your telegram asking me to stay in England came this morning and I shall of course do as you ask and remain with the Millars. They are extremely kind and suggest I should stay for as long as a year if the war lasts that long. But I hope that I shan't have to stay as much as that since such a serious and complicated situation can't go on for very long. I am not getting any letters from you, for communications are almost completely interrupted, but all the same I'll write every three or four days and you should write too. Travel to Russia is completely stopped and there is no way of getting home – the consul and others I have asked have stated this quite categorically. However, I am well situated at the Millars so you shouldn't worry about me. Naturally, I shall use the time to go on learning the language. Don't send any money – if I should need any I'll send a telegram and you can send it by telegraph. The English telegraph works very reliably and still remains as a means of communication between us.

(A)   St. Fillans, 14 September 1914

. . . Mrs Millar is very kind and intelligent (as far as a woman can be!), very energetic and pleasant to talk with. But there are some things which I don't like – for instance, she exaggerates everything and I have been transformed from the son of a colonel to the son of a general! At least, that's how she introduces me to her friends. Naturally, it would be embarrassing for me to correct this kind of nonsense. But, by the way, I hope it won't be long before I do become the son of a general! . . . I am terribly longing for home and I have thought of the following plan. There are freighters which go from Glasgow to Archangel and cross the North Sea where it is fairly free of mines. The journey lasts about ten days and they often take two or three passengers. Let me know what you think – I won't take any action till I hear from you . . .

(A)   17 Beaumont Gate, Glasgow, 3 October 1914

. . . Mr Millar told me that if there were any serious danger, no steamship company would risk its ships, so please send me a telegram authorising me to make my own decision. This would stop me worrying and I would leave at the first opportunity, probably on 16 October.

We have now moved into the town and I am learning English manners. In this respect Mrs Millar is very kind and tells me quite straightforwardly if I am doing anything wrong. She teaches me what and how to eat. For dinner I have to put on a dinner jacket with stiff shirt – Mr Millar teaches me just what to wear. He himself is always dressed to the nines and demands the same of me. I have ordered a new suit from his tailor and he makes sure that everything fits properly. It seems I have the kind of figure that is difficult for the tailor to cope with. I had to try about a dozen collars before I found a suitable one. I don't protest – after all this too is a kind of education and I can always unlearn it if I want to. I'll try to manage all this on the funds I have!

The city and its bustle have cheered me up. The Russian successes [on the Eastern front] and the way they are appreciated here have made me very happy. I have heard it said that the Russian army is the only one in the world that has never been defeated! May God grant that this opinion should still hold after this war.

I have given you the details of my itinerary in a previous letter, but here it is again: Dundee – Göteborg and then through the north of the Bothnian Gulf. The second class fare is about 100 rubles and the steamer leaves every Friday.

(A)   Glasgow, 6 October 1914

. . . It's not possible to say anything definite about the safety of the route home, but I think the most dangerous bit is the North Sea. However, I shall be on a steamer carrying the Norwegian flag so there is no danger of being captured by the Germans and the mines are only in the southern part of the North Sea so that the northern part should be completely safe . . .

Life in the city is peaceful, but I've got tired of idleness. During the weeks that remain I shall try to see something of Glasgow and Edinburgh but I shan't manage to get to London – I haven't enough money and don't feel in the mood. I have got tired of foreign parts and my impressionability has got blunted, so that I don't even want to see London. I hope it will prove possible a year or two from now.

(A)   Glasgow, 14 October 1914

Tomorrow it will be just four months since I left Petrograd – not a
great interval, but how much has happened in these four months! . . .
I'm feeling fine, but am a bit bothered by the news from Galicia where
evidently things are not going well – though so far there are only
hints. But I feel sorry for the Belgians who are getting the worst of it.
Here in England everything is quiet. The women are involved in
charitable work while the men prepare to fight and donate money. But
in general it all seems peaceful and you'd hardly guess that the war
is in full swing. I went to the police today to register as a foreigner.
All I had to do was to give my name and occupation – they didn't
ask to see my passport or other evidence of identity. It's all very simple
here! I continue to study English but not very energetically – I have
got rather tired of it and think I have learned enough for the time being. It only remains not to forget what I've learnt and build on it
later . . .

(A)   Glasgow, 15 October 1914

I have definitely decided – as far as it's possible to decide anything
definitely at present – to set out on 26 October from Dundee by a
Swedish steamer bound for Göteborg, from where I'll go on to
Stockholm and then by way of Tornio to Petrograd. I will telegraph
three times – first just before I leave, then from Sweden and finally
from Finland to give the exact time of my arrival.

I got your telegram yesterday. Many thanks for your trouble and
for the £10 you have sent. I see that you are afraid that I might have
to cross the Baltic, but rest assured I shall be careful and avoid it. I
am now well informed about all the possible routes and I think they
are much the same as regards safety. The route I have chosen is the
cheapest. The 2nd class costs only 100 rubles, but I shall cross the
North Sea in 1st class since if there is any unpleasantness on the way,
1st class passengers are treated incomparably better. I will go 3rd class
in the trains unless circumstances indicate that 2nd class is
necessary . . . I am feeling fine and I think it will be pleasant to have
a [sea] voyage, though I still have to bother about packing my luggage
of which there is an awful lot.

(A)   Glasgow, about 17 October 1914

I shall set out from Dundee on Friday, 23 October at 2 pm travelling
1st class on the "Thorsten", a steamer under the Swedish flag. The
teacher who was recommended by Mrs Millar and gives me English
lessons in Glasgow is very pleasant and I greatly enjoy my lessons with
her. I write essays and read aloud to her, but "on principle" I don't
take dictation. I have learnt a bit of English grammar which is very
easy and there isn't much of it anyway. But I'm still far from perfec-
tion and make many mistakes in speech, especially in the use of the
past tense.

As always, everyone is very kind to me – indeed so kind as to make
my conscience twinge. Mr Millar decided that my dinner jacket
wasn't a good enough fit and gave instructions for it to be altered –
he himself dresses very well and demands the same from his son and
from me. There is real English fog outside today which makes me feel
the difference between the Russian and English climates. Not
everything is going well at the Front – our troops got badly beaten
up near Silesia and there is even a report (from Vienna) that Lublin
has been retaken but I think this is a lie. I have the impression that
our troops are beginning to take up new positions . . .

(A)   Glasgow, 24 October 1914

Well, everything has been packed up and I'm ready to go . . . Tomorrow I leave Dundee at 4 pm. The first few days will be the most
unpleasant, while crossing the North Sea. Mrs Millar will send a
telegram to say I have left. I feel fine though a bit tired after packing.
I had a look at Edinburgh and was very pleasantly impressed by this
fine city, which is very picturesquely situated . . .

P.S.   I tried to take out a life insurance for the crossing, but this
       didn't prove possible.

49. By his ambulance, 1915.

For a few months early in 1915, Kapitza worked with an ambulance unit on the Polish front, part of the time together with his brother Leonid.

(A)   Polish Front, 21 January 1915

I am sending this letter with one of our students, so I can write about things which would not be allowed by the censor. Big battles are going on to our right, close to Zhirardov – perhaps you remember the Zhirardov linen? Our ambulance goes to Zhirardov where the factories have been turned into hospitals to which the wounded are brought.

There have been big battles near Bopimost [this name is not readily decipherable and may be wrong] where the Germans broke through but have now been more or less contained. There are very many wounded and the ambulance units to our right have been very busy, though at present we have not had to do any ambulance work here. I am very glad to have been taken off the job of supervising the motor transport. It is possible that in two or three weeks time Reinwald will send me to Petrograd to get various spare parts for the vehicles. In the meantime, I expect soon to visit Warsaw.

As always, there is a lot to do [repairs, etc.]. Although I am working conscientiously and my work is appreciated, I feel very critical about many organisational questions and that is the most unpleasant. We sleep in a small room holding nine or ten people; it is cold, especially at night but the sleeping bags save us. We share the room with the drivers,* with whom I am on very good terms – much better than with the command. I am keeping well and cheerful. Apart from the cannonade, which goes on day and night, I have seen nothing of the battles so far.

(A)   Zhirardov, 8 February 1915

Dear Mother, Father and Zhenya [a niece of Kapitza's mother]

I drove to Warsaw recently with N.N. Tuturin in his car and we serviced it together. I am still sitting in Zhirardov where I am trying to bring some order into the organisation of the motor transport. In my spare time I play chess. I feel well and spend most of my time in the open air. I go about mostly in a leather jacket and trousers – essential when you have to mess about with motors. I gave my old clothes to one of the mechanics.

Lyonya is preparing to leave the unit, as indeed many are doing and I suppose they are right because there isn't any work to be done. Our unit is badly organised, so that even if there should be some work to do, the unit is hardly likely to function properly. I, too, should like to leave and enrol in either an aviation or a motorised group. There, at least, the duties would be more clearly specified and more productive. Here, I don't get any feeling of satisfaction . . .

P.S.   In my next letter I shall enclose half a dozen live (!) lice, so you should open the envelope carefully or they will crawl over you.

(A)   Zhirardov, 8 March 1915

. . . As usual, disorder and confusion reign in the unit. We shall probably be transferred to Lomzia. I am busy repairing motors. One of our cars collided with a lorry and was broken to bits. It turned over

---

* Kapitza would have ranked as an officer in the unit, so some degree of fraternisation is implied by his sharing a room with the drivers (i.e. "other ranks").

and Varlamov, the driver, was hit so hard that he had to be taken un-
conscious to hospital. He is back with us now but it will be a month
or two before he recovers from his injuries. The body of his vehicle
was completely smashed and I am reconstructing it to my own design.
It will accommodate four lightly and three heavily wounded men, in
spite of being shorter than before, but it's difficult to say if it will be
as good as before. There is very little equipment available for the
reconstruction. The other machine had its chassis broken and also
needs repair, so I am very busy . . . The food is very bad just now
– worse than before.

After returning to Petrograd to resume his studies at the electromechanical
faculty of the Polytechnical Institute, Kapitza became engaged to Nadezhda
Kyrillovna Chernosvitova a few days before she left for China on 23 March
1916, in charge of the young children of her brother who worked in the
Russo-Asiatic Bank in Shanghai. From then until 22 May (shortly before he
left to join her in China) Kapitza wrote to her nearly every day. Some extracts from these letters, and also a few later ones after their marriage, now
follow.

(F)   Petrograd, 25 March 1916

My dear Nadya,

I'm in the lab but I can't work calmly for thinking about you, so I
decided to write. Well, how was your journey? Let me know at once
and tell me how you are going to spend the summer and how you have
settled down. I have thought of a little plan for the summer that I'd
like to discuss with you. It's this: I see nothing to prevent me going
to Shanghai and returning with you. I got very tired during the winter
and need to have a rest one way or another. So can you see any reason
why I shouldn't go to China, spend some time there seeing the sights
and then bring you home to the general rejoicing of our parents?
Nadya, I don't see anything impossible in this plan.
    Let's look at the project from the practical side.
    (1) When should I go? I think I should leave at the end of June or
early in July and reach you at the end of July. We could then be there
together for two or three weeks and come back home at the end of

August, just in time for the start of term. During June I'll get some practical experience somewhere.

(2) Where shall I get the necessary money? There is a simple solution to this. I have succeeded in saving about 600 rubles – intended for buying a yacht and a motorbike – and this should be enough for the journey, travelling third class of course. There is only one possible snag: since I am liable for military service I may not be able to get a passport for foreign travel – I'll find out in a few days.

Anyway, you must let me know how much it costs to live in Shanghai and also what I must take with me. Well, that's my plan – how do you like it? I simply can't bear to think that it will be so long before I see you. During these last two days I have been reading and walking but doing very little work. I'm impatiently waiting for a letter from you and I do hope that you'll use [the familiar] "thou". I want to start using "thou" myself but somehow can't manage it. I'm either afraid or else I feel that it would come out awkwardly, but it's you who should break the ice since you suggested it.

(F)   Petrograd, 28 March 1916

It's just a week ago today since we last met. To me it has seemed a very long week: how did it seem to you? . . . You will probably get this letter after the one in which I suggested making a trip to see you. Reply at once. I have already considered all the obstacles to such a trip and am now quite clear how to carry it through. But remember that what I'm now writing about my trip to Shanghai is, of course, only a provisional plan and I should very much like to know what changes and improvements you may want to suggest . . . We mustn't keep it secret from our parents and so I ask your permission to go directly to your parents and tell them all about the plan. I have already spoken to my mother about the trip and she is completely in favour of it and has even promised to help me carry it out.

But to return to the present – I shall spend these next two months working, mainly on my immediate duties at the Institute. Yesterday and the day before I worked with Abram Fedorovich Joffé, on experiments which were not very successful. He is now off to the Crimea and I must prepare for the exam. I spent today working in the drawing office. I've no inclination to go visiting or to the theatre. Somehow I don't feel like seeing people . . .

(F)   Petrograd, 4 April 1916

I'm still not very happy about my studies . . . and for the next couple
of months before the summer I shall probably concentrate entirely on
studying, so my life will not be very interesting. As I look into the
books I shall have to study, I begin to wonder whether it might not
be better for me to abandon science and technology and escape from
this "culture". But this is only an impossible dream – I can't do it
for a very simple reason. It's just that I know for certain that I'm less
well suited for any other kind of life than for just this "culture". You
know, Nadya, I'm probably not ambitious and up to now I've never
courted fame or anything of that kind. Such hints of success as I have
had, have given me no lasting satisfaction. I work mostly because I enjoy working and if at any time the work no longer satisfied me, I'd
throw it all up and go off somewhere. I think I'd go into the country
and the idea of looking after a small estate seems very attractive (don't
laugh!). You may find this surprising, but that's how I feel.

Nadya, you must write and tell me what you want from life. The
answer I should like to hear is a family. You know, Nadya, I have
never made any plan for my life, but I couldn't think of myself without
a family. I divide myself into two parts – family on the one hand, the
outside world on the other, and I don't know which is more important.

(F)   Petrograd, 11 April 1916

. . . I wrote to my Professor [Joffé] about my call-up and I'm very
curious to know what he thinks about it. Personally I am completely
resigned to the idea that I may have to interrupt my work and get side-
tracked for a few years – heaven knows how long the war will last.
If I can make the trip to Shanghai and if you* won't forget me,
everything will turn out all right. Don't you agree?

---

* Actually, he is now using the more intimate "thou" throughout.

(F)   Petrograd, 12 April 1916

I've an odd story to tell you. I looked at the call-up list today . . . and
I wasn't on it. I was astonished and can't understand how this came
about. Either there is some mistake or else I have miscalculated which
class I am supposed to be in – I don't know! Perhaps the Professor
has done this for me. I thought I'd be considered as second year since
I hadn't yet taken the physics exam (out of silly perfectionism, though
I knew the subject quite well). When I mentioned it to him recently
he indicated that he needn't examine me and would pass me on the
basis of the work I had done . . .

13 April. I still can't believe that I'm not going to be called up –
I had already made all my preparations, and I'd like to know why I'm
not on the list. In any case my conscience is clear, for I made no request and I did nothing to avoid call-up. I consider one should never
do such a thing . . .

14 April. I was at the Institute again and it is now almost certain
that I won't be called up since, quite against my expectations, I have
been considered as in the third year. This came about because I
miscalculated my credits and erred on the low side. So, Nadya, in all
probability I shall now be able to come to you. That will be wonderful . . .

(F)   Petrograd, 17 April 1916

. . . I am looking forward with keen anticipation to my journey, to
crossing the whole of Siberia and then the sea. I'll be able to have a
complete rest which I badly need. You know, Nadya, although I've
done very little this year as regards work and study I've had to endure
a great deal all the same and this has evidently taken a lot out of me
– I hope next year will be more peaceful. Well, we'll think about next
year and what to do with it, together.

Nevertheless, I have achieved something this year. I've not only
made a decision about my personal life but it has also become clear
that my field of activity is going to be science. You know, Nadya, this
was really something very important for me. I was always vacillating
and doubting but now everything points to my future being in science,
though I'm still apprehensive about my powers and abilities. To

achieve something in this field you need to have a brain in your head and the devil alone knows what's in my head. I feel very sceptical about such successes as I've achieved so far – after all, they might be just due to good luck. But I think if there's a chance to try my strength in this field, I should have a go at it. Don't you agree?

18 April. I'm now awaiting the arrival of Professor Joffé because the distribution of my time depends on him. I had first of all to finish the physics work I had started. There was little left to be done but if we were to achieve what we hoped for, this would be very good, so good that I'm even afraid to write about it. If successful our work could become classical. Abram Fedorovich says that if we discover this phenomenon*, which is predicted by the modern theory of electricity, it would precipitate big changes in the whole of contemporary physics, so he is constantly urging me to continue these experiments. However, I don't believe I can get anywhere during this month – that would seem to be just too good to be true. So I'm waiting for Abram Fedorovich's advice on my work. However, I still hope to finish at the Institute as soon as possible, in order to become independent and work on anything I please. I know that Abram Fedorovich will probably be annoyed [at my dropping research for the time being], but what can I do? To me personally, it would be pleasanter working on a pet project of my own. If I am not called up for war service, I should now be able to finish at the Institute in less than two years. Well, I can wait that long before starting on my own work in the lab.

(F)   Petrograd, 25 April 1916

I'm feeling unsettled though I expect no one notices since I try to show the same outward face as usual to those around me. I'm worried mostly about you . . . but it's not worth saying much about my worries – there's nothing that can be done about them. I have only two requests to make: write more often and don't get involved in any kind of entanglement – but I think your parents and your brother will protect

---

* This probably refers to observation of the angular momentum associated with magnetization, often known as the Einstein – de Haas effect. Kapitza reviewed the question in his first scientific publication (2).

you from that. They will do it far better than I, mainly because I'm afraid of restricting your freedom by my pleas.

As for me, I am doing all I can to get out to you in Shanghai or Japan, in whichever you may be. The one thing that could prevent me would be a call-up of students in the third year. So the question can't be finally resolved before the end of May. It is very possible that I shall have to enter the military school in the middle of September. That would guarantee me freedom during the summer. In any case, I reckon I have a 50% chance of joining you in Shanghai and a 95% chance of meeting you in Vladivostok. Incidentally, technicians like us will be taken into the engineer corps.

To get a good idea of my present predicament, Nadya, you should remember the Krylov fable in which a carriage was hitched to a bird, a fish and a crab, all pulling in different directions. I am now being pulled in three directions: by my duty as a citizen in time of war, by my intellectual urge toward science, and by my heart to someone whom I love very much in China. It need hardly be said that I hope my heart will win . . .

Well, that's enough about this topic, though it's my main worry. I confess that it's the first time I've been in such a complicated situation, but I'm holding fast to the view that has always upheld me: look at everything in the simplest possible way. The solution will then also be simple. For instance, what must I do now? – I think I must wait and keep busy. What must I aim at? – reduce the geometrical distance between myself and Nadya on the surface of the globe. How? – Time will tell. In this way everything becomes clear and tranquil. And so, we must patiently await the end of May for clarification. That was what was worrying me, Nadya, and that's how I solve problems . . .

(F)   Petrograd, 28 April 1916

May is drawing near and with it the end of my studies. Then comes summer, but this doesn't excite me . . . I think a lot about the War and about the present situation. I could write much of interest if it weren't for the censorship. I shall say only that my optimism is fading day by day . . . Still, I can't describe my mood as gloomy – on the contrary, I follow events with interest and live very intensively . . .

My present activities are as follows. I wake between six and seven and read, usually some textbook, until nine, when our servant gets up and serves coffee. So about this time I get up and have coffee. The morning post arrives and for two weeks I have been waiting in vain for a letter from you. Then I go to the Institute or work at home. Dinner is at two. After dinner I feel justified in doing some light reading or going for a walk. Tea is at five, after which I again read or go walking. At ten there is supper and at eleven I enter the land of dreams.

For reading I have Rousseau's "Confessions" (as you know, I have read very little all my life). I have never read it before and in places I find it boring, but then for a few pages really delightful. Rousseau was a strange person, basically incomprehensible and foreign to me. Besides, I find him antipathetic, probably because of his unbounded egoism. He never did anything kind to anyone. He is incapable of understanding the joy that can be found in self-sacrifice. I think that only a man who is able and willing to make sacrifices can genuinely fall in love . . .

(F)   Petrograd, 29 April 1916

. . . Abram Fedorovich arrived four days ago but I couldn't catch him till today when I had a very long talk with him. He said a lot of nice things, so I'll give you a full account. In the first place he had been to the Dean of my department . . . to enquire about me. He learned that I am entered for the third year and that is why I wasn't called up. He also learned what exams I have to take in order to qualify for the next year and advised me strongly to take all these exams now. There is little time left and I don't know if I'll succeed in doing this, but perhaps I'll try. Then he asked for my lecture notebook and said that he would now give me a pass in physics. I began to protest because I was embarrassed to get a pass for one of the basic courses without sitting the exam. But he took my book and wrote the pass in it. I then thanked him of course and said that he probably hadn't ever examined anyone so quickly before. He replied: "On the contrary, I have never examined anyone for so long as I have you". This pleased me a lot.

We talked about many things and he told me of his plans for next year, about his students, what tasks he would allot to them, how to set up the projects, and so on. He sees everything so clearly and well. I'm very glad to have such a good teacher and guide. He is a successful mixture of good teacher, distinguished scientist and sensitive person. He has just come from the Crimea . . . where they rent a small house, and he will return there about 20 May. For some time they have had no help in the house and Abram Fedorovich likes to tell that his wife acts as the cook and he is the maid of all work. He has to tidy up the garden, milk the cow and so on, so that scientific work is impossible. We also spoke about my work and were in complete agreement on the questions that have arisen, which was nice . . .

Next year we shall organize a small working group in the Institute consisting of six or seven people who will work on physics in friendly collaboration. In Russia there is very little cooperative work but now, especially in science, cooperative work is almost essential. I hope we shall be successful.

(F)   Petrograd, 20 May 1916

Today I passed my last exam for the third course and so completed my plan*. Naturally I am very glad and I can now take a holiday with an easy conscience. If nothing goes wrong I shall leave on the 26th at 10.30 a.m. by the express train. So my dear Nadya, we'll overcome 10 000 versts!

Kapitza actually left a day or two later and travelled to Shanghai via Dairen. On 23 June the Polytechnical Institute received a telegram from the Russian consul in Nagasaki saying that the student Peter Kapitza required permission to marry and asking for confirmation of his bachelor status. The Director of the Institute replied rather unhelpfully that they had no information regarding Kapitza's marital status, so it was only on 28 July, following their return to Petrograd in mid July that the marriage took place.

In 1918, food was very scarce in Petrograd but the situation was rather better in the country. The next letter was written just after Kapitza had returned to town after visiting his wife and son, then living in a rather far-off village.

---

* His course was officially completed on 1 September 1919 with approval of his diploma thesis on "The mechanism of ferromagnetic phenomena" under Joffé's supervision.

(A)   Petrograd, 14 October 1918

My dear Nadenka,

I got back to "Peter" [Petrograd] on Friday and learnt that Lyonka [his brother] and Natasha have got engaged, so our speculations have been justified. I found the family very exhausted after their move to the new apartment, though it's a very fine one and you and I will have very good rooms. The following day I went to the Polytechnical Institute and met Abram Fedorovich, who was glad to see me, as was I to see him, of course, and he immediately began to tell me about his plans. He has the job of organizing the new Röntgenological Scientific Technical Institute of which he will be the head with eight directors of research and 12 scientific workers under him. The directors will all be professors whom I know and the scientific workers our young people, with a salary of 1100 rubles a month. My job will be to continue my previous work in the lab, with the title of engineer. So, as you can see, the conditions are very attractive – 1100 rubles and doing what I like best. Moreover, I can take on some private work, which will bring in another 500 – 700 rubles.

In spite of all the attractiveness I didn't immediately agree. You would also have to move to Petrograd and earn some money. I have already been to see Olga Konradovna [Nedzvetskaya, a close friend of Nadya] about this and there is a possibility of fixing you up at the Information Bureau of one of the Commissariats. The work would be close to us, daily from 10 to 4 and the pay 700 rubles [monthly]. The work would be to read English, German and French journals and to make summaries. As you can see this would be a very suitable job for you. We would employ a daily nurse. So if we can manage to bring a few things away from Priyutnoe [in the country] for the first months we should be able to have a tolerable existence.

I haven't been too badly off for food lately. We get 1 ½ bottles of milk from the old man and the hospital promises to provide a small ration of milk for Nimka [their year-old son]. I think we should now bring over some of our food reserves and I will come over for them with my father. He is extremely nervous. He can catch up on his eating for a week and then we'll take away what we can. Nadezhda Aleksandrovna [his mother-in-law] may perhaps come with us too. I was over at their place and they were very amiable (they fed me with

50. With his year-old son Ieronim, 1918.

dinner!) . . . Today I am going to Tsarskoye Selo to arrange things about my exemption from military service. I am extremely busy these days so we shan't come before Monday. How are you getting on without me? Well, so long, I kiss you both tenderly.

Your loving tomcat,
Petya

At the end of 1919, when Nadezhda Kapitza was expecting a second child, conditions in Petrograd were even worse than a year previously and epidemics – particularly the Spanish influenza – were rife. Kapitza's father died on 9 December and the Kapitzas' two-year old son Ieronim died of scarlet fever a few days later on 13 December. Peter Kapitza himself was infected by his son's scarlet fever and his illness was complicated by nephritis and by his grief. Nadezhda was just recovering from 'flu when she entered the maternity ward of the Military Medical Academy at the start of 1920 to have her confinement. The following two letters – the last from Peter to Nadezhda – were written while he was gradually recovering, but had to be separated from her because of the risk of infection. The baby, a daughter, christened Nadezhda, was born on 6 January 1920 but her mother, weakened by her illness, died three days later and the baby died three hours after her mother.

(A)   Petrograd, 2 January 1920

My dear little wife,

I think of you constantly, night and day. I hope you didn't catch cold on the way. How was the journey? How are they looking after you? What did the doctors say? What is the food like? It's so trying not to know what's going on. The main thing is that you should write about everything – if it's not too difficult – or else your tomcat will sleep badly and not know what to do with himself.

As for me, my temperature is as before 37.5°C [just above normal], but I think that the glandular swelling has definitely gone down, the painful feeling has almost gone and my throat is much better. I get on well with Nadezhda Aleksandrovna [who was looking after him] – she tries very hard and falls in with my wishes. We talked a lot [last night] and I couldn't get to sleep for a long time. We talked mostly about her favourite topics – how she could go about finding a job, where to go, how to arrange things with the apartment, etc., etc. So life is quite normal, no special happenings and no visitors. I am waiting impatiently for Lyonka's [his brother's] visit to hear how you are. So we all think only of you and the little kitten to come. May God keep you both healthy.

Your very loving little husband,
Petya

P.S.   Let me know if you need anything. We shall send milk as often as possible.

(A)   Petrograd, 5 January 1920

My dear and infinitely beloved Nadya,

I am thinking only of you – my thoughts are constantly with you. I long to be with you and to kiss you. You must try to get well quickly. My mother says that you're not eating anything but that's not good – you should make an effort to eat, so as to keep up your strength. I should be well again in about a week and should be able to come to you. – I long so much to find you in good shape. . . I am definitely getting better, my temperature is down to 37.2°C, the glandular swelling has gone, I am hardly coughing and I can smoke as much as I like. Nephritis is not a particularly painful illness – it's just unpleasant to have your face swell up and to have a headache, but both are disappearing now . . .

I am getting on well with your mother who does everything necessary. The most important is that she is able to get hold of milk, both for you and for me. Make yourself drink as much milk as possible since it's very necessary for your health . . .

Well, so long, I kiss my infinitely dear pussy-cat tenderly and wish you speedy recovery and a successful delivery,

Your infinitely loving Tomcat

51. Olga Kapitza, 1926.

3

Letters to his mother (1921 – 1927)

Kapitza was sent abroad in 1921 as a member of a Soviet commission to renew scientific relations with the West after the Revolution and the Civil War and eventually spent 13 years in Cambridge. This chapter contains extracts from the frequent letters he sent to his mother in Petrograd (mostly also intended to be read by his brother Leonid (Lyonya) and other relatives), starting with his sojourn in Estonia, while waiting for visas to go further, and charting the progress of his scientific career in Cambridge and his relations with Rutherford. This series of extracts finishes in 1927, for after his marriage to his second wife, she replaced his mother to some extent as his confidante and consequently his later letters home are of less general interest. A few extracts from the early letters sent by his mother are interspersed with his letters to her, to illustrate the great affection between mother and son and also to give a hint of the hardships of life in Petrograd in the aftermath of the Civil War. It is indeed rather remarkable that this whole correspondence survived the hardships of the early 1920's and the even worse hardships of the blockade of Leningrad in the Second World War, since such archives often ended up being burnt to keep people from freezing.

Some themes, such as his introspective philosophising, his loneliness and unhappiness away from home and reproaches to his mother for not sufficiently appreciating his difficulties, recur in many of the letters, especially the earlier ones, and only a few typical examples are included. Opening and closing endearments are usually omitted, as are many details of purely family interest. A few letters to and from others during the Cambridge period are also included, but his correspondence with Rutherford is collected in ch. 5.

(C)   Reval*, 3 April 1921

. . . I got here safely after a good journey, with every comfort, and I'm staying in a comfortable room with friends of E.Ya. Laurmann – it's quite difficult to find accommodation.
Tomorrow I shall send you a parcel through the Russian mission and another through the Estonian mission. My departure for Berlin is being held up and goodness knows how long I'll have to sit here in Reval. Next Thursday I have to lecture on ''The nature of electricity'' – fancy that! . . .

(C)   Reval, 6 April 1921

. . . I have been busy trying to get visas but progress is slow. I haven't yet managed to get in touch with Abram Fedorovich (Joffé). I sent him a letter and a telegram but have had no reply yet. I haven't been in too good a mood but am feeling better now. If only I could get the visa I'd soon be in a much better frame of mind. The main thing is that I'm longing to get down to work. Meanwhile I keep intending to write and there is plenty to write about but somehow I don't feel like writing letters . . .

(C)   Reval, 21 April 1921

I have already been waiting three weeks for a visa and the situation is just as indefinite as ever. This is really intolerable . . . Tomorrow I shall send you a parcel [with someone travelling to Petrograd] . . . I am doing everything possible to keep you supplied so that you shouldn't go hungry. The food situation is very good here . . .

    Yesterday I read H.G. Wells' book ''Russia in the shadows'', in which he describes his stay in Russia. The style is very lively and many of his ideas are quite sound. This Englishman has achieved a much better understanding of what is going on at home than have the Russian emigrés. They obviously have a difficult existence, not so much materially as morally. Their animosity towards the Allies on the

---

* The old name for Tallin.

one hand and their inability to be reconciled to what is happening in Russia on the other hand puts them in a totally isolated position . . .

(C)   Reval, 13 May 1921

Today my affairs took a significant turn for the better. I got permission to enter England* and I shall probably leave for London either tomorrow or the 18th . . .

(C, E)   London, 26 May 1921

I'm sitting in the hotel lounge looking out at the Thames which is shrouded in fog and stinks pretty strongly even here in the middle of the city. I have already travelled on the underground and on trams and buses. Life is seething here and the traffic is heavier than in peacetime Petrograd. As you can imagine, all this makes my head swim. Trains that hardly stop at the stations and buses that go every which way. I bought a map of London and yesterday I began to study it, not without results, for I made several journeys on my own and all went well.

But, strangely, none of these new experiences and none of the comforts I have, give me any joy. I don't even want to visit museums, though this would be the right opportunity before becoming absorbed in my official duties. Later I shan't have time to look at anything. I don't feel like buying myself clothes. I found it difficult to part with my cap and as for my suit – it is, after all, from Petrograd and I don't want to be parted from it either. I'll wait a little before I make myself buy a suit.

I'm alone here and that's the worst of it. Of course I haven't lost my energy or my drive but the joy has gone out of life and left only sorrow. How good it would be, my dear, to visit the British Museum with you! And how often my Nadya [his late wife] told me about London and how we longed to be here together!

When I look back at all I have lived through I am astonished and

---

* In an earlier letter Kapitza mentioned that L.B. Krasin, the Soviet representative in London, whom he knew personally, had undertaken to help in getting a visa.

awestruck; have I really been able to bear all this? Sometimes I even think that I'm not a man at all but some sort of machine that continues to function in spite of everything. This is a sad letter, mainly because Nadya loved London so much and I think of her all the time.

(C, D, E)   London, 2 June 1921

Well, I've already been a week in London and, thank heavens, have got settled in – not in a hotel, because of the noise, but in a small apartment with service and board. I have fitted myself out and now look quite respectable as the following incidents show. In my former get-up, when I asked a "bobby" (as they nickname the police here) the way, he took me by the arm and told me where to go. Now they no longer take me by the arm but address me as "Sir". Englishmen, as always, are very strict about dress. When I looked for lodgings wearing a cloth cap there was never a vacancy. But after putting on a good suit, they showed me two rooms in the very same house where only the previous evening they had said there was no vacancy, and I took one of them.

Yesterday I was at King's College and met Professor O.W. Richardson, who is a Fellow of the Royal Society and a scientist of European standing. I got so carried away that I talked with him for an hour and a half. He is a clever chap but I'm afraid I didn't show sufficient respect and rather impertinently got into an argument. Next time I shall be more circumspect. It was only afterwards that I noticed the assistant of this famous man staring in amazement at me. But, anyway, Professor Richardson was very kind and gave me the information I needed; we agreed to meet again tomorrow. He looks quite young . . .

(C, D)   London, 10 June 1921

. . . The purchasing work (of equipment, etc.) is all falling on my shoulders and doing it under Abram Fedorovich is very difficult. He gives me no freedom of initiative and he frowns on whatever I do, though he gives no precise instructions. Our purchases in England should of course fit in with those made in Germany but I don't know what he bought in Berlin for he has no receipts – he says there were so many that he couldn't bring them?! . . . Krylov [whose daughter became Kapitza's second wife in 1927] and Rozhdestvensky should arrive soon and I shall be glad to see them. I am especially attracted to Krylov who has quite a different style [from Joffé] . . .

Tomorrow I am going to a lecture by Einstein on the theory of relativity at King's College. He is highly regarded here and called a second Newton. Now I am off to see a picture gallery. By the way, I have bought a camera and am taking a lot of pictures. I shall try to send you some and hope you will like them . . .

(E)   London, 27 June 1921

. . . Recently I have got better acquainted with Aleksei Nikolaevich Krylov and the more I see of him the better I like him – he is a very good raconteur.

3 July. As you can see, I have interrupted this letter for a few days. This was because I travelled from London to Manchester with Abram Fedorovich in order to visit the famous Professor Bragg and we had some adventures on the way. Just now, because of the coal strike, which is in fact already over, the trains have been running very irregularly, so that when we got to the station we found that the train we were intending to take would go only as far as Derby. We worked out that if we spent a night at Derby and went on the next morning we should still be able to keep our appointment with Professor Bragg on time. But when we arrived in Derby it turned out that an International Agricultural Exhibition was being held there and all the hotels were full. So we went to the police to enquire about accommodation for the night and were given several addresses to try. Eventually, after considerable effort, we found ourselves two small rooms, rather dirty and lit only by gas mantles, since electricity was not laid on. Just as we were getting ready to go to bed the landlady came to tell us that two other young men were looking for a night's lodging and she asked if we would mind sharing a bed in one of the rooms so that she could put them up in the other. However, I turned down this suggestion very firmly and eventually we spent the night more or less comfortably.

The following morning we reached Manchester and were very well received and invited to lunch. Bragg himself is very pleasant, and although at 32 he is still very young, he is very knowledgeable in his field and told us a lot of interesting things. We also met Professor Sedgfield, who speaks Russian well (he spent six years in Russia) and wants to organize an Anglo-Russian Institute. We had tea with him and he discussed his project with Abram Fedorovich. Next week we are invited to dinner by the writer H.G. Wells to meet various English academics. We also plan to go to Cambridge to see Professor Rutherford . . .

(B, D, E, e)   London, 13 July 1921

. . . I have been very busy and that's why I haven't written. I went to a party at the Wells', and also had tea at the Wright's, where I was introduced to Bernard Shaw, Soddy, Lord Haldane and others. You may well be impressed, but I shan't dwell on all this as there is something more important I must tell you about, which is that I shall probably stay in England for the winter, living in Cambridge and working in Professor Rutherford's laboratory. He gave his approval when we visited him yesterday and Krasin is agreeable to my staying here. I don't know whether to be glad or not. My heart is already aching to be with you, my dear ones. How will you manage without me? But, on the other hand, it wouldn't be possible to work in Petrograd this winter and for me all that is left in life is work – and, of course, all of you, my dear ones.

I shall do all I can to help you, but if I don't take advantage of this lucky combination of circumstances I shall probably have to wait a long time and much time has already been lost. I am afraid for myself too, that I shall be terribly lonely among the English. I shall go up to Cambridge in two weeks' time – and then to work. I am in a hurry to finish the purchases here . . .

(B, D, E, e)   London, 15 July 1921

As always, I was glad and excited to get your letter yesterday but, above all, I am worried about your health. You must rest. I am worried, too, about the winter. How will you manage without me?

. . . My dear, don't be too lonely without me. I shall, of course, be very sad here without you but I shall just have to get down to work. Youth passes by in a flash and never returns. Just now I am rather apprehensive about how I shall get down to work in Cambridge and how I shall get on with Rutherford, bearing in mind my poor English and my irreverent manners. I go to him on 21st July . . .

(B, C, D, E, e)   London, 24 July 1921

Today is your "name day" so I send my congratulations and all best wishes. I have been very busy these last few days, having moved from London to Cambridge and started work in the laboratory. On the 22nd and 23rd I worked very hard, but today I came to London because Cambridge is dull and, moreover, I have some business tomorrow connected with the purchases.

Little is clear as yet about my work in Cambridge. I am learning how to measure radioactivity and simply taking the laboratory training course but I don't know what I shall do after that. I can't make any plans, or even guess anything about the future. Time will tell. I am very worried about your affairs and how you will get through the winter without me. I shall do my best to send help if possible . . .

(A) From his mother   Lesnoy [near Leningrad], 25 July 1921

My dear Petya,

Yesterday I got your letter [of 13 July] in which you say that you are going to spend the winter in England working with Professor R[utherford]. I imagine you will appreciate how I feel about this. This is a great piece of good luck for you and I am infinitely happy that you will be living in a country in which not only can you work productively, but under the supervision of such a great scientist as R. Of course you will be lonely away from your near ones but you must come to terms with this and put up with it this year. I won't conceal that for the first few moments I found it very difficult to adjust myself to the thought that I shan't be seeing you for so long, but this was only a momentary weakness and now I rejoice for you more and more. But I must ask you for one thing – do write more about your feelings, your work and your living conditions, since all the trivial details of your future life will be very interesting for me . . .

As for ourselves, don't worry – we shall get by and struggle with the present conditions of life. Probably we shall move to Kameno'ostrovski Street [now Kirov Street] in September and instal heating stoves and build up a stock of firewood. In a word we shan't give in – but after that only God knows. . . .

(B, D, e)   Cambridge, 29 July 1921

I received your letters . . . in which you, and Lyonka too, reproached me for writing too seldom. I have indeed written little recently, but not less than one letter a week. The reason is that I have been very busy working in the laboratory from morning to evening, getting home only at 6 and then having to write and calculate. I am so tired and weary that I think only of getting to bed! When I have got on top of things I shall write more often.

My dear, if I stay here over the winter it is only in order to work. You know quite well that, apart from you and my work, I have nothing left in the world. You are constantly in my thoughts and I am doing all I can for you. Your letters move me and my heart beats more strongly when I get them.I really don't deserve all your reproaches. I hope I shall soon be able to send you a few things from here. I have already sent you spectacles and a lorgnette. Pince-nez wouldn't suit you and would only disfigure your nose. The spectacles I sent are in the latest fashion and should be convenient for reading. I can't send spectacles to other people – they cost too much and sending them is complicated.

Don't forget, my dear, that I am alone here among the English, hearing not a word of Russian all day long, with no one to whom I can unburden my heart or joke and argue with. Only the possibility of working keeps me here and the opportunities for work are good, although as yet I am not doing independent work but only attending the practical class. I now have a better relationship with the other research workers, although my inadequate knowledge of English prevents me from expressing my thoughts clearly . . . But I am bad at explaining my ideas even in Russian . . . Write as often as you can

– your letters are half my existence. I am terribly afraid of being lonely here . . .

(B, D, e)   Cambridge, 6 August 1921

. . . I've already been more than two weeks working in the Cavendish and now comes the riskiest moment – choosing a topic for my research. Not an easy matter. At such a moment I don't like to say a lot so it's difficult for me to write anything definite about my situation and my work. Once I've settled something I'll write to you. . . .

I'm about to set out for London with Müller, who also works in the Cavendish. I have to see Aleksei Nikolayevich Krylov and Anna Bogdanova Feringer before they leave. We are going by motorbike and if all goes well it should take about two and a half hours. Müller is a Swiss from Geneva, 32 years old, and the first person in the lab with whom I have established friendly relations so far. He is a cultivated chap, very pleasant, almost as lively as a Frenchman and as talkative as a Russian. He is working on X-rays. He is very skilled and seems to have a good head. The other people in the lab are quite friendly, but it is not easy to get to know the English at all intimately . . .

(B, C, D, E)   Cambridge, 12 August 1921

Your letters now arrive fairly regularly and this gives me great pleasure. I am worried about how you will get on when Lyonya goes to the North, and my heart aches for you. I have now been in the lab for three weeks and made a little progress but it's upsetting that the lab shuts down a week from now for a three-week holiday from the 20th August. I really don't know how to spend these three weeks. I should like to continue working but, whether you like it or not, you are sent packing for three weeks. I am thinking of going up to Scotland to visit the Millars.

Here in Cambridge I have taken two rooms with board with a family – not very intellectual, lower middle class, but very kind to me. In particular, the lady of the house, who is a great gossip, visits me in the evening and talks at great length. Her conversation is not very interesting but I regard it as an English lesson.

Yesterday for the first time I had a scientific discussion with Professor Rutherford. He was very friendly, took me into his room and showed me some apparatus. This man certainly has something fascinating about him, though at times he can be rude. And so my life runs smoothly, like a river without eddies or waterfalls. I work till six and then read or write letters, or else go for a ride on my motorbike which I greatly enjoy – the roads here are ideal . . .

(A, B, e)   Cambridge, 16 August 1921

I continue to work regularly in the lab and I've made a little progress but I have not yet begun any real scientific work. On 20 August the lab closes for three weeks holiday and after that I'll start some real work. I intend to spend part of the holiday completing the purchase of equipment and the rest in Glasgow with the Millars [see pp. 89 to 94], whom I haven't yet seen, though I've written to them occasionally. I told them I had these three weeks free and they replied that I must be sure to spend the holiday with them. They were even so kind as to send me a postcard to return to them on which I had only to fill in the date of my arrival.

I wanted to go on my motorbike but I had a rather unpleasant accident. On 13 August I set out for a weekend at the seaside, 70 miles from here, with Müller, about whom I've already written to you. Everything was going fine though we weren't going slowly and then, after 30 miles we had a crash. Here all danger spots are marked by a special red triangular sign which warns drivers to reduce speed. But at one very sharp turn for some reason there was no sign and I didn't slow down. The sidecar flew into the air and I had to steer hard to the right to prevent it overturning so that we ran into the ditch at full speed. The collision was pretty violent. My companion was an experienced motor cyclist (he had once broken his leg in two places falling off his motorbike) so he first asked me how I felt and then made me move all my limbs – since I was driving, the whole force of the collision was on my side. Having convinced ourselves that neither of us had broken any bones we burst out laughing, got out onto the road, and decided to have a smoke. I was rather a sorry sight. Half my face was covered in blood and some muck was dribbling from my nose. But the main problem was the motorbike. The front fork and the front of the sidecar were smashed to smithereens and there was no question of riding any further. With some difficulty we started dragging the motorbike out of the ditch; having removed the front fork we discovered that the frame of the bike was also bent. It was an awkward situation since the nearest town was some seven miles away.

We decided to look for help and at that moment along the road came a girl on a motorbike. Motorcycling is very popular here even among girls, who wear trousers and race around just as fast as the men. We stopped her by making the appropriate sign – every contingency has its own special sign here and to stop a rider you have only to lift up your hand. We explained our predicament and asked her to send help from the nearest town. After three quarters of an hour a car arrived – just in time, for I had begun to feel rather groggy. Müller and the driver of the car started to straighten the front wheel of the bike so that it could be dragged to the village and left there. Since it was Saturday there was no way of getting a lorry. I lay on the grass and occasionally got up to help them. After about an hour they took the bike away and I stayed to keep an eye on the car.

At this point a "bobby" arrived. He had been told there was an accident and had come from the nearest town to investigate. He asked me what had happened, took down my address, the registration number of the bike, the number of my driving licence and the name and address of my companion. He asked whether the machine was insured and how great was the damage. Just imagine, I had insured neither myself nor the machine although I had been advised several times to do this. So now the whole cost of the repairs (a considerable sum) will make a very nasty dent in my budget. My first priority now will be to insure myself and the machine.

So eventually we got into the car and were driven to the nearest town, where after considerable difficulty we found a hotel and were able to have a good meal at last. Instead of spending two days in some seaside resort we had to sit out the time in this little provincial town. There were no trains on Sunday so we couldn't get home. Anyway, I didn't want to return home for it would have been embarrassing to show myself. There were six bruises and abrasions on my body. This wasn't too bad but it was painful to move and especially to sit down. My face was the worst. If only you could have seen it! One half was swollen to twice the size of the other, and there were scabs of clotted blood. I was very reluctant to show myself in the lab on Monday. I created quite an impression in the street, with everyone staring at me as if I were a Chinese scarecrow. However, my friend told me that in Cambridge one needn't be ashamed of such an appearance, in fact it was considered rather smart and would arouse only admiration and respect (assuming, of course, it was a consequence of sporting activity and not of a drunken carouse). When I did appear in the lab I caused a minor sensation. Even Aston came to have a look (ask Kolka who Aston is).

Today the swelling has gone down by half and sitting is much easier. But for at least three weeks I shall be deprived of my only pleasure. You can't imagine how frustrating it is to see everyone else riding while I can't. I have to wait till the bike is repaired. But from now on I shall be much more careful.

(C, D)   Cambridge, 20 August 1921

I have just had your letter and I am very upset both by its content and its tone. You know very well that when I am concentrating hard and am very busy, and while my ideas still lack a firm foundation, I find it difficult to write letters. I do still write because I think it is better to write a bad letter than none at all. As regards my letters to others, and your learning things about me from them, that is because I don't like writing twice about the same thing, as I already explained in a letter to Natasha [his sister-in law]. . . . You don't seem to understand my psychology. What I am doing now is, in every respect, a *tour de force* and instead of giving me your support you write such a letter. Do you really know me so little? And I thought that you knew me better than anyone else.

Mother, you know very well that life, for me, has ceased to be joyful. If I say little about myself that does not mean that there is nothing going on. My wound is deep and God knows if it will ever heal. Here, among strangers, I keep busy on the work I love in the hope of feeling better and of recapturing the joy of living. I am not saying that I am unhappy; only that I shall never give up the struggle to the end of my life. To live one must always go forward. Calm or equilibrium means spiritual death. Here in Cambridge I have to begin all over again. At the Polytechnical Institute I had already reached an independent position on a level at least above the average. But here in Cambridge no one knows me. So I've been a month in Cambridge – quite a time. I have already achieved something but very little, not worth writing about. What I should like, and shall use every effort to achieve, is to get into the scientific life of the lab. Only then shall I be able to work at full speed. So far I have not been able to do this, which is quite natural since [up to now, in Petrograd] I was working on topics that don't interest the others here.

I am writing all this to you alone and not for passing on to others. So don't be strict and demanding, just be patient. When things have crystallised a bit I shall write in more detail. But for the present it's too difficult. Just now it is vacation and I find it very frustrating, since everything is completely shut – library, workshop and so on. Life comes to a complete stop and I regret every lost day.

(A) From his mother   Lesnoy, 21 August 1921

I was happy to get your letters describing the details of your everyday life, which are always very dear to me. Knowing how you live, I can better imagine you, far away as you are – it is as if you were brought closer. I quite appreciate that you don't feel like telling about your work while you are still feeling your way . . . I have often heard that the English are very reluctant to admit foreigners into their circle, but I believe in the international nature of science and on that basis I think you will be able to overcome the barriers to personal relations.

You write in every letter that you are worrying about us, but my dear so far there is really no reason to worry. Work, study and try to make use of all the opportunities afforded by your stay abroad. To know that you can give yourself up so completely to your work makes me so happy that I am prepared to put up with everything. As you know, my dear, all my life your scientific work has been my main preoccupation and now that you are at the very source of that science, I would hardly want to spoil your being there in any way. I feel that two or three of the letters I have written to you have been overagitated, but put that down to my quite understandable nervous condition and forgive me. Let me repeat again that apart from the loneliness and sadness of your absence, our material position is in no way affected by it; the parcels you sent from Reval are sustaining us very well. We have enough to eat and I hope we shall manage as regards stoves and firewood as soon as Lyonya returns. He and Natasha are very energetic and good managers. . . . For God's sake don't stint yourself to help us. I shall be infinitely grateful for the spectacles . . .

(D)   Glasgow, 26 August 1921

Here I am in Glasgow, sitting in the Millar home. They have evidently grown a lot richer during the war and live very comfortably. The boys have grown up into men. Mrs Millar herself is at present at their holiday home and I'm going there tomorrow. On 7 September I go to Edinburgh for the British Association meetings, where Krylov is giving a lecture on the Kursk magnetic anomaly. I return to Cambridge only on 26 September and shall then get back to work again.

(D)   St. Fillans, 30 August 1921

I am sitting by the fireplace in the same spot where I used to sit seven years ago. My hosts are very kind. Good Heavens! Seven years and nothing has changed here. The war seems to have had no effect here and this seems rather strange, for so much has happened to me in these seven years. Good Heavens, if I had foreseen all this in a dream at that time I wouldn't have believed it possible to survive all that has happened but apparently it is possible. Yes, one can respond to life with great resilience and the moulds into which fate presses us are fantastic . . .

I'm glad to have this rest as otherwise I'd be overworking. However, the rest from work is only partial because talking constantly about Russia tires me out. Besides, you meet total incomprehension among people here, and this is understandable. The boys in the family have now grown up and I find it amusing to see the enthusiasm of the younger one for the cinema. He is a bit of a musician and American music is all the rage now in the cinema. The man with the big drum in the orchestra plays what is called "jazz". The drummer has to hand a whole variety of instruments such as the xylophone, tambourine, cymbals, motor horns and sirens, hammers, plates, tin boxes, bottles and so on. So, using his judgement and keeping time with the music, a skilful drummer produces sounds from all these things. This gives the music a very peculiar character, typical of the Americans, who care only for this sort of music, which we would describe as caterwauling rather than music. But the role of the drummer is considered the most important one and he sits above the whole orchestra so as to be clearly seen by the audience and his name appears on the programme. The younger son of the Millars is this kind of drummer and performs two or three times a week, for which he gets thirty shillings each time. The family is not in any material need (they have five servants) and he does it just for the love of the "art" (it is considered an art form here) . . .

(D) St. Fillans, 2 September 1921

I am still staying in Scotland on the shore of a loch but I feel the pull of Cambridge and my work in the lab more and more strongly. However, vacation is vacation and no one is allowed to work. Yes, my dear, three days ago it was exactly five months since I left you, almost half a year. Never before have we been separated for so long. During these five months I have worked only one month in Cambridge and I look back on that month with great pleasure and satisfaction. But that, after all, is very little and I want more.

You ask me to write more about myself but I don't find it easy. I am always looking around me and pay little attention to myself. I speak English a lot, I can say almost everything I want to and my vocabulary is increasing but I fear my pronunciation is bad. I dress well, even by local standards. I have two suits, a grey one for everyday use and a blue one one for more formal occasions. I wear a grey felt hat and am able to indulge to the full my passion for clean collars. But the situation as regards my hair is worse. Mrs Millar gives me no peace with her insistence that I should go to the barber. I've had only two haircuts during these five months. But I'm worried about the amount of hair I'm losing and I fear I shan't be able to make fun of Kolka [Semenov] when I come home.

I have no joy in life and I don't know whether I ever shall have. But at times I work in a frenzy and I shall never give up as long as I live. To live or to work – they are becoming one and the same for me.

(B, D, E)   London, 19 September 1921

. . . The weather is getting colder and they say winter here is horrible, more like our autumn. I have already caught a cold and am not feeling too well, with my nerves again in a poor state. I don't know what I shall do when Krylov has left. At the Edinburgh meeting of the British Association I made a new acquaintance – Professor Timoshenko*, a very clever and pleasant person. He is tall with white curly hair, a small beard, a pale face and bright intelligent eyes. There is an aura of books and study about him. He really is clever, with such a calm and deep understanding as I have rarely met. There is something fascinating about him though he speaks little and seldom, always in a calm and quiet voice. I spent the whole week with him at the meeting. He is a friend of Abram Fedorovich.

I am very pleased that Natasha writes to me. You are lucky to have such a good daughter-in-law! You really have done well! Of course, your sons also are not too bad! – except for one who, you say, writes too little; but this son is a poor fellow with a rebellious spirit. When will he find his place and be at peace? I should like to believe this will happen.

This is perhaps one of the most critical moments in my life. If I win through I think I shall find peace. What torments me just now is whether I shall succeed in completing the experiments that I have thought up in the Cavendish. Am I not again spreading my wings too wide? I have had some important ideas but perhaps they will all come to nothing. Moreover, Rutherford himself is a puzzle to me. Shall I be able to solve this puzzle? And on top of everything there is the uncertainty of my financial position . . .

(A) From his mother   Lesnoy, 22 September 1921

. . . There is a lot to do in the library and I am being offered work in a publishing house, which I shall have to take on if it is decently paid, since all my other work gives practically nothing financially speaking. At present it costs 1000 rubles for a tram journey and so far nothing has been done to help professors and teachers in this respect.

---

* In his reminiscences Timoshenko (1968) comments amusingly on Kapitza.

At the 2nd Pedagogical Institute I get about 3000 rubles a month but I have to spend about 16 000 rubles on tram journeys. Of course something will eventually be done, but in the meantime the situation is very difficult and I have to think about increasing my earnings, which can best be achieved at a publishing house – my lectures and library work are more for the soul. . . .

(A)   Cambridge, 3 October 1921

I moved into my new quarters yesterday – only two rooms, a bedroom and a room for eating and studying. As usual, it is clean but there is one snag – the lavatory is outside the house and to reach it I have to go through the landlady's room and she can keep statistics on the frequency of my bowel movements. However, these rooms have electricity rather than gas, as in the former rooms. The landlady and her family are friendly. There is a very special kind of attitude towards university people here. A member of the university considers himself on a somewhat higher plane and looks down on ordinary folk. Although I don't behave in this superior fashion myself, people nevertheless treat me in a special sort of way . . .

I am working flat out all the time which gives me a lot of satisfaction. People are returning from holiday extremely slowly. There are still only six or seven at the lab instead of the forty or fifty who should be here. I often see Rutherford but I've been quite unable to get the measure of him yet. It's not all that easy to understand these English. Has anyone been to see Kustodiev and collected the portrait* and the picture? Have you received the spectacles? . . .

(A) From his mother   Lesnoy, 4 October 1921

Your last letter was from London [19 September] and today, to Lyonya's great delight, there arrived tobacco and a pipe and also cocoa, chocolate and dried eggs. We shall save the last two items for Lyonchik [little Lyonya]. These parcels are a great joy for us. I have

---

* This refers to the portrait of Kapitza and N.N. Semenov painted by Kustodiev at the beginning of 1921 (see p. 203 and fig. 6).

already written to you about the glasses; they are very good and the lorgnette is particularly convenient – every time I use it I think of you with gratitude . . . Thinking of you, I often remember my father who worked in difficult conditions and moreover had a very difficult character so that he had no friends. Yet in spite of this he rose above his difficulties and through his scientific achievements won wide recognition.

I am much worried by your material well-being, and I am always embarrassed when you spend money on us. But your support has been very substantial and thanks to you we have the possibility of installing stoves and laying in reserves of firewood. . . . I have a lot of things to do but that's not bad. I can work well only when my heart is not aching for you and Lyonya – the two of you are my whole joy in life . . .

(B, C, E, e)   Cambridge, 7 October 1921

. . . I work in a large room which I shall be sharing with several others. So far as I know, one Japanese . . . and an American will be coming, so you see what a motley company we are. There is a constant stream of visitors to the Cavendish lab. Today Rutherford introduced me to Professor Perrin, famous for his work on Brownian motion and there was also a lecture on molecular structure by Professor Langmuir from America. This provoked a lively discussion in which Rutherford, [J.J.] Thomson, Darwin and Perrin took part. Langmuir was criticised rather sharply and his theory was not well received. Here they prefer the theory of the Danish physicist Bohr.

Tomorrow I hope to get the motorbike from the repair shop and on Sunday I'll go for a spin on it. My work is going very well. This is a paradise as regards getting things. If you need anything such as platinum foil or wire you have only to go to the Cambridge and Paul Scientific Co. to get all you need. Incidentally, I have met the Director and the engineering staff of this firm and they took me round the factory and showed me some interesting things. They gave me all the little things I needed for nothing, purely out of good nature, because they don't normally sell them. In return, I did them a small service by giving advice about their work.

It is only in the evenings that I sometimes feel lonely. Just now I no longer want to read or work and that's why the evening is dreary. If only I could have the present conditions for work and have you all here as well!!

(B, C, D, e)   Cambridge, 12 October 1921

. . . I'm very glad that you received the spectacles; I was beginning to worry about them since they cost me £4 and they are the smartest you can get here. You can't imagine how happy I am with your letters. All at once I feel more cheerful and I want to write to you as much as I can. I am working as hard as ever and I'm satisfied with my work and also with how I am treated in the lab. Many of the research workers have returned from holiday, among them Müller. My motorbike is repaired and I enjoy riding around on it. If I didn't have this toy I should feel a lot worse.

Professor Rutherford greets me increasingly pleasantly when he sees me and asks how my work is getting on. But I am a little afraid of him. I work almost next to his office, which is bad because I have to be very careful with my smoking – if he should catch you with a pipe in your mouth you're in trouble. But, thank God, he has a heavy tread and I can recognize his footsteps a long way off. Besides, I have a fume cupboard in my room in which I can smoke and that certainly helps to avoid trouble . . .

(A)   Cambridge, 18 October 1921

The day before yesterday I went to London on business and spent a couple of days there. I travelled there and back on my motorbike without any problems. . . . You can hardly imagine the amount of motor traffic on the roads here. Even on the ordinary highways the traffic is astounding. You have to drive well and to know all the rules. When you want to turn or stop you must give the appropriate hand signal. If you want to stop someone there is a special signal for that too and you have to be thoroughly familiar with all the roadside signs.

I think you will be very amused to know that there are two road signs for every school, about 100 metres apart on either side of it. When the pupils come out of school they could be run over, so the signs tell you not to exceed 10 m.p.h. Moreover, any policeman can stop all the traffic by raising his hand. The concern for children here is sometimes very touching. Not long ago I was standing at the gates of Hyde Park and observed the following scene. Some children and

52. "Talk softly" (Rutherford talking to J.A. Ratcliffe by some delicate equipment: made to look as if the sign is specially designed against Rutherford's heavy tread and booming voice) 1933.

nannies with babies were coming out of the gates. A four-year old toddler began to cross the road and as soon as he saw the toddler come out, a policeman immediately put up his hand and stopped all the traffic, while the four-year old grandee proceeded slowly across the road. Nannies with babies are accorded the same privilege . . .

(B, C, D, E, e)   Cambridge, 25 October 1921

. . . My research is making some progress and my relationship with Rutherford or, as I call him, the Crocodile*, is improving. I am working hard and enthusiastically and already have some results but I have chosen a difficult problem and there is a lot of work to be done.

Just now Academician Shcherbatskoy is in Cambridge to examine some Sanskrit manuscripts and I enjoy talking Russian when we see each other. Winter is only just beginning here and the leaves are only now turning yellow. Freezing weather is not expected before January but it is already cold and the cold here is something special, worse than our frosts, because the air is always damp and, of course, there are fogs. I prefer our winter. I haven't had a cold for a long time but this weather has got me down and I have a sore throat so that I can't smoke, which is miserable . . .

(F)   To P.S. Ehrenfest   Cambridge, 26 October 1921
      Professor of Theoretical Physics
      in Leiden

Much respected Pavel Sigismundovich,

It is a great pleasure for me to write to you, not only because it enables me to share my thoughts with you, but also because you know Abram Fedorovich [Joffé], our young physicists and our Russian physics school, so that you are, as it were, no stranger. I heard a lot about you in St. Petersburg and your memory is still fresh there. I was always

---

* This is the first mention in these letters of Kapitza's nickname for Rutherford (see p. 11).

53. P.S. Ehrenfest.

very sorry that I was too late to meet you there. I must confess that I am still a long way from feeling settled in Cambridge and feel very lonely here. I am pleased with the way I am treated by Rutherford and the other physicists I have to deal with, and our relations continue to improve. The possibilities for work are good.

Although at first I had to go through the radioactivity practical class, I am now engaged on a fairly serious experimental project, which even if not in the most difficult category is still very difficult. I don't know how successfully I shall cope with it. The problem is the following: α-particles strike a microradiometer and by putting this at various points along the α-particle trajectory it should be possible to measure how they lose their kinetic energy along the trajectory, and to verify if all the loss of energy goes into ionization. This problem was earlier investigated by Rutherford himself and by Geiger, but they were unable to study the last 40% of the track because of the insufficient sensitivity of their methods.

The microradiometer is a devilishly sensitive instrument. I have already built one which should be sensitive enough to give positive results – it can detect 10^{-11} of a calorie. But, as always with such extreme sensitivities, various background effects begin to show up which interfere with the measurements. However this first successful step has been useful in encouraging people to have confidence in my experimental ability. As a result I have been given permission to use a sizeable quantity of radium and the equipment necessary for further progress has been bought for me.

But somehow or other all this has not brought me closer to the English. I think the basic difference is that I look on the work as a kind of "aesthetic enjoyment", as it were, while the English reduce it all to "business". Because of this, it seems to me that the young physicists here have only a very narrow interest in physics, each knowing only his own special bit. I suppose this is useful for getting things done, but for my Russian nature it is rather dull . . .

(A) From his mother   Petrograd, 31 October 1921

It's already nearly a week since we moved into the city and things are beginning to settle down though there is still a long way to go before everything is in order . . . . We are quite well accommodated and indeed even luxuriously by present standards. We have three rooms with stoves; in my room it is about 10°C without heating and gets to about 15°C with heating – the other rooms are warmer. Today the plumber is at work and we should have a tap connected to the water supply. The main question is whether we shall manage with firewood . . .

Today we had a big event – Lyonya bought a big French roll for tea for which he paid 15 000 rubles. He sold your father's medals for 100 000 rubles and this made it possible to pay off our debts. Recently we have been selling and selling without end and it's getting quite dreadful. I am occupied with lectures nearly every day and these are the best hours of my life – I don't know how I'd manage without my work with young people. It's not only the lectures but the intercourse with young people which has a good effect on me. They treat me in a very friendly way and show interest in the subject. I feel both happiness and pain in this apartment. I constantly think of your father. In

essence, even now, after he has left us, we still live a good deal on the
fruits of his labours. But you needn't worry, my dear, we nevertheless
live better than many, many others. Around us many are freezing and
hungry . . .

(B, C, D, E, e)   Cambridge, 1 November 1921

I got three letters from you today and that is always a great joy, but
from the tone of the letters I feel that you are overworking and very
exhausted. You know, my dear, that having too little to do is bad, but
having too much work is also bad. So, reduce the steam pressure and
don't overload the machine. You are giving lectures in so many places
that you make my hair stand on end. . . .

Don't worry about me – I am "alright", as they say here. My cold
has gone and I feel well. My work is making progress and the next two
weeks will be decisive for me. The results I have obtained so far give
me hope that the final outcome of my experiments* will be successful.
Rutherford is pleased, his assistant tells me, and this shows in his attitude towards me. When he meets me he always says a few friendly
words. He invited me to tea last Sunday and I was able to observe him
at home. He is very charming and unpretentious though he can be
quite fierce. When he is displeased you had better look out for he lays
about him. His intellect is quite unique and he has a remarkable flair
and intuition, such as I could never have imagined. I am attending
a course of his lectures and seminars and find his explanations very
clear. He is a quite exceptional physicist and a very original personality . . .

I have a big request to you but don't hurry to fulfil it as I know you
are very busy. Some time when you have three or four hours free
please visit the Smolenskoe cemetery and see whether our graves there
are in good order. Today is just seven months since I left Petrograd
but I feel as if I hadn't seen you for two or three whole years. Don't
feel too lonely without me, my dear. Remember that I too am a bit
sad here, all alone among total strangers who have a different kind of

---

* These experiments, Kapitza's first research in Cambridge, were designed to
measure more accurately the loss of energy of α-particles near the end of their tracks
[see (6) and p. 132].

temperament and do not speak my native tongue. But the fact that I am able to work here, and to work well, makes up for everything. I do indeed sometimes find the evenings dreary but what can I do about it? I study a bit and write letters to you and this seems to shrink the distance between us. You must know what good fortune I have had to have found a congenial field of work in which I can achieve some success. This makes it possible to put up with a lot.

I have just started a new correspondence but please don't tell anyone about it. It is with Professor Ehrenfest in Leiden who was formerly in Petrograd, and was very popular with the physicists there [see p. 131]. It was Professor Timoshenko who advised me to write to him: he said in a letter that he had met Ehrenfest at a physics meeting in Jena and Ehrenfest had agreed to a correspondence. So I wrote to him and he replied in a very friendly way and promised to reply in future. . . .

(A) From his mother   Petrograd, 12 November 1921

. . . It always grieves me to hear that you are unwell – I should so like to be by you my dear. But one phrase in your letter cheered me up greatly. You write that you are working with enthusiasm and that is the most precious thing of all you said. . . . Here winter has set in quite suddenly – a very unusual thing. . . . Most people haven't got firewood and they freeze in their homes. Everywhere staff is being reduced by 50%, and this is particularly so for lower paid employees. It is said that work in the Pedagogic Institute will stop from 1 December till 1 February because of lack of heating fuel. . . .

(B, C, D, E)   Cambridge, 21 November 1921

I feel a bit guilty for not having written to you last week. In fact I didn't write any letters at all, because I was so busy with experiments in the lab, and I couldn't get the results I hoped for. There was so much work that I couldn't settle down to writing.

The situation is that I have to increase the sensitivity of my apparatus by a factor of at least 10 or so though I have already attained a sensitivity greater than usual for the type of apparatus I am using.

The problem is difficult and demands a lot of skill. The Crocodile often comes in to see what I am doing and last time, after looking at the graphs I had obtained, he commented that I was already getting close to the desired goal. But the closer I get the greater are the difficulties.

My financial position is much better, though life in Cambridge is very expensive, much more so than in London. The town is full of the sons of rich parents and it lives off them. I am worried about your health, my dear. Don't begrudge selling things and sell mine too. Only the pictures should not be disposed of. The important thing is that you should be well fed and warm. There is no point in hoarding things . . .

(A) From his mother   Petrograd, 25 November 1921

. . . Abram Fedorovich got a letter from Rutherford who writes that he is very pleased with you. This gave great satisfaction also to A.F. He said that he attaches great significance to praise by Rutherford and that praise by an Englishman means more than praise by a Russian.

You write about my idea of coming to you, but I'm afraid this will remain only an idea. Only if your stay [in England] is greatly prolonged would I be able to decide on what for me would be such a heroic step. First of all I don't know the language and that is an enormous drawback. Secondly I think that in your work you should be protected from all kinds of worries and bothers such as you would incur if I was with you. My dear, carry on working and devote yourself entirely to science – we shall meet in the summer. . . . Working in such conditions as you enjoy would be the ambition of many people and I would never want to be a hindrance to you. I deeply believe that our attachment to each other and our closeness depends neither on time nor distance.

(A)   [London, about 1 December 1921]

. . . There is a dense fog in London – I simply couldn't imagine that there could be such a fog. You can't even see your own legs as you go along the street. You walk slowly, whistling so as not to collide with anyone. You have to keep close to the wall to feel how the street goes. The fog is sort of black and heavy, irritating your throat and pressing on your chest. All traffic is at a standstill. Policemen gather pedestrians into groups and lead them to the police station where some wait till the fog clears and others are formed into groups and conducted to their homes. Even in the middle of the day the street lamps are alight everywhere and the shops are illuminated as well. There are many accidents in the fog. Cars and buses mount the pavement, collide with one another or run into the Thames. Passers-by are also sometimes killed in this way.

I travelled by train in the fog and the train was several hours late. The most difficult problem for trains is that the engine driver cannot see the signals in the fog so they have to use a different signalling system. Special explosive charges which are set off when the train wheels pass over them are put on the rails. The number of explosions conveys a message and so when you are travelling by train in a fog you keep hearing these fusillades. In one incident a train stopped on a bridge and one of the passengers, thinking it was at a station, jumped out of the train directly into the river below and perished. After a fog the newspapers are full of accounts of tragic events like this and also of comic incidents. The English lose their self-control and morality in a fog. You can kiss a girl in the street and then, by taking a few steps, disappear into the fog – this happens frequently. . . .

(A, B, e)   Cambridge, 5 December 1921

The winter is still not here though the leaves have all fallen but the temperature is only about zero and I go about in a summer coat. But it is winter with you and you will be going about in warm coats with the fur collars turned up . . .

Yesterday I was at the theatre and saw "The Beggar's Opera", a comic opera by the English composer Gay. The remarkable thing is that it was written in 1728 and it was produced here in the old style, following the original theatrical conventions. The language was archaic and I had difficulty in following .The music was very simple and the orchestra included some antique instruments such as a clavichord dating from 1784, according to the programme. This was all very entertaining.

I went with Dr Chadwick, Rutherford's assistant, and this was my first visit to the theatre in Cambridge. The theme of the opera was very silly, of course, even more so than "Vampooka".* There is a ruffian who marries three or more wives for which he is thrown into prison. One of his wives frees him from prison but he is caught again and condemned to death. The noose is already around his neck and he is surrounded by his weeping wives and children when suddenly a pardon arrives and all join in a dance – the ruffian himself, his wives and the executioner. This completes the action. The audience was delighted with all this and nearly every song had to be repeated once or twice. I have never seen applause so readily given by an audience as here in England – on every appropriate and inappropriate occasion.

I am working flat out again and so I am feeling fine. There is now hope of a satisfactory outcome to my experiments though there is still a long way to go. Rutherford often comes to see me in the lab and has twice paid me compliments and though our relationship is not yet as I should like it to be, it is far better than before. You know, a visit from the Professor is considered to be quite an event here and in the last three weeks the Crocodile has come to see me five or six times. . .

(B, D, e)   Cambridge, 16 December 1921

. . . The holidays will soon be here and the lab closes for two weeks. I asked the Crocodile to allow me to work but he said he wanted me to have a rest, because everyone has to have a rest. There has been a striking improvement in his attitude towards me. I now work in a separate room, which is a great honour here. I have obtained some results but they are not yet final; however, I now have a good chance of success.

I must tell you about the dinner of the Cavendish Physical Society, which was very amusing. The members of this society are all the research workers in the laboratory. The dinner is an annual event and Professors Thomson and Rutherford are invited as guests, with a few professors from other universities – this year Professors Barkla and

---

* This is a parody, making fun of the conventions and traditions of the opera (music and libretto by V. Ehrenberg, first performed in St. Petersburg in 1909).

Richardson. Some 30 to 35 men were there, seated around a U-shaped table with one of the young physicists presiding and the guests on either side of him. First of all we ate and drank, and though the drinking was only moderate, the English get drunk very quickly. You can soon see this from their faces which lose their stiffness and become lively and animated.

After coffee the port was circulated and toasts were proposed, first of all to the King and then to the Cavendish Laboratory, proposed by a young scientist and replied to by one of the professors. The toasts were largely in humorous vein for the English are very fond of jokes and witticisms. The third toast was to "Old Students" and the fourth to "The Guests". Between the toasts songs from a collection written by the physicists themselves were sung, in which the lab, physics and the professors were all serenaded in comical terms. Everyone joined in singing to tunes borrowed from operettas. These customs date from Clerk Maxwell's time.

You could do anything you liked at the table – squeal, yell, and so on – so the general picture was rather wild and quite unique. After the toasts everyone stood on their chairs, crossed arms and sang a song recalling old friends and so on [Auld Lang Syne]. It was very funny to see such world famous luminaries as J.J. Thomson and Rutherford standing on their chairs and singing at the top of their voices. Finally we sang "God save the King" and at midnight we broke up, but I didn't get home till three in the morning. This was because some of the diners had to be taken home, but I can assure you that I was among the takers and not the taken. The latter were doubtless the happier but my Russian belly seems better adapted to alcohol than the English one. There were no ladies at the dinner.

(B, C, D, E, e)   Cambridge, 22 December 1921

I am writing to you sitting by the fire. It is still not freezing outside and there have been night frosts on only two occasions . . . Today I have at last obtained the long-awaited deflection in my apparatus and the Crocodile was very pleased. The success of the experiments now seems almost certain. There are still difficulties, but I think I shall overcome them.

I think I already told you I have been given a separate room for my work and this is very convenient. This not only flatters my vanity, for it is a great honour here, but it makes it easier to work. If the experiments succeed I shall have solved a problem that the Crocodile himself and another good physicist, Geiger, failed to solve in 1911. There is no point in trying to describe these experiments to you because in any case you won't understand. I shall only say that the apparatus I have built is called a microradiometer, and I have perfected it to such a degree that it can detect the flame of a candle placed one and a half miles away – it can detect a rise of a millionth of a degree. By means of this apparatus I am able to measure the energy of the α-rays emitted by radium.

Tomorrow the Christmas holidays begin and the lab shuts down, so I am going to London and perhaps I shall go to Paris for a few days to visit Madame Curie's laboratory, but this is rather uncertain. I have never seen Paris and I shall enjoy flying there, which takes only two and a half hours or so. Well, so long! and my very best wishes for the New Year. I shall be thinking of you at exactly midnight on New Year's Eve and I suppose you will be doing the same . . .

(B)   Cambridge, 3 January 1922

. . . I have got so used to working that the break doesn't give me any joy. But Rutherford noticed that I have been overworking and he has advised me to go off for a holiday . . .

(B, D, e)   Cambridge, 17 January 1922

I'm afraid you'll have been very worried not to have heard from me for so long. I have been to Paris and from there to the Riviera. It was curious to see gardens blooming in bright sunshine in the middle of winter. I wasn't able to write from France, apart from one letter, though I'm not sure if it reached you. I have many impressions from my journey, enough for several letters but, though I can't describe them all at once, I'll make a start. In Paris I visited the laboratories of Langevin and de Broglie and this was indeed the main purpose of my journey. Monte Carlo was just for a little entertainment but surely one needs that too.

Before I write about Paris, I must first tell you about something that happened to me five weeks earlier, which was unpleasant for me and which, I'm afraid, will upset you too, though all is well again now. One Sunday I went for a spin on my motorbike, taking Chadwick, one of the young Cavendish scientists, with me. I was stupid enough to let him drive, and he managed to upset the machine while going at a good speed and sent us both flying. I landed awkwardly, directly on my chin and split it open. Chadwick landed on his side and bruised it badly.

The doctor in the village sewed up my chin, but as the wound wasn't clean it suppurated later. When I got back to Cambridge I went to a doctor who is a specialist for this kind of injury. Such injuries at sports are very frequent with the English. I am indebted to this doctor, who took a lot of trouble, for the repair of my chin. Moreover, as soon as he saw me he immediately gave me an anti-tetanus inoculation as is always done here in such circumstances. The wound had almost healed about ten days ago but there was some inflammation and pus, and when they removed the stitches the wound opened a little, so I have an ugly scar on my chin, three and a half centimetres long. I don't think you will love me the less for it and so it doesn't bother me much – if necessary I could always grow a beard. That's why I couldn't satisfy your request for a photograph up to now, but I'll have a photo taken within the next few weeks and send it to you. The bike was not damaged. . . . Although I had a slight temperature and my head was so swathed in bandages that only my nose stuck out, I didn't interrupt my work in the lab. The Crocodile pressed me to go to bed but I didn't go. He was much concerned about me, asked me which doctor I went to and so on. Altogether, this has earned me a plus.

So, when I had finished in the lab and had got a special plaster on my chin, I decided to go to Paris. The French scientists got me a visa. I now carry on an extensive and expanding scientific correspondence. I decided to take the plane to Paris as I had long wanted to fly but, as luck would have it, there was a storm on the day of my flight. I had not forgotten about you and had insured my life, leaving the receipt for the premium with A.N. Krylov, to be passed on to you. Because of the storm they brought out a big 15-passenger plane of the type they use for transatlantic flights. I was taken to the aerodrome by car and my luggage was weighed. If you want to take more than 20 kg there is a big supplement to pay. I was offered a place in front of the pilot, where there are seats for two, rather than in the main cabin and I accepted with enthusiasm. They fitted me out with a flying suit – helmet, fur lined gloves and breeches and so on. They said it would be cold up there.

Boarding of the plane started after the passport examination and I climbed into my seat next to a young Englishwoman. The pilot and mechanic sat behind us and the other passengers sat in the cabin further back. Because of the bad weather there were only four or five in the cabin. The propellers started to turn and the engines made a loud noise. While we squeezed ourselves into our places they tested the two engines, each of 700 H.P. Finally, they took away the chocks and the plane rolled slowly across the aerodrome to the opposite end where it turned into the wind and, steadily accelerating across the airfield, took off.

As the ground dropped away beneath us the plane swung a little to the right and then to the left. I must admit – I felt scared. After all, neither I nor any of my forbears had ever flown, except perhaps in dreams in a flying trunk, and the fear that I felt must have been simply because of unfamiliarity with flying. But the result of my fear was that, like a fool, I clung instinctively and tightly to the side of the little box in which I sat. Of course, if the plane had crashed that wouldn't have helped me at all, but I continued to cling to the side of the box for the first half hour of the flight, even though I understood how foolish this was.

We flew at no great height since the clouds were low. You don't feel any sense of height, and indeed it is much more unpleasant to look down from the balcony of a four-storey house than from a plane at a height of about 500 feet. On the contrary, you wish the plane would rise higher to avoid the feeling that it might strike a tree or a church tower. For the first 45 minutes we flew over England and gradually approached the Channel.

Then we were over the sea and my realisation that we should be getting a bath if the plane were to come down here also somewhat disturbed my equanimity. But soon we were over France, with firm ground below us again. The weather was worse here, with even lower cloud, and after half an hour's flight we were right in the cloud. All around was grey and the rain lashed us. In the plane you don't feel the speed, though it must have been at least 80 to 100 miles an hour, but when the rain started I realised just how fast we were moving since the raindrops hit the unprotected part of my face with such force that I thought they would cut the skin.

I had already come to terms with the insecurity of our situation and had overcome the fear that we might crash at any moment but now a new anxiety arose. I had made a hearty breakfast in the morning and drunk a lot of coffee. It was cold up in the air so I was suffering intolerably from a desire to satisfy a call of nature, a well-known effect of cold. I was beginning to think I should perish from something bursting inside me and as the plane began to descend towards the Paris aerodrome, I looked around for a small building where I might find relief. The plane had barely touched down before I jumped out and made a beeline for the hangar. The police at the passport control tried to detain me but when I explained the problem the French police laughed loudly and showed me the way. The next letter will be devoted to Paris, and so, bit by bit, I shall describe the whole journey*.

(B, D)   Cambridge, 5 February 1922

Yesterday I was glad to hear from the American committee that Aunt Sasha had received the food parcel I sent her. You too should have received the one I sent to you at the same time. Just now I am a bit short of funds but as soon as my finances improve, I shall send some more to you (for the third time) and to Aunt Sasha (for the second time). But first I want to know if the food arrived in good condition?

I'm living quietly and working a bit slowly, but I hope to get back to full speed soon. However, I'm now doing a lot compared with what I did in Petrograd. Last term I worked about 14 hours a day, but now I find about 8 to 10 hours is sufficient. I've started reading some literature. I know what I'm like – when I don't feel like work it's no good going against Nature. Moreover, the weather here is enough to make you peg out. Just imagine, it can be so warm that you go out without a coat, and then suddenly it freezes. I sleep with an open window all the year round, as is the custom here, and the bedroom is never heated, so that sometimes the water freezes. And you know, my dear, how much I hate being cold. It is also damp, so that occasionally

---

* This intention was not, in fact, carried out.

I have rheumatic pains in my right knee and left shoulder towards the morning.

Today is Sunday and I have been reading *Thais* by Anatole France with enjoyment all day. Anatole France is not much admired in England and Maupassant is considered pornographic, so it's "not done" to discuss these authors in polite society.

(B, D, e)   Cambridge, 16 February 1922

I had a talk with Rutherford today. The Crocodile looked at me very fiercely. You can't imagine what an expressive face he has, it is simply fascinating. He asked me into his office and we sat down. I looked at his fierce expression and somehow felt amused and began to smile. Just imagine, the Crocodile started to smile too and I nearly burst out laughing, but I remembered in time that I must behave respectfully, so I began to talk about one of my experiments and the difficulty caused by the need for a large quantity of radium. He became very genial and, seeing that he was in a good mood, I told him about one of my ideas concerning delta radiation*, the theory of which is still not at all clear. I presented my interpretation which is supported by a fairly complicated mathematical calculation and clarifies a whole lot of experimental results.

So far everyone I have spoken to thought my hypothesis too daring and expressed considerable scepticism. But the Crocodile grasped the essential point with his usual lightning speed and, just imagine, approved of my idea. He is a very direct person and if he doesn't like something, he curses so vigorously that you don't know where to hide. But in this case he commended my idea warmly and advised me to take up confirmatory experiments at once.

He has a devilish intuition and Ehrenfest, in his last letter to me, calls him simply God. His favourable opinion has greatly encouraged me, and it's amusing to find that as soon as the professor smiles on you, this immediately affects everyone else in the laboratory – they too begin to pay more attention to you. Yes, Mother dear, the Crocodile is really unique and I only wish you could somehow see his face. I am not usually timid but with him I feel shy. My experiments

---

* For Kapitza's theory of delta rays see (7).

are going fairly well and it looks as if they will end up successfully. I have overcome many obstacles and very few remain. But now, for some reason, my head feels empty, and I have no inclination to do theoretical work. My chin is very ugly. I haven't ridden my motorbike lately.

(B, D) Cambridge, 6 March 1922

. . . My work is still going satisfactorily but, to judge from your letters, you greatly exaggerate my successes, for so far, I have done nothing remarkable. And, my dear, please don't talk to others about my successes or else I shall be uneasy in my mind: God knows what people may think. At present I am only one of the rank and file and up to now all that I have done has been achieved by someone who is neither better nor worse than 30 others working in the Cavendish. . . .

(B, D, e) Cambridge, 28 March 1922

. . . Things have moved forward and I have got almost final results. The Crocodile is pleased and we have discussed plans for further work. A funny thing happened today. As I told you, the work I have done was started several years ago by the Crocodile himself and then continued by the German physicist Geiger, but neither of them was able to complete the research because their apparatus was too insensitive and I have now succeeded. When I compared my results with theirs it turned out that my data agreed better with Geiger's than with Rutherford's. When I pointed this out he said quite calmly: ''That's how it should be. Geiger's work was done later than mine and he worked under more favourable conditions.''. This was very generous of him. On the whole he has a good relationship with me now. I am pleased.

In three days time it will be just a year since I left you, my dear ones. A year of wandering, living alone, intensive work, many new impressions but not without results. And so, my dear, here I am, lost among strangers, often and often thinking of you all, and longing above all for you to be happy, well fed and warm . . .

(B, C, D, E, e)   Cambridge, 7 April 1922

It is 10 days since I wrote you, but I have been working like an ox. I finished work in the laboratory today and am going to London tomorrow for the holidays . . . Lately I have worked to the following schedule: arrive at the laboratory at 10, prepare the experiment until 3, take a nap between 3 and 4. Then between 6 and 9 I do the experiment (I have special permission from the Crocodile to work after normal hours). Finally I go home and work out my results till 4 or 5 in the morning, so as to be ready to start again next day. I am a bit tired but I already have definitive results and I can now say with confidence that my experiment has been crowned with success.

In the meantime I have had three hour-long conversations with the Crocodile and I think he is well disposed to me now. I am even a little worried at the way he pays me compliments. He invites me for tea in his room, just the two of us. The worry is that this man has a strong and uncontrollable temper and with such people a sudden change is always possible. But his head, Mother, is really staggering. He never indulges in destructive criticism, is bold and easily moved to enthusiasm. With such a temperament he has no difficulty in keeping 30 others hard at work.

If only you could see him when he is angry! Some samples of his scolding: "Whenever are you going to get results?", "How long are you going to mess about senselessly?", "I want results from you, results and not chatter", and the like. In intellectual power he is reckoned to be on a level with Faraday, some even put him higher. Ehrenfest wrote me that Bohr, Einstein and Rutherford must have been sent by God to the physicists, among whom they take the first place. I shall rest a little and must then get back to work. I have so many ideas to work on now – my own and the Crocodile's – that I really don't know how I shall cope with them all . . .

(D)   Margate, 14 April 1922

Well, at last I've got time to write you a long letter. I finished work in the lab last Friday and on Saturday I went to London on my motorbike. From Sunday to Tuesday I was busy in connection with my work but on Wednesday I went to the seaside where I intend to take a badly needed rest. I am writing this by the fire in the parlour of the boarding house. Fyodor Ippolitovich Shcherbatskoy, who is here with me, is playing *Faust* on the piano. He often makes mistakes, and hesitates even more frequently, but I enjoy listening to his strumming. There isn't much good music in England so I am glad to have anything.

I got here yesterday on my motorbike (100 miles). I had intended to take Shcherbatskoy with me but he cried off on account of the rain and I came alone. I got thoroughly soaked and today I am sneezing and coughing. I covered the 100 miles quite successfully, cutting right across London. For the first time in my life I found myself in this world-famous traffic but, if you know the rules, it is not at all difficult to drive. Just imagine, Mother – your son rode on his own motorbike past Westminster Abbey and the British Parliament. Could you have imagined this a year and a half ago?

This is a very crowded and noisy place, which I don't like, but I decided to come with Shcherbatskoy so as not to be lonely and he, for some reason, insisted on this resort. Of course, I can't regard him as a close friend. Imagine someone a head and a half taller than me, as stout as Father, with a broad clean-shaven face, pince-nez and almost totally bald, though today he bought some hair cream (probably just to flatter himself). He is about 60, knows Tibetan and Sanskrit and studies Indian philosophy and Buddhism. His face is smooth and round, very reminiscent of the Buddha. He is very cultivated, a member of the Academy of Sciences and a bachelor who likes to pay court to the ladies – just imagine, not without success.

We have a curious relationship with one another. He is rather condescending not only to me but towards science in general. When he visited me in the lab in Cambridge, he noticed that I was soldering something and, after looking at my work, he declared that the blacksmith on his estate soldered much better than me. Our conversations are rather odd. Whatever you tell him, he assures you that this was known long ago in Indian philosophy. But what he really knows about is politics. Thanks to his excellent knowledge of languages he is able to read all the newspapers and can follow events infinitely better than I can, so I have been able to learn a lot from him. As you see, we suit each other like a saddle suits a cow, but we have spent quite a lot of time together. He pokes fun at me but I usually give as good as I get. He returns to Petrograd soon and has promised to call on you

54. Second-hand American lathe; a present to himself in 1922.

and give you my news.

As I have already told you, my research has finally proved successful. But during the last two days I have stumbled on an effect that I shall have to study in more detail and this will take another two weeks. At any rate, I expect to finish by the beginning of May. I felt some satisfaction and decided to give myself a little present so I bought a small lathe which is very necessary for my work [see fig. 54]. The weather here is remarkably miserable – it has been wet and cold for about a month and I miss the sun terribly . . .

(D)   Cambridge, 24 April 1922

. . . The weather is bad. I don't like spring and it still hasn't got properly established. I feel sad and melancholy and memories keep going round and round in my head. I think of the past and though it is gone for ever, I still remember it. Nadya, Nimka – it is two and a half years since I kissed them for the last time. Two and a half years since I heard Nimka's laughter and chatter for the last time – "Papa – tphoo", and this has gone for ever. How much I would give to bring all this back! What are my successes and the comfort I now enjoy really worth? It is terrible to be so dependent on others. At times I am overcome by grief and dejection. I feel I am changing into some kind of soulless machine, without any real content, able only to demonstrate that the α-rays emitted by radium cause certain effects when they undergo collisions. And what is all this for, if there is no joy such as I had two and a half years ago?

I am indeed like a machine that works and works without knowing what for and why. That is my fate but I shall not give up and shall continue to be this machine to the end. Anguish and melancholy are weaknesses and it's easy to succumb to them here. But thank goodness, at least as a machine I work smoothly. I am writing all this to you, my dear, because I know that you too are suffering. But in my loneliness here I need much greater courage to bear it and I feel that my whole life will continue like this. It is as if fate had teased me by showing me that there is something more in life, indeed the best part, which can be appreciated only after it has been lost. It is a cruel lesson, but a powerful one and I now understand much that I didn't understand before. But all this has gone and in the future I shan't see the like again. I don't complain but I am sad.

Here I am surrounded by people who are striving for happiness, seeking it in money, in their careers and in prestige, but now it is very clear to me that one can be happy quite simply, without any of this trumpery. We read all this in books and especially in the Bible but to experience it oneself and to live it out, is perhaps not so simple. Well, enough of this philosophy, which doesn't lead anywhere . . .

(B, D)   Cambridge, 15 June 1922

I have shaken off some of my work and feel appreciably freer. First of all I have finished preparing my paper for publication and it has turned out quite lengthy – 25 to 30 pages with many drawings and diagrams. I am told it is a good bit of work. It has been translated and is now being typed. It will be ready tomorrow and I shall perhaps give it to the Crocodile the next day. It will be very interesting to see what he thinks of it and I am a bit anxious. . . .

I shall now continue the research, the first part of which is about to be published and I have also started a new project* in collaboration with a young physicist here. If this is successful it promises to yield a lot of information. The Crocodile is keen on my idea and thinks we shall succeed.

The following anecdote about me is circulating here. When I was explaining my idea to the Crocodile I got rather carried away and to demonstrate the size of one part of the apparatus I flung out my arms and nearly struck my esteemed professor. He jumped from his stool and said "But please don't hit me!" I am told that he laughed a lot when he described this incident.

I am really delighted that he has approved my plan. He has the devil of a nose for experiment and if he thinks that something will come of it that is a very good omen. He treats me better and better which makes me very happy. But he is very temperamental and if the pendulum swings too far to the right there is always the fear that it may swing to the left. . . .

(B, D, E, e)   Cambridge, 19 June 1922

. . . Today the Crocodile summoned me twice into his room to discuss my paper. He had read it, made a few changes, and consulted me about each of them. It will be published in the Proceedings of the Royal Society – the greatest honour that a piece of research can achieve here. I cannot recall any Russian scientist having published his work in this journal for the last 10 or 15 years.

---

* This was the idea of using high impulsive magnetic fields to bend α-tracks in a cloud chamber (10).

The paper [see (6)] turned out to be very long (23 to 24 pages) and contains a lot of material. I am the first to have observed some of the phenomena that I describe. The Crocodile wanted me to state this explicitly in the paper but I declined his suggestion. I have never felt so agitated as on this occasion. Somewhat tentatively, I have advanced two hypotheses and I am very worried about how they will be received. When you chat among friends you don't have any feeling of responsibility but it is sobering when you step on to the European stage.

The Crocodile told me to write an abstract of my paper which will be read at a meeting of the Royal Society. Today I took it to him but he didn't like it and rewrote it himself. The attention that he has devoted to my paper has moved me to the depth of my soul. I don't like writing about my successes, so do please keep what I say within the family – I really mean this. But I want so much to share my own joy, and there is no one to share it with here. I know that you will be glad for me, my dear. And so, the first step has been taken, but a long road now lies ahead and much work . . .

Just now I am working with a young physicist called Blackett. On the one hand, I enjoy this work because the subject is very interesting and the results that may be expected are very important. But, on the other hand, Blackett himself has the reputation of being not very likeable, so this collaboration goes against the grain*. I couldn't undertake this work alone because he had already started it. I proposed a radical change of method but, inevitably, I had to start the work with him. If it were not for the great interest of the project I would never have done this. The ability to get on with people and to understand them is always important, even when working with inanimate nature. Perhaps I shall be able to get on even with this young man but it adds one more difficulty to the problem, which is not easy anyhow.

Yes, Mother, my whole life is a kind of struggle, as if fate had decided to test me. For nine years now I can't remember a time when I could work calmly and not think about many things far removed from my work. I often wonder where I get my strength from and what the future holds but the way ahead is dark. I've had no reply yet from

---

* Kapitza's attitude to Blackett soon changed and they became good friends, as can be seen from the letter of 20 June 1923 (p. 169).

Abram Fedorovich to the letter I wrote him. I can't possibly return at present just when my work is in full swing and it's only now that I have really become part of the Crocodile's school and am getting on friendly terms with the young physicists. They keep coming to my room for a chat, for advice and to discuss their ideas. I know about the progress of nearly all the research going on and I have a good relationship with everyone.

Kolya Semenov has written urging me to return but to do so at present seems wrong to me, for I am only just beginning to do real work and I feel myself at the centre of this school of young physicists presided over by the Crocodile. This is undoubtedly the leading school in the world and Rutherford is the greatest physicist and organiser. To return to Petrograd and torment myself with the absence of gas, electricity, water and apparatus is simply impossible. It is only now that I have felt my strength. Success gives me wings and I am carried along by my work. After all, that is all I have left after the death of my family. . . .

(B, C, D, E, e)   Cambridge, 6 July 1922

In the last letters from you and Lyonya you seem to be dissatisfied with my letters and I sense a reproach. I think you have some justification for I have certainly not been a good correspondent recently. But, as I imagine you know, if you are not in the mood for writing there is nothing to be done and I have merely fulfilled your request that I should write once a week. So, my dear, you mustn't be too hard on me. . . . I shall try to give you a general idea of my situation. Imagine a young man arriving in a world-famous laboratory, set in the most aristocratic and conservative university in England, where the royal children are educated. And this young man, who is quite unknown, who speaks English badly and carries a Soviet passport is accepted by the university. Why did they admit him? To this day I don't know. When I put this question to Rutherford once he laughed and said: "I was surprised myself when I agreed to take you, but anyhow, I am very glad I did so.".

And the first thing this young man encountered here was a warning from Rutherford: "If instead of doing scientific work you spread communist propaganda I shall not tolerate it.". Everyone shunned this young man. They were all afraid of compromising themselves if they befriended him. Seeing that I must stake everything in this game, I took on a very difficult problem, hardly believing that I could make a success of it and very often thinking that it would all end in failure. But luck was with me. True, I often worked almost to exhaustion but I have now forced the breach. That, after all, is a happy ending but it cost me a lot of effort.

Don't imagine that I live here in luxury and in a state of bliss. The mental suffering I've had to endure during this time has prevented any possibility of an easy life. I've had the sensation all the time that I'm becoming more and more like some drilling machine, which is boring a tunnel through rock. This is only one of the difficulties. It's not worth describing all the others, mostly to do with the security of my position here, and so I can't often reply to the questions that Lyonya puts to me in his letters – the situation is such that I have difficulty in understanding it myself.

As regards my physicist friends I am not at all surprised at their attitude to me. I expected it from Semenov too. But it will pass. I now await waves of dissatisfaction from them all and also from Abram Fedorovich. The grounds for this seem to me psychological rather than rational and I feel I know how to handle the situation. The best way is to become like a reed that bends before the wind only to stand erect again when the wind drops. If the reed did not bend its slender stem would break before the lightest wind. Semenov writes: ''You're becoming estranged from us and you'll never get close to the English, so you'll be neither Russian nor English.'' He goes too far, for I'll never desert Russia, but all the same he is partly right. Breaking with our own physicists [in Petrograd] is inevitable, and I'm not afraid of it.

The fact is that I'm now subject to a different authority and have a different point of view from my friends in Petrograd because I have joined a different school. The working methods are also different. In Russia everything is cut according to the German pattern, which has little in common with the English scientific world. I can't think of a single Russian physicist who worked for long in England. But England has produced the most outstanding physicists and I'm now beginning to understand why. The English school lays great stress on the development of individuality and gives infinite scope for personal initiative. The absence of rigid patterns and routines is one of its basic characteristics.

Rutherford does not bear down on a man and is less insistent on the precision and elegant presentation of results than Abram Fedorovich. For instance, they often do experiments here that, in my opinion, are so absurd that they would be laughed out of court at home. When I enquired why these were undertaken it turned out that they had simply been thought up by the young people themselves and the Crocodile judges that to let a man prove his worth, he should not only permit but should even encourage him to work on his own idea, while at the same time trying to put some sense into projects which are sometimes absurd. The absence of negative criticism, of a kind that kills individuality and is used too much by Joffé, is one of the characteristic features of the Crocodile's school.

The second feature is the emphasis on obtaining results. Rutherford is insistent that a man must not go on working without getting results, because he knows that this can kill the urge to work. So he does not like to propose difficult projects and when he does it is simply because he wants to get rid of the man. The kind of experience that I had when I spent three years on a single piece of work, struggling with enormous difficulties, just couldn't happen in his laboratory. . . .

I'm now working on a very bold project*. In a few days Emil Yanovich Laurmann is due to arrive, but I'm very anxious that our physicists in Petrograd do not hear about this. The best proof of the Crocodile's high regard for me is his permission to bring Laurmann over.

I'm very worried about your situation, about the health of Natasha [his sister-in-law] and about the general commotion at home. You are evidently overworking, I once said to the Crocodile that I should like to work during the holidays. He flew at me with the remark: ''Everyone must have a rest and I want you, too, to relax or else I shan't allow you to work'' and so on. He was right. The English are sensible in this respect – three months work and then ten days of relaxation, repeated four times a year. I think you too should adopt this rule.

---

* This was the production of very high magnetic fields lasting only a fraction of a second by discharging a specially designed accumulator through a coil; the high fields were first used to extend his earlier study on α-particles (see footnote to letter of 15 June, p. 150).

(B, D) Cambridge, 30 July 1922

I haven't written to you for two weeks or so because I've had many things to do besides my usual work in the lab. First, as you'll see from the new address, I've moved to new lodgings. Second, Abram Fedorovich came to London and I had to visit him there twice. Then he came to Cambridge to visit the lab and Rutherford welcomed him warmly. He invited him to dine in college and I was invited too. After dinner we played bowls – Rutherford, Fowler and I against [G.I.] Taylor, Aston and Joffé and we won. . . .

As regards my stay here, it is essential that I should prolong it because just now my research has assumed a very wide scope. I have achieved considerable success in the new project and the results should be very interesting. Judging from what Joffé has told me about the Röntgen Institute, I gather that the temperature there is barely above freezing point though he has tried to present things in a more attractive light. His personal view is that I should work here for about another year. But funds are now very tight and I don't know what the outcome will be. All this creates an uneasy atmosphere for my work . . .

(B, D) Cambridge, 17 August 1922

. . . Preliminary tests of my new experimental set-up have been completely successful. I hear that the Crocodile can talk of nothing else. I've been given a large amount of space in addition to the room in which I work and for the full scale experiment I have permission to spend a fairly large sum of money [£150!]. Nevertheless I sometimes feel very unsettled, worrying about you, my dear ones, and everything seems precarious. But I think I now have great opportunities. It looks as if I've actually succeeded in opening up an interesting untouched area of research and that doesn't happen often in a lifetime. In all probability I shan't return home before the winter but I'm planning to come and see you all at Christmas, or failing that at Easter. This will be very difficult but perhaps not impossible . . .

(C, D, E).   France, 2 September 1922

I left Cambridge two weeks ago and I'm now in France. I was invited
to go on a yacht trip with a good friend [G.I. Taylor] who had just
returned from a trip to Scandinavia and Finland. He has a beautiful
two-masted yacht and we enjoyed two excellent days at sea. I then
went on to Paris and spent two days there. I'm now in the South of
France and I intend to relax for a couple of weeks here – a rest won't
do me any harm. I have missed the sun in England but here there is
more than enough . . .

(C, D, E, e)   Nice, 14 September 1922

For you alone

Today I very much want to have a heart-to-heart talk with you – but
with you alone. You know that I don't often write fully about all that
is going on inside me. Dear mother, you alone understand my
rebellious nature which finds no peace. I should have stopped rushing
around long ago – after all, I am nearly 30.

I had a letter from Rutherford yesterday, saying that my financial
support is almost arranged. It seems that the funds will come from the
Royal Society*, although I don't know the details. I shall remember
the last conversation I had with Rutherford for the whole of my life.
After paying me a whole series of compliments he said: "I should be
very glad if I had the possibility of providing you with a special
laboratory here in which you could work with your own students".
I now have two Englishmen working with me [P.M.S. Blackett and
H.W.B. Skinner]. Judging by the generous way he provides for my
work, and the attention he pays me, these are perhaps no idle words.
He has already given me two rooms, one of which occupies almost the
whole of the attic and is half the size of our apartment in Petrograd.

Am I really so gifted? I feel overawed. Can I cope? Is it perhaps only luck? I have certainly had luck, but so much I never expected in my
whole life. After all, I am still inexperienced – a young and stubborn

---

* The Russian text says Royal Institution, probably a mistake by Kapitza (see next
letter).

boy. Someone recently took me for an 18-year-old. There is already talk of my presenting my work at the Royal Society. The prospects ahead are wide but frightening. So much depends on good fortune and success. Mother, pity me!

You know, I'm nearly in tears as I write this – I don't know why. But one thing I do know: that I would give all this up if only Nimka and Nadya could come back to me. I now have enough money. I travel first class and am staying in Nice in a hotel room looking out to sea, with all comforts such as bath and so on. Through the window I can see palms and the endless blue of the Mediterranean. I have everything, yet I am as lonely as the boat I can see out there. It knows when it will return to shore, but it will be a long time before I find an anchorage. I shall have many a struggle against storms and bad weather.

What is happiness in this life – where can it be found? I seem to be losing it. I thought that if I accomplished my scientific projects I should be happy, and I have indeed achieved more than I had hoped for. But for what, for whom do I have to achieve all this – these magnetic fields of enormous strength? They may open up a new field in physics – perhaps they will. But what is it all for? It will only swell the number of envious eyes, of which quite a few are already directed at me. Perhaps a personal life is essential for me? I don't know, but any kind of personal relationship takes a lot of effort which, now as never before, I need for my work. Whatever is all this for? Outwardly I am as calm as ever, I have the urge to work and I look well. But sometimes I want to tell you what is going on inside me. For you understand me better than anyone and you alone know how deeply I experience everything and how terribly difficult it is for me to forget the past.

You mustn't, of course, say a word to anyone about my successes. After all, these are only plans, but do remember, my dear, that if these plans come to fruition, then I shall take you to Italy, to Rome, Naples, and Florence next summer as I promised long ago. I still remember this promise and shall fulfil it if possible*. And the thought of this journey gives me greater joy than the discoveries I hope to make this year. Perhaps nothing will come of all this, but we shall see . . .

---

* This journey did not in fact prove possible.

(D)   Cambridge, 10 October 1922

I've come back much refreshed from my travels and have taken up work again. I must tell you about the state of my affairs. From now on I shall receive a grant from the Royal Society to cover my expenses (£450 a year). This is flattering but, considering the high cost of living here, it isn't all that much. . . .

In Paris I often went to the theatre and in Nice I was warmed by the sun and got a good tan. I went to the Comédie Française and the Palais Royal and recalled a little of my French. The Louvre and the Musée Rodin made an enormous impression on me. But by now that's all hazy and I'm thinking only about work. Am I going to fail this time? . . .

(B, C, D, e)   Cambridge, 22 October 1922

Yesterday I lectured in English for the first time to a fairly large audience in a scientific club. I spoke about one of my theoretical studies that I am thinking of publishing in the near future. Everything went very well. About 30 people were present, all established scientists, some of them mathematicians and some physicists.

My research is going very well and the definitive experiments have given satisfactory results. However, I made a small error in the technical details of the apparatus design and when I told the Crocodile about it he said: "I am very glad that even you can make a mistake!". You see how skilled he is at paying compliments, for in actual fact I very often make mistakes (after all, if you never make a mistake you never do anything). During the next two weeks I shall make further tests on my apparatus. . . .

Rutherford is really exceptionally good to me. On one occasion he was not in a good mood and he told me I must economize. I showed him that I was doing everything very cheaply and in the end he couldn't deny it and said: "Yes, yes, that's all true, but it is part of my duty to talk to you like this. Bear in mind that I am spending more on your experiments than on all the other work of the laboratory put together". And, you know, that is true, for our equipment fairly runs off with his pennies. Although everything is going well at present, it is always easy to blunder in a new field, so great care is needed and I, my dear, am still very young and inexperienced . . .

(B, D, E, e)   Cambridge, 3 November 1922

. . . I've had some success in getting to know people. I've organised a scientific club [this came to be known as the Kapitza Club, see p. 40] and I am giving lectures on magnetism. This takes up a lot of time but though the scheme still has many snags it is going quite well.

Now about the Crocodile – at the moment I have good relations with him, indeed very good. He not only gives me all sorts of support and shows great interest in my work but he also frequently chats with me about scientific matters. I find it funny that, like Joffé, when he has given a lecture he calls me in (when no one else is about, of course) and asks: "Well, what did you think of that?". He loves to be praised and, while he is always brilliant, I try to give some criticism as well, but in such a way that he won't be vexed. You know, mother, he is the very greatest physicist in the world! Yesterday we talked together for nearly two hours about an idea he had put forward in his previous lecture. Fortunately it touched on a matter that I know a lot about. As you know, my dear, I'm not very clear when I speak, because my thoughts jump ahead of the logical argument and there are few people who can understand me quickly. Joffé was one and Kolka Semenov, too. But if you allow for my poor knowledge of English, the Crocodile unquestionably holds the record.

He has called me in quite a few times to advise on experimental apparatus. He has told me of his clashes with various eminent scientists such as Lord Kelvin, and he has given me his private opinion of various scientists in the lab and outside. But, in spite of all this, I am not sure that his good relationship with me will last. He is extremely temperamental and can swing far to one side and then bounce back again. By now I know his character fairly well. Because his room is opposite mine I can hear him closing his door and from the way he does it I can almost infallibly assess his mood.

Now about my work: at present I am working on the production of magnetic fields of great intensity. These are needed to study certain phenomena in radioactivity. I began this work, as I have already told you, with a young English physicist called Blackett. I proposed three methods for obtaining such fields. The first of these had to be rejected on theoretical grounds, leaving two others. We began work on the second method and straightaway struck almost insurmountable technical difficulties. While my collaborator continued with this approach I tried the third method and immediately obtained positive results. After that Laurmann, Blackett and I worked on this third method for six weeks and succeeded in firmly establishing its suitability. It was then only necessary to go from the small-scale experiment to a larger one. Since these high fields open up a new field of research and since the Crocodile believes we shall be successful, he has decided to fund the large-scale experiment and has allocated me a large room, almost a whole floor, though in the attic. More than a thousand gold rubles [about £100] have been laid out for the purchase of equipment and the work is already in full swing. What the outcome will be is hard to say. The apparatus is nearly ready and we shall be trying it out next week. Much depends on the success of the test. . . . Wish me success!

(B, C, D, E)   Cambridge, 29 November 1922

For me this day is somewhat historic, for today I obtained the result I had been hoping for. In front of me is a photograph [see fig. 55] on which there are just three curved lines. But these three curved lines are the paths of α-particles in a magnetic field of enormous strength. These three lines have cost Professor Rutherford £150 and myself and Laurmann three and a half months of very hard work. But here they are and everyone in the university is talking about them. Strange! Only three curved lines. The Crocodile is very pleased with these three curved lines. Of course, this is only the start of the research but even from this first picture one can draw a whole lot of conclusions about matters hitherto not known at all or only guessed at from indirect data. Lots of people have come to my room in the lab to look at these curved lines and admire them. We must now go further and there is much work ahead. The Crocodile called me into his office today to discuss plans for further work. And so, this time I have not failed. That's good!

In a few days I shall send off another paper for publication, a theoretical one, entitled "On the theory of delta-rays" [see (7)]. This paper is not particularly remarkable but there are some positive

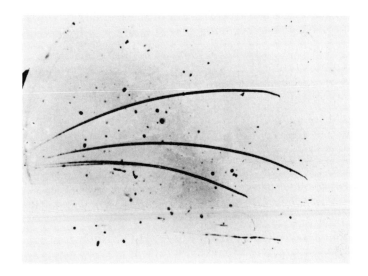

55. Three curved lines (α-particle tracks bent in a momentary high magnetic field).

results. So, my dear, you can see that I'm working flat out and not without success. But how I miss you all! I'm very tired, but the holidays are in two or three weeks time and I hope to have a rest . . .

(B, C, D, e)   Cambridge, 4 December 1922

For some reason or other there have been no letters from you for a long time and that always worries me. My work is going well, indeed, at the moment, excellently. I wrote in my last letter about the success of my experiments and I have now obtained even better results so that for these last few days I have become a kind of celebrity. On Saturday, 2nd December, there was a reception at Professor J.J. Thomson's home on the occasion of a visit by the Dutch physicist Zeeman. I was invited, as was everyone in the lab. About 70 people were there and, of course, I had to squeeze into my dinner jacket.

Your name is announced as you enter the drawing room and you shake hands with your host. I then slipped away into a corner but J.J. Thomson at once came after me, took me by the arm and said: "I want to introduce you to Zeeman who is very interested in your experiments'', and so on. I talked with Zeeman, having been introduced as "the kind of physicist who solves problems that are considered impossible". These generalissimos put me through the hoop for 20 minutes or so, after which I again crept into a corner. But J.J. Thomson found me again and invited me into his study with a few young physicists. There he entertained us for about an hour with various stories about the inventions he was involved in during the war.

Today Zeeman and Lord Rayleigh (the son) visited me in the lab and looked at my work. At the Thomsons', Newall, the astronomer (Aunt Sasha will know his name) also came up to me and said that he had heard so much about my experiment that he wanted very much to see it. All this is, of course, very pleasant since it will make my position here more secure.

I have had an invitation from Ehrenfest to visit him in Leiden and lecture to his seminar and I'll go there at the end of the week. This is a very welcome invitation for I'm so overtired that I shall be very glad to interrupt my work and have a complete rest for a month or so. From Holland I am thinking of going on to Berlin to arrange for my visit to Petrograd at Easter. My success elates me mainly because it sets me more firmly on my feet here. I wanted to tell you all this because I know that you will be happy for me, my dear, more than anyone else in the world. This time I've had more luck than I could have expected, but I'm terribly tired, as never before. I have decided to do nothing this week and moreover, I've caught a slight cold. . . .

(D)   Leiden, 13 December 1922

I have already been here three days and shall stay for two more. I am staying with Professor P.S. Ehrenfest and tomorrow I have to lecture at the University. Today I have been to see Professor Zeeman in Amsterdam at his invitation and I spent the whole morning in his famous laboratory and had lunch with him. Afterwards I looked at the Rembrandts in the museum and thought of you. The Night Watch, Group Portrait of Elders of the Guild of Clothiers, and The Anatomy Lesson are the best exhibits they have here and are almost the only things worth looking at. Frans Hals is very good, of course, but they haven't much of his work here. . . .

Ehrenfest is a brilliant man, and very original. Although intelligent and quick on the uptake, he is somewhat irresolute and doesn't concentrate long enough on a problem to get deep into it and force a way through. His special gift is for discussion and a conversation with him is very rewarding . . .

(B, D, E)   Cambridge, 27 January 1923

I am working hard and have given two lectures. The Crocodile invited me to lecture to the [Cavendish] Physical Society but I asked him for a postponement till the end of term. The work is going well, though it seems to me rather too slowly. However, I think I'll be able to get numerical results within a month.

I decided to join a College and become a member of the University. On Wednesday I was made a member of the University and on Friday admitted to [Trinity] College. I have been given some exemptions and it seems that in five months time I shall be able to obtain the degree of Doctor of Philosophy. This, I think, will be useful. The dissertation is a trifling matter and the examination even simpler. But the most amusing thing is that I have been registered for the degree of Bachelor [B.A.] (all this, of course, was arranged by the Crocodile, whose kindness to me really knows no bounds), and these five or six months I have to be under the supervision of a tutor and wear a uniform. The uniform is very amusing, having been handed down from olden times – it consists of a square black cap with a tassel and a black gown.

As you probably know, the colleges are descended from mediaeval monasteries and much of the ceremonial has been preserved to this day. I shall be promoted to Doctor in a mediaeval ceremony of laying on of hands, rather like what we saw in *Le Malade Imaginaire* at the [Moscow] Arts Theatre. Meanwhile I have to be home by midnight and am not allowed to go out after 10 p.m. But the main advantage is that as a member of the University and of a College I enjoy their protection. I'll have my photograph taken in cap and gown and send you a print in a day or two. I made an Englishman rather angry when I told him that this uniform would do for a fancy dress ball in Russia . . .

(D)   Cambridge, 1 February 1923

. . . I matriculated on Monday. This is not at all dangerous. You are
led into the Senate House and sign your full name in a thick book.
After this you are considered a member of the University for life . . .
I'm a bit worried whether I'll be able to come at Easter. On the Continent things are boiling up just now and an explosion seems likely*.
London is now cut off from direct communication with Berlin (by way
of Dover and Ostend). It looks very much like revolution and war. My
work has made some progress and I have sent off a small paper on the
theory of delta radiation [see (7)]. In about a month Ehrenfest is coming here at my invitation and will give three lectures to our club
(where I, too, lecture). . . .

(B, D, E)   Cambridge, 18 February 1923

. . . By the time you get this letter it will be just two years since I left.
Good Heavens! – two years of separation! But how much has happened in the meantime. These years have been perhaps the most difficult
and most testing of my life. I was thrown into the air and managed
to fly with my own wings. The flight was daring but I think I can now
say confidently that I didn't crash, as might easily have happened. I
came here and worked for a year and a half; at first I was treated with
suspicion and mistrust but now I'm playing one of the first fiddles.
People talk about my work and come specially from London to see my
equipment. I have been invited to work in other universities. The
Crocodile is almost like a father to me. People value my opinion and
come to me for advice.

How everything has changed! How strange it is looking back. Some
of the moves I made could almost be called crazy, for instance, asking
for Laurmann, but the Crocodile, too, appreciates him now and has
increased his salary. He is regarded as my personal assistant. I
sometimes feel awestruck when I realise how much better the present
situation is than I could ever have dreamt of. And so, my dear, these
two years have so far yielded results. What of the future? God alone

---

* This refers to the crisis arising from the occupation of the Ruhr by Belgian and
French forces.

knows. After sunshine comes rain, but then the clouds clear again. But how much, at times, I long for home! I'm extremely unhappy that I'm unable to describe to you all that I'm doing – what my projects are and what I propose to do. It's difficult to write about this because my projects are often bold and oftener still I act intuitively not knowing where my action will lead me.

One of my main problems is where to make my base – here or in the [Petrograd] Polytechnic? If I compare my chiefs, then the support and encouragement I get from the Crocodile is more than I have had from any other chief. The funding and in general the opportunities for work are also much better here. But what should I put first – the possibility of doing scientific work calmly and successfully, or to give up scientific work and return to my former situation? If I returned I would, of course, be with my nearest and dearest, my friends and, moreover, have my mother tongue. How can I decide? However, like an ostrich, I make no decision at all, and put it off for the future. Were it not for the sad memories of what I lived through, I might perhaps have returned long ago. But here, in the excitement of my work, I am so carried away that I forget everything. . . .

(D)   Cambridge, 9 March 1923

. . . Professor Ehrenfest* from Leiden came over for a week and he stayed with me, so I had to spend all my time with him. Then there was my lecture to the Cambridge Philosophical Society . . . I have never felt so nervous as I did before this lecture, because I was putting forward a rather bold hypothesis with as yet little foundation. This was a preliminary account and I wasn't very happy with it. It was not as rounded as I would have liked but it will be published. I have heaps of things to do and Ehrenfest's arrival has put me off my stroke.

I'm terribly sorry to say that my trip to Petrograd will have to be

---

* Ehrenfest was very enthusiastic about his visit and in a letter to Joffé on 16 May 1923 (see *Ehrenfest –Joffé: Scientific correspondence*, 1973, Nauka, Leningrad, p. 167) he wrote: "I spent a week with Kapitza in March. He is a phenomenal chap. He has managed to win the respect and sympathy of all those around him and occupies a unique kind of position in Cambridge. Rutherford rates him highly and is exceptionally fond of him".

postponed until the autumn. However much I long to go, I cannot just now absent myself from Cambridge for more than about ten days or so, which is hardly enough for a visit to Petrograd. In autumn there will be a whole month's holiday, or even more, and then I'll be able to visit you. The situation, my dear, is such that it isn't possible just now, as I hope you will understand. Ehrenfest is of the same opinion. All my friends are trying to talk me out of going and I myself understand why. I only now obtained results from my experiments and during the next two or three months I must confirm what I have already found and, perhaps, go further. My experiments have now assumed a large scale * and cost a lot (5000 gold rubles or so) so it is difficult to interrupt them in view of my moral obligation to the Crocodile. Secondly, the award of my doctorate will be finally decided in June but I'll receive it only if I do not absent myself from Cambridge for more than two weeks, that is, as they say here, I must "keep term". Third, I am expecting an improvement in my financial situation at the end of March and it would be awkward if I were not here. Finally, my current account is such that I cannot spend more than about £15 on holidays.

I know how sad you will be that I am not keeping my promise, but by making this trip I could destroy at a stroke all that I have built up in the course of two years. Autumn is much more favourable and I think there is a 90% chance that I can be with you then for two or three weeks. And so, my dear, don't be sad. I am sure you will understand that I shouldn't stake everything on a single card. Remember that your son's fate is in the balance. It has been so difficult for me to write this letter because I know it will grieve you.

(B, D)   Cambridge, 18 March 1923

. . . I am afraid that you have a false impression of my situation here. The fact is that my life is not all that sweet, but full of worries, struggles and work. I am about to send another paper for publication and I plan to publish three more papers this year. I have been allocated £1200 for my experiments (all this must remain strictly within the

---

* The reference is to the study of the Zeeman effect in the very high magnetic fields developed earlier [see (11)].

family). You can see that the scale of my research is growing larger and larger. The club I have organised [see p. 40] takes a lot of effort. But the one thing that eases my work is the Crocodile's kindness which bears comparison only with that of a father. I am up to my ears in work and this is one of the most critical moments. I am very tired and dream of the time when I can relax. I shall probably leave Cambridge for a week.

I'm afraid that you, my dear ones, do not appreciate that I if I don't get my feet on firm ground now there will be no second chance for me in life. What am I and what am I living for? The most valuable thing I have is my work, and if I were to tear myself away from it, my life would be darker than a black cloud. But you know my opinion: science is only worthwhile if you play first fiddle. As I have written more than once, I don't know what I am aiming at, but one thing is clear to me, it is not the tranquillity of the family hearth. Such comforts as I allow myself are very modest but they are essential. For dinner I take only one course, and I live in the lower class part of the town in a very small room. All my surplus money is spent on my work – books, instruments, Laurmann, and on the one luxury that I allow myself – relaxation during the holidays. I travel to the Continent, to Paris, where among new acquaintances I can escape from work and its worries.

My dear, I have an infinite longing to come to you, but this is so difficult that I have to say frankly that it's impossible without destroying everything I have achieved during these two years. In the autumn it will be easier. You know that I am not afraid of difficulties in life, and if I say that a thing is difficult then it really is. But I fear you do not fully appreciate what I have actually achieved – it is quite difficult for you to have any idea of this. I didn't mention it before, but I've had four fainting fits brought on by overwork. So your reproaches that I am drifting away from you hurt me deeply, very deeply. No, everything near and dear to me in the world is with you.

People are always sceptical about my initiatives and my projects and every new thing I undertake is thought impossible. And I have often heard complaints, for instance from Joffé, about my bringing Laurmann over. Now that I have succeeded in getting him well established, his wife is also coming. But all these complaints are as nothing compared with your writing that I am forgetting you.

(B)   Cambridge, 14 April 1923

. . . The main work has been done and the results make my head spin. The scale of my work is now very large and this always alarms me, but the fact that the Crocodile is behind me gives me courage and confidence. You can't imagine, my dear, what a great and remarkable man he is . . .

(B, D)   Cambridge, 15 May 1923

. . . I finished my dissertation today and tomorrow I'll have any mistakes corrected. I'll take it to the typist the following day and submit it on Saturday. It has turned out rather short but I think it will do. The exam will be in June, presumably the last exam in my life. However, life itself is the best exam that you can think of. Laurmann is getting somewhat excited as his wife and daughter are due to arrive tomorrow. The Crocodile made it possible for them to come here – he has a heart of gold. There are clouds on the political horizon*. I don't understand what is going on.

(B, C, D, E, e)   Cambridge, 20 June 1923

The degree of Doctor of Philosophy was conferred on me yesterday. The ceremony in the Senate House was very traditional. The [Vice] Chancellor of the University, in a scarlet gown with an ermine collar, sat on a chair like a throne on a dais. Various officials of the University stood around him: the Proctors, the Public Orator and others. I'll send you a list of the recipients and you'll find my name there.

Wearing evening dress with a white bow tie, small black cap and red silk gown, I was led by the hand to the [Vice] Chancellor. The whole ceremony was conducted in Latin. I was presented to the [Vice] Chancellor in a fairly long speech in Latin of which I understood only two words – Pierre Kapitza. Then I knelt on a small red velvet

----

* On 8 May 1923 the English Foreign Minister, Lord Curzon, sent a memorandum to the Soviet Government which amounted to an ultimatum, and threatened to revoke the Soviet – British trade agreement.

cushion at his feet, placed my hands together and held them out before me. He took my hands between his own and began to speak in Latin as if praying. After that I arose with the title of Doctor. The whole ceremony and the costumes have, of course, frankly mediaeval origins and bear the stamp of the era when Cambridge colleges were monasteries. It is only 75 or 100 years ago* that Fellows of colleges were first allowed to marry.

As a doctor I have acquired a lifelong right to a vote in the Senate, the right to participate in all festivities, to wear a black and red silk doctoral gown and a velvet cap shaped like a pancake with a golden tassel. In addition, I have the right to use the University Library and to have four free dinners a year in my College. Nevertheless, this degree has cost me so much that I am left almost penniless, though thanks to a loan from the Crocodile I shall be able to go on holiday. I shall probably accept Ehrenfest's invitation to visit him and I may go for a few days to Paris.

I have been involved in the following affair. This year the Clerk Maxwell Scholarship becomes available. It is given for three years to the best research worker in the laboratory and the award is considered a great honour. It is, moreover, a fairly large sum, £750 spread over three years. I gave no thought to it, of course, but several colleagues asked me whether I was going to apply for it. I said no. On Monday, the last day for submitting applications, the Crocodile sent for me and asked why I was not applying. I said that I considered my present salary entirely adequate and, as a foreign guest, I ought to be modest and remain content with what I had. He told me that my foreign origin would in no way prevent my receiving the award and then asked, in strict confidence, whether I knew that Blackett, one of the most capable of the young physicists here, was also applying for the scholarship. I replied that I thought Blackett should receive it, and considered he had more need of it than I, as he was going to get married and could hardly get by on the income he had now.

Once I knew that Blackett had applied for it I definitely decided not to apply, as it seemed to me that I shouldn't stand in the way of a friend. Besides, the receipt of this award is much more important for an Englishman since it represents a valuable qualification. For me, as

---

* In fact, it was in 1878 that Fellows were first permitted to marry without losing their Fellowship.

a bird of passage, it doesn't play any role. But, evidently, the Crocodile couldn't understand my psychology and we parted rather coolly.

Then I looked into the matter more closely and discovered that the Crocodile thought I was the right candidate and when others had proposed applying he had dissuaded them, saying he had earmarked the scholarship for me. No one had told me this before my talk with him. Evidently he hadn't thought I would decline and my refusal rather perplexed and offended him. Nevertheless, I feel I did the right thing, though in my heart I suspect I may have offended the Crocodile, who has been so infinitely kind to me and has taken so much trouble on my behalf. I am afraid he can't understand the psychological reason for my refusal.

However, it seems it will all end happily. Before his departure for a month's holiday I met him in the corridor. I had just returned from the Ph.D. ceremony and I asked him directly: "Don't you think, Professor Rutherford, that I look more intelligent?" "Why should you look more intelligent?" he replied to this somewhat unusual question. "I have just been created a Doctor" I said. He at once congratulated me and laughed, saying: "Yes, yes, you look much more intelligent. And, moreover, I see you've had your hair cut!"

Such levity with the Crocodile can be very risky, because usually he tells you to go to the devil, and it seems I am the only one in the whole lab who takes such risks. But when it comes off, I know that all is well between us. I have chaffed him a few times with this sort of whimsy and at first he is at a loss and then says: " Go to the devil!" He is not at all accustomed to such an approach from a younger person, and I have had such compliments as: "Fool!", "Ass!" and so on from him half a dozen times, though he is a little more used to it now. Most of the people in the lab can't conceive how such flippancy is possible but I find it very funny when the Crocodile is so taken aback that at first he is at a loss for words to reply. However, on the whole we get on well . . .

After all these events I am very tired, although I don't know why, for I haven't worked all that much. Perhaps it's a reaction to all the last six years . . . Looking back at what I've had to live through in these six years since my marriage it seems more than most people have to put up with in a whole lifetime. And the final result is: Doctor of Philosophy of Cambridge University, widower, wanderer and scientist.

(D, E) Paris, 7 July 1923

. . . I shall spend a few days here before going on to Leiden for a further few days and I expect to return to Cambridge in about ten days time. You reproach me for writing little about myself. The fact is simply that I know very little about myself. You say that I am losing touch with you, but I think you are quite simply mistaken. When I'm working and I feel my strength and have enough energy, I don't have the possibility of thinking about the past, and then I suppose I'm happy. But during the holidays I sometimes feel awful, like a lost puppy, miserable and lonely. Others value me just for my work – without that I am simply an unwanted lump of flesh. That's why I can never forget you, my dear, because I am precious to you in myself, as your Peter. I am often faced with the painful question: Is it right that I should behave as if I had completely given up any aspiration to a personal life and turned into some sort of machine, which goes forward without knowing whither? I really don't know, but if I get my teeth into something I keep at it till it's finished. That's the rule I try to follow.

Two stars shine brightly as a background to my daily worries and troubles – how I can help you, and how we can meet. Both are difficult problems, and I don't know whether I shall be able to resolve them satisfactorily this year. But believe me, my dear, I think of them all the time and if I don't fulfil them it is only because I can't. When I'm with other people, I appear cheerful, but inwardly I often feel wretched.

(D) Leiden, 18 July 1923

I am staying with the Ehrenfests for a few days. . . . Tatiana Alekseyevna, Ehrenfest's wife, is planning to spend a couple of months in Russia – most of the time in Moscow, but she will also visit Petrograd. She is very interested in pedagogical questions and went through the Bestuzhev courses*. It would be very nice if you could invite her to stay in our home. The Ehrenfests, as I have already told

---

* These were pioneer higher education courses for women in Russia (see p. 2).

you, have been very hospitable to me and I stay with them when I'm in Leiden. I think too that she will be able to tell you a good deal about me and you will enjoy getting to know her, for she is a very likeable person . . .

(B, C, D, E, e)   Cambridge, 23 July 1923

I'm back at work in Cambridge again . . . When I got here the Crocodile again suggested I should apply for that scholarship, saying that he didn't think any of the others deserved it, so I gave in and wrote an application. I'm very pleased it has turned out like this and the scholarship will be most welcome. My material circumstances will now improve considerably so that at last I shall again be able to send you some help, and I think Lyonka's trip abroad will now be completely assured. . . .

(A, B, e)   Cambridge, 23 August 1923

I got the Clerk Maxwell Scholarship and have had many congratulations. This scholarship is awarded by the University once every three years. There are others, awarded by the Colleges, but they are not so highly regarded. This is the only scholarship specifically for physics. I'm sending you the University Reporter [of 13 August] in which you can read in one place that I've been awarded the scholarship and in another my name appears in a list establishing the seniority of those recently created Doctors of Philosophy.

(B, D)   Cambridge, 30 August 1923

. . . Tomorrow I leave Cambridge for the vacation. The lab has already been closed for a week. I have been busy ordering some equipment of a rather unusual type for my work and have had to make drawings and calculations. I'm attempting new experiments according to a very bold scheme. It will be very good if my luck holds out once more. Yesterday evening I visited the Crocodile and we discussed some of the problems. I stayed to supper and we talked a lot on a variety of subjects. He was very friendly and much interested in my projected experiments. I spent about five hours with him. He gave me a picture of himself and I'll make a copy to send to you. You asked me for this long ago and now I can fulfil your request [see fig. 70].

In your letters I often detect a note of sadness. Yes, it will indeed soon be three years since we saw one another. Never before have we been parted for so long. But I think you will be pleased with what I have achieved during these three years. Only recall how helpless and miserable I was when I arrived here and compare this with the way I'm treated now. But it's alarming that everything I've achieved could be destroyed by a careless move and a second chance in life is hardly likely to be granted. Beginnings, they say, are always difficult. . . .

(D, e)   Annecy, 15 September 1923

I've arrived at Annecy in Savoy . . . I'll stay here for about a week, go up on the funicular to Mont Blanc, and then return to Cambridge. The cost of living in France is so much lower than in England that the holiday, including travel, costs me just about as much as if I had stayed in Cambridge. A young Englishman called Skinner [see fig. 56] came with me so I'm not alone, and we shall go walking together during the holiday. He, too, works in the Cavendish but is much younger than me, having just turned 23. We shall probably collaborate in an experiment this year. I've stayed with his family a couple of times. They are fairly well-to-do and have been very kind to me. On the last occasion they took me to the theatre to see Pavlova, who is on tour at present.

Today it is raining so heavily that it is hardly possible to go walking. This trip to the Alps has reminded me of my visit to Switzerland with you many years ago. The same mountains, the same blue lakes, and the same crowds of tourists. It is not the high season just now, so there are fewer people and prices are lower.

I must tell you about my voyage with G.I. Taylor in his yacht. We sailed for eight days, and encountered a storm which swept waves over the yacht. Sailing day and night, we covered about 400 miles, mostly in the open sea out of sight of land. There were three of us and we sailed by the chart, using compass and log. We started from the Isle of Wight and got nearly as far as Hull before turning back to London.

56. H.W.B. Skinner.

The yacht rolled frightfully almost the whole time. The first day I was seasick, but after that I got used to it. I prepared the meals, using a Primus stove. It was a large yacht with three cabins and bunks for seven persons. Three people could only just manage the sails. During the storm, which lasted two days, we were thoroughly soaked and there was no point in changing clothes only to be soaked again. The rain poured down and the waves lashed us. At night it was rather frightening. Sitting at the helm you glance at the compass and then look out for the lighthouse. When you see the light in the distance your spirits rise. I shall soon be back to take up my work again.

(B, D)   Cambridge, 4 October 1923

Today I got your letter and Lyonka's, telling of T.A. Ehrenfest's arrival. Very many thanks to you all for giving her such a warm welcome. It's quite a time since I wrote. I simply haven't felt like writing, my dear. It's about a week since I got back to Cambridge and started work again. I was one of the first to arrive in the lab, and I've been assigned a large room divided into three parts and specially renovated for me. . . . It's a splendid room. I've not yet got going at full speed but I hope to soon. I enclose the photo of the Crocodile, a copy of the one he gave me. I'm waiting impatiently for your book – I am always glad when I hear you have published something.

(A)   Cambridge, 12 October 1923

Life is gradually getting back on the rails again after the holidays and the town is assuming its term time character. Undergraduates in gowns stroll along the streets and the lab is full of them too. There are lots of eye-catching notices of lectures and other activities.

I'm gradually getting drawn into my work too and wonder what this year will bring me. I must hurry up and finish the work I'm doing and publish it. There's not a lot left to do but the last step is always longer than you anticipate. I had expected to finish everything during the summer but now I shall be glad if I get it done by Christmas. I haven't felt much like writing recently and I've neglected my cor-respondence . . . Above all, I can't bear writing to English people since I don't understand English spelling at all . . .

(B, D)   Cambridge, 21 October 1923

. . . I'm working every day but nothing special has happened. My life flows along steadily and monotonously. I've joined a chess club for relaxation. Laurmann is very keen on chess and solves problems given in the newspapers. There are competitions for problem solving and we each won a book. I have got bored with it but he still goes on com-peting. Then there are meetings of the club I founded and these too provide relaxation. Things seem to be going well and our discussions are very free. The Crocodile, too, has started a colloquium in the Cavendish and I have been elected to the five-member committee which selects material and organizes things.

Two young men will be working under my guidance*. I have got

---

* This refers to P.M.S. Blackett and H.W.B. Skinner.

a good deal involved in the work going on here and I get badgered a lot. But I'm doing little work myself just now, mostly reading. I ought to give an optional course of lectures, but somehow it hasn't been possible to arrange this. There's a lot to be done but I feel rather washed out . . .

(D)   Cambridge, 3 November 1923

. . . I feel so lonely! English people are kind but I am a stranger to them and they to me. I am a bird of passage, who flew in here for a while and will fly away again, and I have no nest. This room in which I live is small and crammed with books and papers, so there is nowhere to pace around as I usually do when I'm thinking. I detest it, and the horrible wallpaper, which is such as we might stick on the wall of a passage . . . A portrait of Edward VII with legs spread wide and Queen Alexandra with a long train aggravates my gloomy mood. I long for our drawing room and our pictures.

I have been here for two and a half years. Scientifically I have grown stronger and matured, but what have I gained in spirit? I have no family circle to visit, to chat or to joke with (our family weakness). I hardly even know the language. People here are aloof, only interested in sport and the weather. Conversation is boring, insincere and limited. It's as if everyone is afraid of saying something unconventional or stupid and so prefers to say nothing at all. I ought to get busy learning the language, but I have no time. I don't feel comfortable here.

I am well fed and well clad but bored, Mother, bored to tears. But what can I do? I have splendid possibilities for work here, and nowadays one has to become someone and to know something. I am absorbed in my work and find happiness in it but my character will suffer if I stay here too long. But if I were to go home that would be equivalent to stopping, or almost stopping, my research since I should have to teach a lot to earn enough. And the working conditions would ensure that for the same investment of time and labour I'd get less in the way of results. Would I feel better in myself? I think not, because my work is the centre of my life and as for the rest – I suppose it is more or less unavoidable and I can bear it for another couple of years. But, my dear, write to me more often! You don't know how much I need your letters . . .

(D)   Cambridge, 25 November 1923

I'm working terribly hard all the time. The experiments have moved forward considerably and I also have to give two lectures soon. Then I must begin to write up my work and this is a lengthy and tedious task. There is also a small theoretical calculation which I plan to publish jointly with one of the physicists here . . . While working I forget about what is happening in the world at large. During the intervals between tasks in my work I dip into your stories in "Living water" and, although they are written for readers somewhat younger than myself, I find them amusing for they are so well chosen . . .

(B, D, e)   Cambridge, 18 December 1923

. . . I'm in a strange state of mind and I feel rather irresolute. I often wonder what is my real aim in life and when I shall cease to be a wanderer over the face of the earth. Living in furnished rooms, often eating in restaurants and solitary evenings are likely to spoil my character in the long run.

The letter from Joffé asking whether he could count on me to fill the Chair vacated by [V.V.]. Skobeltsyn, and his hints about the Radium Institute, sound like an offer of the Directorship. At first all this upset my equilibrium. The offer is very flattering and in normal circumstances it would be hard to wish for anything better. I have discussed the question several times with the Crocodile. However, it is rather early for me to take a professorship because it would take away too much time from my scientific work. Moreover, in the present situation at home, it is difficult to carry on scientific work. What could I do in a laboratory where there is no gas, for instance, and no contact with West European scientists? Here I have organized a small circle for free discussion [the Kapitza Club] and although there are only 10 of us, we invite the most distinguished physicists: Franck from Germany, Bohr from Copenhagen, Ehrenfest [from Leiden], G.N. Lewis from America, and others. I am able to travel to the Continent, to Paris, three times a year, to visit the Sorbonne and Marie Curie's Radium Institute, and I enjoy the lively exchanges and close contacts.

In Petrograd it would take two or three months of effort to get a passport and when I got it, and had collected all the necessary visas, where would I get the funds? My salary here is quite sufficient to cover all my needs, including travel, with a surplus to send to you. My moving to Petrograd would really amount to scientific suicide. But, on the other hand, if I stay here for another two years the following difficulties will arise. My experiments are now on such a scale and require such considerable sums of money that, if I begin to develop them, I have to reckon that it will take me two years to get everything organized before I begin to get results. Until I have gained enough confidence, I am working on a small scale and I am afraid to go over to a larger scale. But I feel that some decision must be made. The Crocodile says that I should work here for another five years or so, and then I could dictate my own terms if I wished to move elsewhere. This, of course, was well meant but I am afraid he may have exaggerated a bit. However, some decision must be made.

The opportunities for work here are such as I never dreamt of in Petrograd, even in peacetime. But the pull of home is sometimes so strong that I long to throw up everything and return. I must come to a decision in the next six months or so. I want to try to visit you this summer. Maybe the political horizon will soon become clearer. There are high hopes of this, especially in connection with the new Parliament here. In any case I shall see Abram Fedorovich and Kolka at the Solvay Congress and I'll clear up many things then. The Crocodile will be at the Congress and has promised to be my advocate . . .

(D, e)   Hindhead, 26 December 1923

. . . I spent Wednesday night in London and on Thursday I got visas for my trip to the Continent. The French sent me a visa for a whole year – unprecedented on a Soviet passport. The Dutch gave me a six-months visa, also unique. On Thursday I bought some expensive optical equipment for my experiment and then went off to my friend Skinner, at whose home I stayed over the weekend, after which I went to stay with the Blacketts here in the South of England. I also visited the National Physical Laboratory in Teddington on Thursday and looked thoroughly at everything. On Saturday I went to the theatre and on Sunday to a concert. So you see, my dear, I have started my holidays and am having a complete rest from work. This evening I shall be at the Skinners again for a large dinner party. The English eat very heartily at Christmas time – plum pudding, turkey, ham and so on. And I don't lag behind! . . .

(D) Leiden, 9 January 1924

. . .I spent four days with my friends in the south of France and then two days in Paris. I return to Cambridge the day after tomorrow. It was very intriguing to come almost directly from the Riviera and the Côte d'Azur, where everything was green and the sun shone warmly from blue skies, into Holland with its – 5°C, snow and ice, black tree trunks sticking up and people walking about in fur coats. In two days' time I shall be back in Cambridge and getting down to work again. By the end of the holidays I always look forward eagerly to the laboratory.

I feel at home here with the Ehrenfests and enjoy the days which I can manage to spare for Holland. Children and family are what I miss most in Cambridge. I am coming to the conclusion that I must be a family creature. The Ehrenfests have strict rules – smoking is permitted in only two rooms, and I am now a heavy smoker. But there's nothing to be done – even Einstein has to abide by these strict rules when he stays here . . .

(D) Cambridge, 30 January 1924

. . . I've been very busy and I haven't had a single evening sitting at home. Work is piling up and my lecture course has started. I had an audience of three for the first lecture and ten for the second. I am much concerned about your health and your work, and about Aunt Sasha, too. If for any reason you have to give up work you can always come and stay with me. The only snag that I can see is that you would be lonely here. I spend the whole day at the lab, the only language spoken here is English and I think you would find it hard to get along without your usual work. I myself would of course be very glad. As for my coming over, this is still very possible but it is difficult to encourage any definite hopes. I am doing everything possible but it is far from easy to arrange it.

I'm very distressed to hear that you are overworking, You really mustn't do this. Just remember that you should never overload any kind of machine. I understand the difficulties that have arisen in your teaching activities especially after my talk with Tatiana Alekseyevna Ehrenfest but, of course, it is not easy to appreciate the situation from a distance. Russia will probably be recognized by Britain in the near future. This is very good news and I think it will make many things much easier. I'm still rather undecided what path I should choose for my future career. I'm a little tired today and this makes everything look gloomy. There's so much to be done and I keep worrying that I shan't be able to cope with it. . . .

(D, e)   Cambridge, 9 March 1924

I am impatiently waiting for a letter from you answering the invitation in my last letter for you to come and stay with me. I have just finished writing up the first part of my work for publication. It has been passed by the Crocodile and will appear in print very soon. A great deal of material had accumulated – enough for one lengthy paper and two small ones.* I should like to complete all this before Easter but I am feeling a bit tired and overstrained. On Tuesday I give my last lecture to the undergraduates. On Wednesday I present a paper to the [Cavendish] Physical Society and next week I have to give another lecture. All this has to be done and then I can relax.

I have recently made an interesting acquaintance – J.M. Keynes, whose name you probably know. He was at the Peace Conference and has written a book about it. He is considered to be the most eminent economist in England and is still only in his late forties. He is very quick-witted, lively and talkative, with a sharp tongue – not at all a typical Englishman. I have met him twice. The first time I had breakfast with him and he then came to the lab to look at my experiments. The second time was at a feast in College. After dinner we

---

* This is the outcome of the "very bold" project mentioned in the letter of 6 July 1922 [see p. 154 and (9) and (10)].

played cards and Keynes invited me to play at his table. He knows a great many people, has seen a great deal in his time and is a good raconteur.

My Crocodile is on the best of terms with me just now, despite a row we had not long ago. That's how it goes between us, but as a rule, we are great friends. I am being given a small bronze crocodile which I shall fix on the bonnet of my car . . .

(D, E, e)   Cambridge, 9 April 1924

. . . Once again I have grandiose plans* and I have been extremely busy, so I haven't written for some time. When you have an ambitious plan it is important to discuss it with other people and this is indeed a very large and difficult undertaking. Moreover, one must strike while the iron is hot.

I have been in Manchester to see Professor Miles Walker, the well-known designer of electric dynamos. I called on him on Saturday at his home in Buxton, in the Peak District, 230 [?] miles from Cambridge. I went by car and thought I would need only an hour and a half of consultation with him, but he became so interested in our project that I spent three days with him. He is a most intelligent and likeable person. Afterwards some engineers came to see me and then I went to London.

Yesterday I got your letters and I'm very glad that the question of your journey has been settled. People in Cambridge know that you are coming and you already have invitations to tea and for evening occasions. To answer your questions, I have so many acquaintances in the University that I can arrange for you to meet anyone you like from bishops to financiers . . .

(D)   Cambridge, 25 April 1924

. . . I got back from Paris only the day before yesterday. I was also in Leiden and saw Abram Fedorovich, who tried to persuade me to

---

* These plans were for a new technique of producing momentary high magnetic fields by using most of the kinetic energy of rotation of a large dynamo to create magnetic energy in a coil. This technique took about three years to develop [see (12)].

57. A.F. Joffé, P.L. Kapitza and A.N. Krylov, Leiden 1924.

return to Petrograd, but in the end nothing was decided. We shall discuss this matter further when he comes to England. I hope that you will be here too by then, and we can come to a decision together. I'm impatient to know the result of your efforts. In a day or two I'll send you £12, probably by telegraph. . . . I should like you to come to me as quickly as possible. I have already started work. The Crocodile is in Brussels at a conference and so is Joffé. They will be discussing my future . . . I eagerly await your letter and Lyonka's telegram.

(D, E)   London, 4 June 1924

I'm sending this letter from London with the Kostenkos who are my very good friends. M.P. Kostenko and I are working on the design of an electric dynamo. I'm still waiting impatiently for news of your departure date. I have to lecture tomorrow to the Royal Society about my experiments – that's why I am in London, as well as to see Kostenko off. I do want to hear soon that you and Natasha have set out. My research is making some progress but I'm up to my neck in a lot of organisational work. Everyone in Cambridge knows that you and Natasha are coming and you will have to pay a lot of visits. . . .

In June 1924 Kapitza's mother, Olga Ieronimovna, his brother's wife, Natalya Konstantinovna, and her son, Lyonya arrived in Cambridge. They stayed with Kapitza until the beginning of April 1925. There are, therefore, no letters for that period.

(D, e)   Cambridge, 27 April 1925

. . . I left Paris on Friday the 17th, having got fed up with the theatres and with my own idleness. I stayed with the Krylovs in London and on Saturday morning I visited the factory and later over lunch at the Krylovs, I met some meteorologists who had come to attend a congress. On Monday I got back to Cambridge and went to the lab where I found a huge pile of work awaiting me. At five o'clock the Crocodile invited me to tea.

When I called on Mrs Gray I found she was moving to a different house and could offer me only two rooms there. I didn't like this at all and moreover she wanted a very high rent for the two rooms, explaining this by the rising cost of living. I said I would think about it and went to stay in my old room at 84, de Freville Avenue. In the morning I went to the "Baron's"* for coffee but it was sad and dreary there without you. I was so busy that it was Wednesday before I began to look for permanent quarters, but then I was in luck. Not far away I found three rooms with an elderly landlady. She is very pleasant and looks after me very well. If it goes on like this I couldn't wish for anything better. At first she took me for an undergraduate but when she discovered I was a Doctor, she showed great respect. We had discussed the rent while she was still under the impression that I was an undergraduate so she had set a modest figure of 31 shillings per week. I moved in on Saturday, the Baron giving me a lot of help. Yesterday I tidied up the previous house and I shall give it up tomorrow. "Uncle Skinner"** also helped to move the furniture – the

---

* This was Kapitza's nickname for Laurmann (see p. 14).
** H.W.B. Skinner was called "Uncle" by Kapitza's nephew Lyonya.

tables, piano and so on. My new landlady has no objection to my installing a wireless set.

My machines and all the accessories are working well and on Wednesday, the day after tomorrow, I shall go to Manchester to draw up conditions for testing and acceptance. The tests will be in the middle of May and will last a week. I hope to goodness that all turns out well. The Crocodile is quite cordial and I had lunch at his home on Saturday. Life goes on as usual but my heart is heavy without you. I feel apathetic and work automatically, doing only what has to be done . . .

(D)   Cambridge, 5 May 1925

I got your letter . . . You seem to be thoroughly immersed in your work already – see that you don't overwork. I have been to Manchester and all is going well with my machine. All is well with the Crocodile too. It is now finally decided that I shall put in for a Fellowship. Skinner is upset since this somewhat reduces his chances. My finances are low but they should recover by the end of June. Next week I shall go to London and on the 21st to Manchester for the tests on the machine. I hope to God that all goes well . . .

(A)   Cambridge, 24 May 1925

My dynamo will be tested next week, probably on the 26th and 27th. I am going to Manchester with Cockcroft and, naturally, I'm rather excited so I can't concentrate on anything else but the impending tests. The fitting out of the laboratory is coming along. We shall finish the wiring next week and everything will be ready for installing the machine. The Crocodile, too, is noticeably excited. Well, come what may! . . . By the way, Ellis announced his engagement this week and I met his fiancée. She is not bad looking, of Polish Jewish origin . . .

(D)   Cambridge, 9 June 1925

I feel guilty for not having written for so long. I've been very busy testing the dynamo. I spent eight days in Manchester and returned here exhausted and then went to London and visited the factories. But, thank God, the machine is more than satisfactory – the results were excellent . . . But I'm extremely tired.

Just now I am desperately busy but as soon as I have any free time I'll write you a long letter describing the tests on the machine. At the moment work has piled up. Tomorrow it's my turn to present a talk to the club. Then I have to prepare a lecture for Göttingen, where I am going at the end of June, and there is a lot to do in the laboratory.

Please excuse this short letter but I can hardly keep my eyes open and must go to bed. In the factory we had to work under very difficult and tiring conditions . . . I have made the acquaintance of Ethel L. Voynich [aunt of G.I. Taylor and author of "The Gadfly", a well-known left-wing novel]. She is a charming old lady and speaks Russian very well. [G.I.] Taylor is also getting married – there seems to be a kind of epidemic of weddings!. . .

(D, E, e)   Cambridge, 17 June 1925

. . . As I wrote previously, I tested the machine in Manchester with complete success. This took about ten days and the work was very tiring. We carried out the tests during the day and the engineers worked on changes and repairs during the night. The working conditions were difficult, with a terrible noise level in the factory which I found very tiring because I wasn't used to it. In the test department lots of machines were all being tested simultaneously. Short circuits and loud bangs were quite frequent. All this wasn't particularly dangerous but the unexpected bangs were unpleasant. The noise was such that you could hardly hear your own voice. Instructions had to be given by shouting into the ear.

The frightful nervous tension was increased by the fact that I was responsible for the tests, and the factory would not accept responsibility for any breakdown of the machine. Consequently, I conducted the tests extremely carefully. I increased the load gradually and after each test made accurate measurements on the components. The factory gave me a great deal of help and support. After finishing work each day I usually went to Walker or Kuyser to discuss the results. So I worked up to 14 hours a day and after all this testing I went about for a couple of days as if I had been hit over the head. Thank God, the tests were more than satisfactory but I was so tired that I could hardly even rejoice. The Crocodile was very pleased. So the most arduous and responsible part of the job has gone well but a lot more work lies ahead. . . .

(D, E, e)   Cambridge, 26 June 1925

. . . These last few days have been very busy because my machine arrived in Cambridge and was unloaded and placed on its foundation. It weighs about 12 tons so you can imagine what a considerable task this was. They began to unload the machine last Saturday at four in the afternoon and finished only at two next morning. Special men were sent from London and they worked with great skill. The whole job was done by only six men with steel rollers, a jack, timber beams and so on which they brought with them. It was so interesting to watch them at work that the Crocodile was present from the start until 11 p.m. The machine is now in position with the bolts cemented in and as soon as I return from holiday it will be tried out. I hope to God that all will continue to go well.

Today I had a long session with the Crocodile and we talked about everyday things as well as about science. He is very well-disposed towards me, since he is pleased with the results of the tests . . . I'm feeling well but very tired – mainly nervous exhaustion, obviously because of the tests on the machine. The nervous tension was great but now I can relax. After the trip to Germany I'll probably rest a bit more . . . G.I. Taylor visited me – very friendly, busy with buying his house and setting up his household. His fiancée is the same age as he is, 39, and a teacher at a very good school. I have not yet seen her. Ellis is already married and has gone off on his honeymoon. Chadwick is up to his ears in love and the Crocodile growls that he is not working enough . . .

(D)   Göttingen, 6 July 1925

Getting a letter from you here was a great surprise and a very pleasant one. I am staying with the Francks. They are a very charming family and live comfortably, with a cultured lifestyle – everything is clean – nice furniture and so on. His wife is also very likeable. On Friday I have to give my lecture. I conveyed your greetings to Franck and he asked me to send you his best regards. I am sitting at the writing desk in his study just now while he is lying on a sofa reading and his wife is mending linen. She has promised to play the piano for us.

I stayed three days with the Ehrenfests in Leiden. They were very hospitable and I had a good rest there. I'll stay here till Friday and then return to Cambridge and get down to work. There is something else to tell you about. Chadwick has asked me to be best man at his wedding. I've never yet been to an English wedding and here the best man has to wear a morning coat and top hat. This is obligatory though it's a useless waste of money, but it would be impolite to refuse . . .

(D) London, 19 July 1925

I've come to London for the weekend and have been to the factory to see my circuit breaker. The Director of the factory invited me to his home in the suburbs and I'm killing time here. Just now I'm lying on the lawn writing this letter. I gave my lecture on Friday, my last day in Göttingen. It so happened that Abram Fedorovich arrived that very day, so he, too, was present at the lecture. There was a large audience to hear me murder the German language . . . When I came back to Cambridge I had to see to lots of things – tests on the machine in situ, with the lab full of crocodiles! Everything went very well. The Crocodile is in a most friendly mood. He invited me twice to his home this week . . .

(D, E, e) Liverpool, 10 August 1925

I'm writing from Liverpool where I've come to get Chadwick married. The wedding is tomorrow and I shall send you a photograph of myself in top hat and tails. Up to now I've been kept busy. I keep having to dress up, either in dinner jacket or morning dress, and to attend lunches and dinners. On Wednesday I return to Cambridge.

I've had no luck with this wedding. First, there's the outlay of money and then the loss of time – both very inconvenient. As best

58. As Chadwick's best man, 1925.

man – and here they have only one – I have to undertake a whole lot of responsibilities and to represent the bridegroom after his departure. The custom at an English wedding is for the bride and bridegroom to depart immediately after the ceremony while I have to remain to entertain the guests. The number of invitations is enormous – 140 people. The reception is in the garden and in marquees. I don't know how it will all turn out but I hope I'll enjoy it. I have had to put up at the very smartest hotel and it's rather hard on my pocket. But, thank God, I did not have to buy a top hat – I borrowed one from Fowler whose head is the same size as mine . . .

(D, e)   Cambridge, 25 August 1925

At last the tide of work has ebbed. Yesterday I handed in my dissertation for a Fellowship and the result will be known after the examination on the 2nd or 3rd October, and the elections about a week later, but I don't rate my chances very highly. However, we shall see. I am not worried. . . .

I returned safely from the Chadwick wedding. The English style of wedding and my part in it were very interesting. I acted as the bridegroom's "best man" and accompanied him all the time. In the church before the ceremony I sat in front on the right-hand side next to the groom until the bride arrived. The bride enters the church and, with the organ playing, she processes up the central aisle. She is preceded by the choir and followed by her entourage. She goes up to the altar where she waits for the clergyman. The groom stands beside the bride and the bride's father stands a little behind, facing the altar, on her side, while I stand on the groom's side.

During all this one must stay very serious, but I couldn't keep it up and started to grin. During the ceremony, when the clergyman asks: "Who gives this woman to be married?" her father takes her by the hand and says: "I do". Then the clergyman asks the groom for the wedding ring. I take the ring from my pocket and give it to him. He asks some further questions, there is some singing and the organ plays. My duties are over and I sit down. There are some further ceremonies and the clergyman exhorts the man and wife about the purpose of marriage as if they themselves had no idea. When the ceremony is over we all go into the vestry behind the altar where the register is signed and I sign as a witness. There the bride and bridegroom receive congratulations. According to custom, the best man has the right to kiss the bride and the bridesmaids. I claimed the former right but not the latter as the bridesmaids are not very well-favoured and it wasn't worthwhile.

With the organ playing we all leave the church, I taking some elderly aunt in tow. In front, of course, go the bride and bridegroom. Photographs are taken as they leave the church. Then the whole company gets into cars and is driven to the bride's home for a large reception. There was a marquee in the garden for 200 people, refreshments, champagne, wedding cake and so on. After general congratulations, group photographs were taken and I shall send you my picture in top

59. Chadwick wedding group, 1925.

hat and tails as soon as I get it. I then accompany the bridegroom to
the room where he changes into travelling clothes. My final duty is to
witness his last will and testament and take charge of it. Then the
newly-weds depart and the guests celebrate and dance until late into
the night . . .

(A)   Cambridge, 30 September 1925

I have got back safely after a very successful trip. On the way back
I came through Marseilles, Avignon and Nîmes and I saw a genuine
bullfight. Once back in Cambridge I immediately started work. I want
to write you a long letter describing all that I've seen but it's not possible just now as I have an exam on Saturday and the election of Fellows
takes place on the 12th. I haven't much hope. I'm completely immersed in work just now because I'm testing the circuit breaker. If the test
is successful all will be well. This is a very important matter . . .

(A) Cambridge, 12 October 1925

Today I was elected a Fellow [of Trinity College] and tomorrow I'll be formally admitted. The whole procedure will, of course, be in Chapel, as is customary here. The other nice thing that happened to-day was a letter from Lyonya in Archangel telling me that he had arrived safely and had successfully completed his assignment. I confess I was anxious about him as I knew that this enterprise involved some risk. His letter came in the morning and I was elected a Fellow between three and four in the afternoon.

Skinner arrived first, very upset that he had not been elected, for this was his last chance. Then came Ellis who had supported me in the committee. Poor Skinner, I am very, very sorry for him. His disappointment has spoilt much of my pleasure. I hope our relationship will not be affected, but he is very distressed. He always has a melancholy look but today he looked particularly sad. Nevertheless, however sorry I am for him, I think he needs to learn something about life. After all, we all have many disappointments, but he had all the advantages in this competition, while everything was against me. . . .

(A, D, E, e) Cambridge, 26 October 1925

. . . I wrote to tell you I was elected a Fellow of Trinity College on 12 October and you have probably already had that letter. I shall now describe the admission ceremony. On the following day, the 13th, I had to present myself to the Master of the College, Sir J.J. Thomson, wearing my gown, a red hood, a white tie and two white bands, like those of a Catholic priest.

For my sins, I had lost my gown the previous evening and I had to spend the whole morning running around to collect these things. When I and three other newly elected Fellows presented ourselves to the Master, he greeted us and handed each of us a copy of the College Statutes. Then we all went into the Chapel, the Master leading and we following two by two. On entering, we remained in the porch and the whole Electoral Commission was waiting in the Chapel. There was some reading and speaking and then they called us forward. One after another we took the oath of loyalty to the College, promising that we would observe its rules and promote its prosperity. After this we had to sign in the old book that contains the signatures of all the Fellows that have ever been elected. I don't know how old it is. It is a thick book with leaves of parchment. Just think, even Newton's signature is there! Isn't that impressive! After signing you go up to the Master, who stands in a special stall with a reading desk before him. You kneel, stretching out your hands like a swimmer preparing to dive, extending them to the Master who takes them between his own and reads some Latin prayer, of which, of course, I understood not a word. One, two, three . . . the Holy Spirit descends on me and I become a Fellow – not simply the first Russian, that's certainly true, but, I think, only the third foreigner. In the evening there was a feast in the College in honour of the newly elected Fellows. I had to put on all my regalia – which meant a dinner jacket for want of tails (all the same, I have had to order tails – £13, scandalous!).

In his welcoming speech the Master, Professor J.J. Thomson, referred to my election in the following words: "Now I have to welcome Dr. Peter Kapitza as newly elected Fellow (loud applause). Here we are establishing a new record in the annals of our College – he is the first Russian we have elected" (two or three weak claps, whether from heaven or earth I can't say). He said that in Oxford they had once elected a Russian as Fellow of a College – this was Professor Vinogradoff – so the two universities were now level again. Then he reminded us of Russians who had worked in the Cavendish Laboratory: [V.I.] Pavlov and Papalexi.* He said he was sure that all would unite in wishing me success in the difficult and fundamental experiments on which I am now engaged (sparse applause). After the speeches everyone except the four new Fellows stood up and drank our health, while we remained sitting. I received many congratulations and some were very sincere and warm.

I sent a telegram to the Crocodile saying: "Elected Fellow. Very happy. Very grateful, experiments going well. Best wishes". It required 20 words and cost 11 shillings. I sent it as a weekend cable. On a weekday it would have cost about £4. Now I dine in College practically every day. Everyone is very cordial and I feel much more at home. Next term I shall move into rooms in College.

Poor Skinner, he still hasn't calmed down. He wanted so much to

---

* The text of the letter says Pokrovski but probably Papalexi was meant. Kapitza often muddled names.

be elected. I am very sorry for him, though, in fact, he is not in any real need . . . For him election to a Fellowship was more a question of prestige than of need. For me it is the closer link with University circles that matters, and I attach little importance to the fact that it is considered a great honour here. In particular, it gave my landlady a great thrill. She gave me congratulations from all the family and was very surprised that I was taking my election so calmly.

Skinner is very upset but my relationship with him is still good and he has breakfast with me once a week as before. But I feel that he is trying to maintain good relations more from the head than from the heart. Time, I hope, will assuage his disappointment. I have often asked myself whether I had stood in his way but I think my conscience is clear. In these elections everything was against me – my age and the fact that I was not English. Besides, if you compare our paths, mine has been cobbled with much larger stones than his. Why stand aside, when it was not even certain that he would be elected? Maybe this spoilt lad should learn to struggle in life. But he is not even able to show a bold face in adversity. He is more bitter than ever. Poor chap!

It is difficult to discover what went on in the election but a certain amount has leaked out. I know that the Master himself was against my election. I know also that the philosophical papers I wrote in the examinations (on religion, on the reality of our existence) were so short and in such English (as regards both orthography and syntax) that no one could decipher them. Perhaps this was taken for such deep philosophy that they all prostrated themselves before it.

I am often asked now whether I shall stay in Cambridge. Evidently I had caused some alarm by telling various people that if I were not elected I would at once leave Cambridge. I also said that people here are so conservative and narrow-minded that they wouldn't dare elect a person holding a Soviet passport. However, the touchy English were so proud of their freedom and independence that they reacted in a manner favourable to me. But, after all, that wasn't the most important consideration. The main thing was that the opinions given about my work by the experts must have been very favourable.

I have not yet begun to work at full speed. Evidently, in spite of my outward calm, these elections have cost me a lot of nervous energy. I now have to face a lot of expense for furnishing my college rooms. I should be very grateful if you could send me some of my goods and chattels. Besides, I don't want to spend too much just now as I'm intending to go to Leningrad at Easter and I need to save up for that. And so, my dear, I have written you a long letter and you should be pleased with your prodigal son.

(D)   Trinity College, Cambridge, 14 November 1925

. . . It is very probable that I shall come over [to Leningrad] next year but the time of my trip depends upon the Crocodile's travel plans and on the state of my work and when it will be most convenient to interrupt it. I hope that Easter will be a convenient time but it is very difficult to foresee how the work will go.

Until mid-December I shall stay in my rooms in Chesterton and only move into college on the 15th; there I have been given a splendid set of three rooms plus lavatory, entrance hall and a small room for washing up dishes. It will cost quite a lot to furnish this apartment. Would you be so kind as to send me some household linen (such as hand-embroidered tablecloths), to liven up my solitude. . . . The weather is very cold just now with freezing temperatures almost the whole week; this is very unusual. I am writing this letter in the college reading room. After 8 December, you can address your letters with only my name and Trinity College; nothing more is necessary – the same for a telegram.

I have visited Keynes, the distinguished economist, who is married to Lydia Lopukhova [the famous ballerina]. He told me about his visit to Russia and he has also given a lecture about it. . . . Skinner is still dejected but apparently consoles himself with work. Taylor, Ellis and Chadwick are enjoying family life, busy with organizing servants and all that.

I had a letter recently from the Crocodile who was very amused by the photograph of me with Chadwick in top hats [fig. 58]. I sent you two copies of this (one for you and the other for Lyonya and Natasha) . . . Today I tested a very important component of my circuit breaker and it proved to be satisfactory. There are plenty of troubles ahead but, all the same, it's good that there's now one less.

At Christmas I shall go to Paris as I have to go somewhere. It's a long time – almost a month – since I was in London. Now that I am in College I see a lot of people and talk a lot, so life in Cambridge has become far more interesting to me than it used to be. But just now I'm concentrating so much on my work that I can't think of anything else until I get my machine going . . .

(D, E, e)   Cambridge, 16 December 1925

Today was the critical day in the series of tests on the machine. Everything went well and now I can say with a clear conscience that the basic idea I had for these experiments has been proved correct, and I have come out on top. I'll write in more detail later. There are still some difficulties but the principle is established and that is the main thing. Today a new record was set for magnetic field strength. I would have gone higher but the coil burst. It was an impressive explosion. But this, too, was all to the good because it gave me a complete picture of what happens when the coil bursts and clears up many points of detail. Everything is even better than I had supposed. Now I can relax though some important visitors are coming tomorrow and the day after.

I moved into College on Monday and this is my third day of living there. It's the first time in four and a half years that I have comfortable rooms, two large rooms – a sitting room comparable with Father's study and a bedroom like yours – with a lavatory and a place for washing dishes. The rooms, thank Heaven, are warm and the service, I think, is good. I have not bought any furniture yet for lack of money, so I am renting some for the time being . . .

(D)   Cambridge, 26 December 1925

Season's greetings to you all! For me these festive days always bring sad memories. It is already six years since I was widowed. I spent Christmas in Cambridge. After the tests on the machine and the experiments, I wanted a few days of complete rest and I'll go to London tomorrow where I'll spend two or three days. Then on the 29th or 30th I'll go to Paris for a week and stay, as usual, with the Krylovs . . . I am getting used to life in College; the rooms are very good, and what is most important, they are warm and large. I had to spend some money on blankets, pillows and some crockery, but now everything is in order. I have been rather busy in the lab these last few days with important visitors from London. They all seemed very well satisfied. I have now calmed down a bit but I still feel some reaction after the nervous strain . . .

(D, E, e)　　　　　　　　　　　Cambridge, 12 February 1926

I am very upset about your not feeling well and, especially, about your working far too hard. You really must, once and for all, moderate your activities. Go more slowly and you'll get further. Lyonya writes that you are hard up. Let me know at once if things are very tight and I'll send something. But in any case, I shall be with you in Leningrad in six weeks and bring whatever is needed for your support.

Now as regards my own affairs, as you can appreciate, I have a busy time ahead, with the laboratory due to be opened in March. Balfour is coming and a whole crowd of crocodiles. I shall obviously have to make a speech, then give a demonstration, and you know how awkward it is to do that in front of such a mob of people. There will be a dinner in the evening and all the rest of it. I'm now constantly besieged by visitors, often very important ones, which prevents me from getting on with my work. Numerous receptions and dinners in Cambridge also take up a lot of time. Moreover, I have to give various seminars on my work (two this term) and a course of lectures to undergraduates.

I shall leave for Leningrad by 30 March at the latest. I have to be back in Cambridge early in May since I can't desert the lab any longer than that. My Crocodile is very busy as he is now President of the Royal Society (the equivalent of the Academy of Sciences) and has heaps of work in that connection . . .

(D, E, e)　　　　　　　　　　　Cambridge, 11 March 1926

My laboratory was opened on the 9th and it was a busy and anxious day. Among the big-wigs was Lord Balfour, a former Prime Minister. He made a speech and I had to speak after him. Wasn't it unfortunate that my broken English should have come immediately after one of the best orators in England! In the evening there was a large feast in

60. P.M.S. Blackett, P.L. Kapitza, P. Langevin, E. Rutherford, C.T.R. Wilson, outside Cavendish Laboratory, 1929.

Trinity College with about 50 to 60 guests and it all went off very well. I'll tell you all about it when I see you . . . I am very busy and can't write a lot but we shall see each other very soon anyway . . .

(E, e)   Cambridge, 22 May 1926

I returned to Cambridge [from the Soviet Union] on the 20th and was immediately immersed in work . . . I spent a day in Göttingen where I was very well received. Langevin [see figs. 60 and 79] is due to arrive from Paris today and will stay for a week. Indeed, there are so many visitors in Cambridge just now that it interferes with work. It looks as if I'll have to go to Zürich on 18th June to attend a small Congress on magnetism to which I've been invited.

The Crocodile welcomed me very pleasantly. I am being asked many questions and have to give an account of my travels. It feels very strange to find myself back in Cambridge again but I missed my work

so much that I'm glad to take it up again. I had no problems on the
journey though our ship became icebound and we had to be escorted
by an icebreaker for part of the way . . . How pleasant it is to recollect
my visit to my native land! . . .

(E)   Strasbourg, 18 June 1926

I'm in Strasbourg, which I once visited with you [in 1912], and I still
remember being in the Cathedral and looking at the famous clock.
Tomorrow the principal clockmaker will show me the details of the
clock mechanism. I have been very pleasantly received here. I gave a
lecture to the Physical Society today and everyone laughed at my
distortions of the French language. I then remarked that I was very
pleased to see them laugh because it encouraged me and this produced
general applause. The lecture went off quite well and they said they
had understood me. On the whole they treated me well and were very
kind.

Tomorrow there is to be a banquet in a restaurant and on Sunday
I go to Zürich, where I shall stay a week for the meeting on
magnetism. Twelve scientists will be there – French, German, Swiss
and Dutch – a very select company. I too shall be giving a lecture,
though I don't yet know in which language. But it doesn't matter, I'll
muddle through somehow. . . . It's now 1 a.m. and I'm tired after
speaking French all day but it was useful practice. I hope all is well
with you. . . .

(E)   Cambridge, 23 August 1926

All this time I have been thinking of you and your health but you
don't write much about this. In my last letter I wrote that you must
give up most of your lectures and adopt a very quiet lifestyle and I beg
you to do this. Let me know in more detail what the doctor said and
you must consult some good specialist without delay. Remember, too,
that you can come here whenever you like and I shall arrange
everything necessary for you to have a quiet life. In the meantime tell
me how much I should send regularly to enable you to do less
work . . .

The lab has emptied and only a few are left of the usual 40. It's quiet and peaceful and a good atmosphere for working. . . . I've received the Kustodiev portrait* but somehow I feel a bit embarrassed about hanging it in my own room. I shall continue working till about 7 or 10 September and then go to France where I'll see Semenov again. I believe that Abram Fedorovich is coming later . . . Do write more often. You can't imagine how unhappy I am not to be able to see you. But I shall certainly try to come again next year . . .

(E, e)   Cambridge, 21 October 1926

Forgive me for not writing for so long, but I have been completely immersed in my work which, alas, has not been successful – the coil has burst again. The forces which have to be overcome are really enormous and I'm spending all my time trying to find a way out of the difficulty. I haven't lost hope, rather the reverse; the failures stimulate me to continue the struggle. We are now making yet another coil and if nothing comes of it we'll make two more. So you see I have three possibilities in hand and I hope one of them will work. Because of these worries I haven't been able to get down to writing and I'm afraid this letter is rather incoherent. . . .

Here in Cambridge life is getting into full swing now that the undergraduates have returned from vacation. I am often invited to parties. My new rooms are very comfortable, and the bedmaker is pleasant and obliging. The surroundings are very quiet and convenient for studying . . .

(E, e)   Cambridge, 10 November 1926

It's ages since last I wrote and my conscience is troubling me, but I've been chasing after new ideas recently. The work is not going well yet, with the coils still bursting, and though I'm almost sure the difficulties will be overcome, it is taking a lot of time and the experiments are not going as fast as I would like . . .

---

* This is the portrait of Kapitza painted by B.M. Kustodiev when Kapitza was in Leningrad (see letter of 7 April 1927, p. 203 and fig. 1).

Khariton arrived three days ago and I have been helping him to settle in. Today he has started work and I think he is pleased with the facilities provided. Many thanks for your book. How is your second book* getting on? I may perhaps start writing a book about magnetism – at least I have an invitation to do this – but it would take a lot of time and I have not yet replied to the publisher . . .**

(E, e)   Cambridge, 7 December 1926

. . . Yesterday I at last succeeded in what I have been trying to do for the last few months. The coil withstood the load which it had to and I have at last obtained the magnetic field I was aiming at. Now I can turn to purely scientific work which makes me very glad. The struggle was very fierce. The Crocodile was hovering about the lab the whole day and he was pleased too. This success shows that all the efforts of the last few years have not been for nothing. I know you will be happy for me and will forgive my irregularity in writing to you. The holidays will soon be here and I'll have a rest, which won't come amiss . . .

(E, e)   Cambridge, 16 January 1927

I feel I am a bad son for writing so rarely. I came back from Paris on the 10th after an enjoyable visit and a good rest. In Paris I met Aleksei Nikolayevich Krylov's daughter a few times and she was a jolly and pleasant companion at the theatre. I went a lot to the theatre and saw many interesting pieces. Of course the French are best of all in farce and they can even turn comedy into farce.

Now I'm back in Cambridge immersed once again in work and I think this term will be interesting. Your letters worry me and I find it painful to be so far from you. As far as work and living conditions are concerned it would be difficult to wish for anything better but I do miss you all. However, I firmly hope that we shall see each other in the course of this year . . .

---

* The books in question by O.I. Kapitza are ''Children's folk jokes, rounds, sayings and songs'' and ''Folk riddles'' published by Mirimanov.
** He did not in fact write this book.

(E, e)   Cambridge, 6 March 1927

All this time I have been working very hard and things are going well. You can imagine how hard I have been working from the fact that I haven't been to London for two months; but I did go on Thursday this week to a session of the House of Commons which was devoted to Anglo – Russian affairs. My friends got a ticket for me and I couldn't refuse to attend the debates of the oldest parliament in Europe.

The debates were very interesting and among the famous orators who took part were Lloyd George, Ramsay Macdonald and Austen Chamberlain. Only rather poor places are reserved for the general public and since I came late I got the very last place at the back of the gallery. But the biggest impression I got was not from the speeches but from the general picture and spirit of parliament. The walls, floor and ceiling of the large hall are finished in carved oak, though part of the ceiling is of matt glass to provide diffuse illumination. Above the chamber are galleries for the public, the press, diplomats, etc., and there is also a gallery where ladies used to be allowed to sit behind a screen. The screen has been removed but the ladies gallery still remains. Below is the parliament, with an elongated rectangular area left empty at the centre, from which the benches rise up. There are only 300 – 400 seats though there are more than 600 members of parliament. But most of the members spend their time outside the chamber, where there is a good restaurant, a chess club, etc., and they come into the chamber only to vote.

The benches, in fact, occupy only three sides of the rectangle, the fourth side being left empty except for the Speaker's chair, which is of oak, stands on a raised platform, and has an oak canopy under which sits the Speaker, dressed in a black gown and wearing a large white wig. Since the light comes from above, the Speaker can't be seen clearly, only his wig being visible. In front of the Speaker sit three clerks on plain chairs at a writing table; they also have black gowns but wear only small wigs. The table is big and occupies something like a third of the whole rectangle. On it lie a gold mace and various books, presumably containing the code of rules or something of the kind. Government ministers sit on the front bench on one side of the table and to the Speaker's right, and leaders of the Opposition sit directly opposite. Thus it is across the table that they hurl insults at one another.

The Ministers' seats are just as plain as those of ordinary members and they sit in a very free and easy way, some even putting their feet on the table (I'm not exaggerating!) so that the Opposition can admire the soles of their boots. There is no special rostrum for speeches and members speak from their places in a very unconstrained way. When he is speaking, a member addresses himself to the Speaker, rather as if he were trying to convince him. He is often interrupted during his speech and is liable to get involved in an argument with whoever interrupts him from the opposite benches. If the altercation goes too far, however, the Speaker restores order, but there is a great deal of noise most of the time and the debate which I witnessed often became very heated. During a speech the member's neighbours often make remarks in an undertone and the orator may interrupt his own speech to make an aside to a neighbour. In general you get the impression that it's all a game rather than a discussion of serious business. But the speeches are concise, serious and almost wholly to the point. There is no pathos but a lot of wit – which, after all, is a weakness of the English.

Of the orators I mentioned, the one who spoke best was Lloyd George in spite of being old, white haired, small and not at all impressive to look at. The tall, thin [Austen] Chamberlain with his formal frock coat was quite another matter. Slightly leaning on one of the despatch boxes on the table and waving his monocle, he would from time to time pause to put the monocle in his eye and look around him. This really created an impression. He looked just like the kind of diplomat portrayed in novels and moreover he kept up the parliamentary tradition, already abandoned by other members, of wearing a top hat throughout the session. But he doesn't speak all that well. Macdonald was, I think, the weakest speaker of all and I didn't care for his vague way of expressing himself and his lack of both precision and simplicity. There was a lot of noise and squabbling. I didn't stay until the voting but went back to Cambridge – not that the voting matters, since the government has a guaranteed majority.

I have talked to Rutherford several times during the last few days and we had a row on one occasion, but today we made peace again . . .

(E, e)   Cambridge, 16 March 1927

. . . I'm working hard and get tired, but things are going well and I am thinking of writing two papers soon, though I have had to interrupt my work a bit just now. Recently Aleksei Nikolayevich's daughter, who is a friend of Kolka's wife, has come to England and I had to arrange a visa for her. She is an archaeologist and is working in the British Museum. She is fond of art and understands pictures very well and I think you would find her interesting . . .

(A) To B.M. Kustodiev   Cambridge, 7 April 1927

. . . By the way, everyone here likes the portrait and I have already been asked to present it to the Fitzwilliam Museum. But since it gives me great pleasure to look at it in my room, I replied to this request with a smile . . .

This letter refers to the portrait (see fig. 1) painted in 1926 during a visit by Kapitza to Leningrad. Kapitza's friendship with Kustodiev dated from 1921 when Semenov introduced him to the painter. Legend has it that Kapitza asked Kustodiev "why do you paint only famous people? Why not paint some who will *become* famous – us for instance?" Kustodiev was tickled by Kapitza's cheek and eventually produced the oil painting reproduced in fig. 6. There were great food shortages at that time and Kapitza paind for the joint portrait with a bag of flour and a cock, which he had just received as payment for mending the wheel of a water mill. The joint portrait is in Moscow in Anna Kapitza's home.

(E, e)   Cambridge, 7 April 1927

This is the last day of work and tomorrow morning I shall go off on holiday, first running around in my car and then for a week to Paris . . . Today I had a row with the Crocodile – quite a row. He was unkind and I was tired. Really it was all about a trifle, but it can't be helped, we'll make it up after the holiday . . . I have decided to tour England since I know it least of all and I haven't spent any of my holidays here. I am pleased with the results of my work. I've discovered something new, though I'm not sure how important it is.

But in any case it couldn't have been done by the usual techniques, so it will serve as a justification of my work . . .

I have met Anna Alekseyevna Krylova a few times during her visits to Cambridge. She is a wonderful girl and clever too, very interested in and knowledgeable about art. I'm sure you would like her and I know your taste very well . . .

(E, e)   Paris, 23 April 1927

It looks as if I shall get married to Rat [his nickname for Anna] Krylova next week. You will love her.

(E, e)   Deauville, about 3 May 1927

We are at the seaside in Deauville resting after all the complications and in a week we shall be back in Cambridge and then to work! The complications were many. The main one was the difficulty of formalising the civil marriage. My wife is an émigrée and is considered stateless by the Soviet government. At first they wanted her to get a Soviet passport before the marriage could be legally formalised. But this would have taken at least six months. However, Rakovsky [the Soviet ambassador in Paris] was very helpful and succeeded in arranging all the necessary formalities, so since 28 April I am a married man. I have already written to the Crocodile [see p. 260 and fig. 61] and he sent a congratulatory telegram. In a week's time we shall be in Cambridge and have to set about equipping ourselves for our new status.

My wife has written to you. I don't know exactly what she has written, but it will be best if you judge her intuitively. I find it difficult to talk about her. The most important thing is that she doesn't wear high heels nor does she use powder. She speaks Russian, French and English. She has short hair and walks well. I am sending you her photograph. If everything works out well I'll bring her to you in "Peter" [Leningrad] and then you'll be able to see what she really looks like. So, my dear, you see that after seven years I am married again and I'm sure you will love my wife. Well, my dear, I kiss you tenderly; kiss the others for me.

27/4 27

c/o Mme E.D. Kriloff.
12 Villa Stendhal
Paris XX

My dear Professor Rutherford,

I am going to be married
tomorow and. Saturday. I
meen to say that tomorow
at the consulat and Saturday
in the Russian Church in Paris
When I shall be back in
Cambridge I do not know.
What do you think about it?.??.!!
I fear you are rather angrey
This why I propose to have no
honeymoon and bring my wife
in few day's time after my
wedding to Cambridge. I hope
you shall be so kind & help

61. Letter announcing his marriage, 1927 (see also p. 260).

to get a British visa for
my wife. Kindly ask the ~~Forein~~
Office to give the visa at once.
as having no visa I shall not
be able to get back to the Lab.
I got engaged about a week
ago & were very busy all
this time in aregeing the
formal part of my wedding

It was very difficult because
my future wife is a Russian
emigrant & the Soviet Consolet
were making great difficulties
in regestering the marriage.
But now after much talking they
agreed.

I hope you understend that
I am a victime of my own
~~soooo~~ goosse and I have to
anfece that the dose which
I received is ~~rather~~ a strong one.

I presume you are interested
to know how is the Lady
I shall refer to Robertson or to
Simpson, who met my future
wife during her visit to Cambridge
recently.

I hope you are not more
cross with me for my last
token, but you see that even
in more important questions
I have a quick decision and
great speed of action.

Yours <u>very</u> sincerly

Peter Kapitza

62. Peter and Anna Kapitza after their marriage, Paris 1927.

From A.F. Joffé in reply to Kapitza's announcement of his marriage.

(F)   New York, 28 May 1927

Dear Piotr Leonidovich,

Your letter made me happier than I ever could have expected – I hadn't realized how closely I feel attached to you and how much I take your fate to heart. In fact I love you very much, and I think your present success is much more important than your million Gauss. I know your wife only slightly, but I cannot think of anyone who could have been better. I am also glad for you that you are not marrying an Englishwoman who would always be foreign to you. I can hardly formulate exactly why, but I am terribly glad for you, all the more because I think your wife will also be happy.

Unfortunately I shall not be able to visit you on my way back to Russia . . . but I should be very glad if you and your wife could come to Russia in the autumn . . .

Your A. Joffé

4

Letters from Moscow to Anna Kapitza in Cambridge (1934 – 1935)

After Kapitza's detention in the Soviet Union and Anna Kapitza's return to her children in Cambridge, he wrote her long letters almost every day. This chapter consists of extracts from just a few of these letters, which give some idea of his inner thoughts, and his emotional and at times even hysterical reactions to his very difficult situation. Some of these letters were translated from Russian by Anna to provide ammunition for Rutherford in his efforts to help Kapitza. These translations are in the Rutherford Archive (a) and have been reproduced, in 1985, almost verbatim by Badash (c); some of our own extracts are also based on Anna's translations (those marked (a)) but the English has been rendered rather more colloquially. A few letters from Anna to Peter are also included.

In reading this chapter and parts of the next two chapters it may help to bear in mind the following synopsis of the sequence of events:

Early October 1934	Kapitza detained in the Soviet Union; Anna returned to Cambridge.
23 December 1934	Decree establishing the Institute for Physical Problems of the Academy of Sciences with Kapitza as Director.
May 1935	Construction of the Institute started.
August 1935	Visit of E.D. Adrian to Moscow and subsequent confidential report to Rutherford on Kapitza's position.
October 1935	Formal agreement to the purchase of Kapitza's equipment by the Soviet Government.
Late 1935	Transfer of equipment started.
Early 1936	Building of Institute for Physical Problems completed.

63. Anna Kapitza with Andrei (l.) and Sergei (r.), Cambridge 1934.

(I)   Leningrad, 5 October 1934

Dear Rat [his nickname for Anna],

This is the third day since you left – I wanted to write yesterday but
Leipunski called and then Lyonya [his brother] took me to the circus.
. . . After your departure I . . . came home and felt very blue. The
following morning (the 3rd) I went walking in the morning and got
as far as the Strelka. Then Nikolay Nikolayevich [Semenov], who had
just returned from Moscow, came at five and stayed for an hour and
a half. . . . He seemed to be sorry that he wasn't in my shoes – just
let him build a new Institute and he'll go without bread! I am simply

astonished how someone like Kolya can be a serious scientist, though in fact he is without doubt our most outstanding scientist. Then he took me to see your father [A.N. Krylov] on the Vasilevski Island and I spent the evening with him.

On the 4th I started the day with a walk in the Botanic Garden and was shown over the greenhouses by an old man who explained everything very clearly. After lunch I started studying Pavlov's book on conditioned reflexes which I have just bought. I am feeling a good deal more cheerful though still rather melancholy – I even feel a bit happy! The point is that I got very tired during the last few months in Cambridge setting up the helium experiments and then there was the car trip through Scandinavia and everything else, so my enforced vacation is actually pleasant.

. . . Well, my dear Rat, kisses to you and the piglets . . . Have the weeds grown terribly in the garden? How did you find the Crocodile? Hurry them up with printing my article and sort out my mail . . .

Your Petya

(A)   Leningrad, 2 November 1934

. . . Yesterday I visited Ivan Petrovich [Pavlov]. He teased me a bit and swore at me, but on the whole we got on very well together. He willingly agreed to let me work in his lab and as soon as I have prepared the ground I shall start experiments on the mechanism of muscles along lines similar to those of A.V. Hill . . . please ask him to send reprints of papers by his group.

So, my dear Anya, I am about to become a serious semi-physiologist, which makes me very happy. I have wanted to work in this field for a long time and now I have the possibility of doing it in collaboration with such a great man as Ivan Petrovich . . . It will be very convenient working almost next door to our home and of course the most important thing is that the lab is very well equipped. Aleksei Nikolayevich [Bakh] also approves of my research project . . . You can't imagine how little we know about how muscles work. Such direct conversion of chemical into mechanical energy is observed only in living nature. Hill was the first to take up this problem – as you know, he used to be a mathematician . . . It's not a problem that can be solved by physiologists. All you have to know is the structure of the muscle and that can be learnt very quickly. Ivan Petrovich thinks that a physicist would need only two or three months preparation to tackle the problem and welcomes the idea. Moreover, I have had several discussions with Hill and am familiar with his techniques. An even greater advantage is that there is no need for a great deal of laboratory space and I can start out on my own. As Maupassant said . . . ''from small beginnings''* . . .

(a)	Leningrad, 9 November 1934

. . . You remember that in Cambridge it took me a long time to decide to build a laboratory, and the year it was being built, while I was still working in the old one, was not the happiest time of my life. Rutherford is quite right, I am not made for administrative work; it is a torture for me and I accept it only as a sad necessity. In Russian conditions the building of a laboratory like mine will mean colossal administrative difficulties and will spoil my life. Above all, I am convinced that the difficulties would be so great that with my lack of experience of all the very complicated bureaucratic processes, it would be quite impossible to overcome them. You know how cross and worried I got and how I cursed at every misunderstanding we experienced with our car . . . the building and administration of a laboratory would be a continuous cursing and there would be a 99% chance that I should end in the lunatic asylum . . .

(A) From Anna Kapitza	Cambridge, 11 November 1934

. . . I have got very independent and have quite lost my fear of people, which makes me very happy. True, I wasn't all that afraid before, but the Crocodile has taught me how to be brave. If I'm not afraid to talk to him, then I'm afraid of nothing.

---

* This refers to a story Kapitza was fond of quoting, about a successful brothel keeper who boasts ''I started from small beginnings – just my wife and her sister''.

I haven't seen Aleksandr Ilich [Leipunsky] for some time and don't know what he's doing . . . John [Cockcroft] is as usual full of life and dealing with many things simultaneously – everything seems to come easy to him. I shall be lunching with the Cockcrofts tomorrow.

(A)   Leningrad, 1 December 1934

Dear Ratlings,

Today was full of events. "Sage" [Bernal] arrived with his wife and we chatted for a couple of hours. It was pleasant to hear some English gossip and I was cheered up by our conversation . . . Later I got a telegram from Volgin [Secretary of the Academy of Sciences] "Come to meeting 4 December. Wire your acceptance." I don't know what it's about. Over evening tea we heard the sad news of the assassination of Comrade Kirov and were very shocked. Kirov was very popular here; he respected science and was full of good will to the academic world. His loss will be keenly felt in Leningrad.*

(A)   Leningrad, 4 December 1934

Ivan Petrovich [Pavlov] told me "You know, Pyotr Leonidovich, I am the only one here who says what he thinks and when I die you must continue to do this. Speaking out is essential for our country, which I have come to love especially strongly in these difficult times . . . I should like to stay alive for at least another ten years to see what happens to my country and my conditioned reflexes. And, you know I'll make myself live that long" [this was not to be – he died in 1936].

He treats me well, but there are differences between us. As regards speaking out, I don't think I would be afraid, but the potential for doing it is very different for him and for me. He has long been the head of an important scientific school recognised by everyone, while I am

---

* In fact, Kirov's assassination was the forerunner of the great purges and terror of the later 1930's.

alone, without support and not trusted. Perhaps I am recognised in the West, but as for the clique of Abram Fedorovich and Co – you know what they think of me. Now I risk damaging my relations with people in Cambridge too – they always disapproved of my left tendencies and I'm sure many are saying "That's what he deserves" . . . Ivan Petrovich [Pavlov] immediately believed in my plans and was much more optimistic about my work in biophysics than were A.V. Hill and Adrian. I have already been working a month and I'm sure they are mistaken. I talked to Bernal and got him interested in my theory of muscular mechanism and I am planning several experiments. Apparently, A.V. [Hill] and Adrian are unable to see any approach other than the one they follow themselves. It's always like that with me – if I were to listen to all the critics and sceptics I should never have accomplished anything in my life . . .

(A)   Solyanka 14, Moscow, 11 December 1934

. . . So far the discussion has been about the suggestion that I should participate in the work of the Academy of Sciences and the planning of its move to Moscow [from Leningrad] – I am of course perfectly willing. For the first time they also offered to find us living quarters . . . Aunt Lisa [a cousin of A.N. Krylov] has offered to look at possible places and advise me. It will probably be near the Academy and the Neskuchni Sad [Joyful Garden, a public park] which her husband [G.N. Speransky, a well-known pediatrician] says is the healthiest part of the town . . . Well, little Rat, I love you all very much and if it wasn't for you, I could of course find another way out. It seems that G[amow] has made a lot of mischief for everyone here. They say that his passport has not been prolonged. It will be difficult to build everything up again from scratch, but the main thing is that I'm not trusted and I'm not at all sure they will help in deeds, as well as in words. It's all words, words and words – quite foreign to my nature. But Russians are born to talk, talk and talk. My kisses to my dear little Rat – be brave and cheerful and hope for a better future. Kiss the little Ratlings too.

64. In the Metropole Hotel, Moscow 1935.

(A)   Hotel Metropole, Room 490, 26 December 1934

I moved here from the Novo-Moskovski Hotel because it was very bad there. Here I have two rooms and a bathroom. Although they are a bit cold I am glad to be able to enjoy a little comfort. . . . I am glad some understanding has been established about what is needed for my work. Kr[zhizhanovski] has been very helpful in formulating the question. Without a lab I can't work in my own field and to build a lab without the right people and know-how isn't possible. Therefore, the right people must be sent and the [Cambridge] lab bought. M[aisky] must talk to Rutherford and may already have done so, while I clarify the conditions here. If the lab is indeed acquired it will be assigned to the Academy and the Praesidium has already made an appropriate order.

Without my special lab I can't continue my former work and will have to switch to something else. We'll see – perhaps I'm wrong but I'm doing everything possible to help. People here are really rather pathetic – they are full of enthusiasm and even if they have treated me so swinishly, it is not on personal grounds or to settle some sort of petty score, but because they think it's for the good of the country and of socialist reconstruction. I understand them very well and in spite of everything I don't bear them the slightest grudge . . .

Joffé had asked Kapitza to lead a discussion on superconductivity at the next session of the Academy.

(a)   Moscow, 1 February 1935

. . . but this would be like the tortures of Tantalus for me. On this topic I disagree with many people including of course the Kharkov school, which will take part in the discussion. My work in this field is at a halfway house. I have some results but they are not yet verified. So of course I shall feel like a Goliath with his arms tied. And if we arrive at some conclusions after the discussion, then the Kharkov people will be able to verify them and I will be left sitting like a fool. It was wicked of Joffé to have asked me and my refusal may be interpreted very badly. To the devil with it! But here you have an example of Joffé's good will to me; superficially it seems as if he is especially concerned for me but it is difficult to believe that he does not really realise how painful it is to me to discuss scientific questions . . .

When I was in Leningrad, I talked to Pavlov about the position of scientists and he very correctly discussed the freedom necessary for a scientist. With a stick over him a man can only be made to dig the ground but not to make scientific discoveries . . .

2 February. Yesterday's talk with Mezhlauk [Deputy chairman of the Council of Peoples Commissars] was a very long one (two hours) and it was very important. For the first time, I really felt that they were interested in me as a man, and are actually prepared to do everything to help me start my work here. You may be astonished that we talked least of all about everyday life and technical conditions. We mainly discussed the question of trust. This took a long time, and the fact that Mezhlauk regarded it as seriously as I did, and did not want to pass over it lightly, but on the contrary considered every point separately, made a very good impression on me. I talked quite freely, but not harshly. Mezhlauk has a winning manner, so that harsh words do not come to the lips. The frankness was complete on my side, and I also felt it in much of what Mezhlauk said. This conversation can be

described as follows. They no longer mistrust me; all that has been left behind. If they do not want to give me complete freedom, it is only because they are afraid that after getting back to my old working surroundings, I should not have enough courage to tear myself away again.

My arguments amounted to the following. If for thirteen years I remained a son of my country, in spite of all the difficulties, I should continue to do so in future. And the time spent would not be lost. It is quite impossible to make a man as well-known as I am sit here by force. If our "characters" do not agree, and if I shall not be "happy" here, nothing will make me stop here, and since it is impossible to guard a man forever, my position will be one of unstable equilibrium, from which it can always slip. So they must throw away all the methods of pressure, and try to find a way by which we shall all be satisfied. This is easy and simple. First, I have no feeling of anger or offence, because in times of great and historic events such as are now going on, it is foolish to talk about personal offence. Misunderstandings are always possible, but they must be liquidated at once. Second, I want to help and take part in the development of science in the Soviet Union, and have already done everything I could. I am sure that as soon as I shall be fully trusted, I shall easily be able to arrange that I and my laboratory will be here in the shortest possible time. And after all, if I am not sincere, and only pretend to be friendly, then in any case nothing will come of it . . .

So you see everything is clear and simple. Of course, if they want to have some formal guarantees, I am quite prepared to accept them, but I shall not produce them myself – I consider it degrading. I felt that Mezhlauk understood that he could talk to me as to someone reliable, that I really am a friend of the Soviet Union and that the main thing is to show that they really care seriously for my scientific work here – seriously, and not merely to justify the words of Maisky. Of course, Mezhlauk is not alone, and I should very much like to talk to his colleagues as I told him explicitly. In any case, we shall meet on the 14th.

This conversation cheered me up and improved my mood which had been pretty bad. Of course, I am working a bit here all the time and I give consultations. If only I could have my work here I should be quite happy . . . I had your recent letter and it is very good that you write about things so openly; go on doing so. Do not worry

Rutherford, I love him dearly . . . yes, Maisky, according to your letter, cannot get a prize for brains.

P.S.   At the moment we are all impressed by the new Soviet constitu-
        tion. Well done! How many vicious mouths will be shut by it.
        Do send me newspaper cuttings with the comments. The im-
        pression in Western Europe must be terrific.

(a, A)   Moscow, 5 February 1935

. . . I want to write more fully about O[lbert] . . . He is, of course,
"non-Aryan", small, energetic and even educated – he finished the
faculty of economics. He was Director of the Institute of Aluminium
and afterwards the Optical Institute and now he has to arrange the
transfer and starting of my laboratory. He has little to do at the moment and still works at the Optical Institute. He is also supposed to
look after my welfare, but this, it seems, he finds beneath his dignity.
He is not in the least modest and when he wants something he tries
every means to get it. So when I was critical about the flat, he tried
to persuade me of the contrary, at first by flattery and then by
frightening me. You remember – first Leipunsky and Semenov –
and now O. But O. did it much more rudely than the others.

This system of "intimidation" is really rather amusing, since it has
only frightened away all my so-called friends, and as far as I know they
will be frightened of me for a long time yet. It is a curious feeling to
realise that one must not visit them – first friendship, then fear. This
is the reason why I don't go to see Aunt Lisa. I talked to Mezhlauk,
and I pointed out that without confidence in me it will be quite impossible for me to have normal relations with people. Of course, when
O. said that it would be a pity if such a good scientist were to be shot,
it did sound a bit old-fashioned to me. As for his care for me, he does
nothing and always gives the excuse of being too busy. But he ar-
ranges his own affairs very well. Sometimes I feel very jealous of his
excellent rooms in the National Hotel which are miles better than
mine. O. uses his "friends" very skilfully indeed. It is strange, but I
am sure I am much more idealistically on the side of the present ideas
and new construction than he is, but at the same time he is there to
bring them to life.

But he does have some good qualities – he is always very cheerful and his cheerfulness is infectious, so that when I am sad, and nowadays I am often sad, I look at his impudence and cannot refrain from smiling. Also, he has no malice towards people. He is very cunning and I think he knows that malice in the end always tells against oneself, so I think I needn't expect any harm from him. I am, of course, very glad that this is only temporary – he was nominated without my being consulted . . . Of course, even if I had been asked, I would have had to give my consent, since all are alike before you get to know them. At present I behave very calmly and peacefully. But I don't despair that when such questions as trust and buying the laboratory, etc., are cleared up, I shall make the same O. work as I want him to, and his energy can then be used in the organisation of scientific work. But always one comes to the same question of confidence – that is the centre of everything . . . Do remember that even if I am sad, I am still clever (in any case I think so).

(a)   Moscow, 20 February 1935

. . . Today I am tired – it is thawing and this sudden change of temperature always tells on my mood. Life is monotonous and dull . . . At the moment the great event is the chess tournament, which is really an all-Moscow event and receives a lot of attention. The other event is the opening of the Metro, which is to happen in a few days time.

21 February. . . . My darling, day and night I think of you and the children and of the present situation. The question of whether I go to work with Pavlov and quietly do biophysics or whether something will really come out of the transfer of the laboratory is likely to be decided during the next three weeks. I cannot understand why Maisky did not talk to Rutherford. I was definitely told that it would be done . . .

Some day I shall write to you about the Academy [see fig. 65] – it is in an awful muddle . . . It is most entertaining when the new building is discussed. Those of an elderly age picture the Academy building as a temple of science, so, for instance, the centre (the altar) is devoted to the earth and the aisles to animals, machines, etc. This is the plan of Fersman who is a geologist. Krzhizhanovsky also wants a temple, but insists on "energy" at the centre (he is an engineer).

65. Academy of Sciences, 1936.

Bakh puts the Praesidium and the library at the centre, which is a little better, but to enumerate everything is too difficult. Suppose I have to build a factory, for instance for making bricks. I can make a plan and it will be easy to criticise it, since everybody knows how a brick factory works. But nobody knows how an Academy works and what it produces and even what part it takes in socialist construction. So how can one build a factory of pure science if no one knows how to make the product? Yet at the same time everyone knows that scientists and science are necessary to the country. This is more an intuitive feeling and sincere too, but these thoughts have not acquired any definite shape.

I read the stenographic reports of the 17th Party conference and also almost all the speeches at the Soviet conference, but nothing was said about pure science. About applied sciences yes, but nothing about pure – I wrote about it to Mezhlauk, but have not yet had a reply. On the other historical occasions of reconstruction of the state, attempts were made to encourage science, as for instance when Peter the Great imported scientists from abroad and created the Academy. Napoleon brought in Laplace, etc., and encouraged science, so that in his time French physics started to flourish, unlike today when

France has almost no science. We also are drawn towards science, but for the present the results do not justify the efforts in the domain of pure science. It is said that when Molotov received the Academicians (I was not present, of course) he said that the Academy would have unlimited help if they present a clear and intelligent programme and plan. I am afraid the Academy won't get very much in this way, but in the end it is not their fault, since they weren't told clearly enough what they are supposed to ask for.

I myself think that it is completely wrong to plan production of a product which nobody seems to understand. Nobody has a clear idea of how science is to participate in socialist reconstruction. This is not because there is no room for science in the socialist state – on the contrary it should have a very grand place – but because the time is not ripe. All efforts are at present directed to accumulating the material basis on which socialist society will be built. This accumulation is going on at a terrific pace, at a pace such as nobody could have predicted, but it is only going so smoothly because its basis is imitation – almost nothing is spent on the creation of new technical forms. Research is all directed to unravelling secrets and mastering various processes already mastered in Western Europe. No special depth of thought or qualification are needed for this, but the results are very spectacular, and the country is indeed growing at a devilish pace.

How long this phase will continue I cannot say, but it is clear that the position of pure science, if not completely nil, is not far from it. A scientist such as Pavlov is treated like a museum piece, of which great care is certainly taken, and of which one is proud, but it is clear to everyone that if Pavlov and all his school should disappear tomorrow, it would certainly be a great pity, but it would have no effect on the life of the country . . . Not only Pavlov, but the whole purely scientific part of the Academy is not linked to the life of the country and if it all disappeared, planning would not be at all affected, because at the moment there is no room for pure science in the active part of life. During the coarse imitative period there is nothing for a speculative and searching mind to do. The tragedy of all this is that they do not know what to ask of a scientist and he does not know what to give. And the best the Academy can do to justify its existence is to provide a mystical awe of science. They want to build a temple for themselves, and carry poor Karpinsky like a live relic in a 12-cylinder Lincoln. The position is tragi-comical – they destroy the Cathedral

66. New Academy of Sciences building, 1989.

of the Saviour by the Moscow river to build another one by the Kaluga Gate.*

I am sure that when we enter the period of original thought in our socialist development, everything will change completely. After acquiring a sufficient material basis, industry organised on a socialist

---

* The Cathedral was demolished as part of a grandiose plan for the reconstruction of Moscow, but much of the plan was abandoned and the site is now occupied by a huge open air swimming pool! The proposed new Academy building (fig. 66), near Kapitza's Institute, has taken over 50 years to materialize and was not yet complete in 1989.

base will want not only to overtake the capitalist economy, but will follow its own original paths; this is already noticeable now. Then the inventive mind and creativity will become free and originality of mind will be valued more highly than organising gifts . . . Academicians carefully selected from among the scientists and freed from the cares of everyday life, will then easily be able to tell what sort of Academy they require to give the most productive work and to be the guiding star of the scientific life of the country. I picture to myself the Academicians as young and gay, and the building of the Academy as a collection of Institutes, healthy and business-like, built for the individual requirements of each scientist and such as can be easily replaced by others as need arises. And the "temple" which is going to be built will be considered only as a memorial to the spirit of the old Academy, which is, alas, still revered by the leading men of our country.

What concrete deduction do I make from all this? First, science has not got a real place in our country at present. Second, once our regime is firmly established, science will become the leader. But when this happens, as with everything else in life, it will happen gradually. If there is war, everything will be delayed, but I think that at the end of the second five-year plan, our science should be well advanced. What ought the scientists to do now? They must train themselves for the leading role and not let any opportunity pass to develop pure science, nor be upset by any temporary separation from reconstruction. Our builders of socialism must reconcile themselves to a temporary breach between pure science and life, since any interference will only damage the true face of science. Let life itself develop the tie between them – then the face of the Academy, both external and internal, will be discovered by the Academy itself. The Academy must and will find its place.

For the moment only one thing is necessary and that is that the scientists should not forget that they are part of a country where a colossal reconstruction is going on, and that very soon they will be asked to play the leading role in it. Do you see what a heresy I have arrived at? I picture the Academy as a society of young, gifted and modern scientists, put together to be trained in the spirit of the time, without being asked anything in return. The building of temples will only turn our scientists into priests who will fool the devotees – as many are already doing with success. I am talking too much – it is 3 a.m. I did not want to write so much about the Academy, and you may not be interested. But when you have no one to talk to, you want to say what you think. And in this I am almost sure I am not mistaken, but, alas, I feel that my point of view does sound paradoxical, even if it may in the end be true. And living relics like Karpinsky will be replaced by some other living relics; the Academy is full of them. I shall not be surprised if the Academy of which I am dreaming, and which without doubt will be required by life, will eventually be created independently. Life will show, and perhaps even soon.

. . . Except for Mezhlauk, nobody talks to me, and I feel completely cut off. The only person who shows interest in me is Shura*; she is a nice creature, and very kind, but she has no broad interests in life and science. The Academicians shrink from talking about the Academy. Your father only curses, but he, too, does not know what he wants. Never in my life was I so *lonely* as I am now. I feel my complete uselessness. Since my visit to the factories I have absolutely nothing to do. I have not refused any work. I was quite ready to take part in the reconstruction of the Academy, but I was not asked. Just one or two consultations, and that is all – what an absurd use of a man! If this goes on for another few weeks, I shall send everyone to the devil and go to work with Pavlov. There I shall have enough space in a corner of the laboratory, and I shall be happier than here with such an impossible perspective. The only things that are well arranged for me here are theatre tickets and cars. I can't complain about that, but I owe it to Mezhlauk directly and not to O. The theatre is my principal support . . .

(a, I)   Moscow, 23 February 1935

. . . I was very pleased with your letter and I was touched by Rutherford's care – he was always very kind to me. Yes, the people in Cambridge spoiled me over the last few years with their care and attention, so the attitude of people here is all the more painfully noticeable. Never before was I so badly treated. I don't think they do it maliciously – rather from stupidity and because they don't know any better,

---

* Aleksandra Nikolaevna Klushina, formerly wife of V.V. Kuibyshev, an important political figure.

but mainly, as I wrote to you before, because they have no understanding or respect for the importance of pure science . . . *Never* in my life have I felt as *alone* as I feel now. Not only Mezhlauk's colleagues, but no one else involved in building up the country talks to me. The Academy of Sciences does not ask me to help in the organisation, as had originally been intended, apparently for fear that something bad might happen. Krzhizhanovski, who at first discussed various questions about the organisation of education, etc., with me and who seemed willing to continue such discussions, also does not ask me to see him. And all my so-called comrades, the scientists, also do not want me. But I don't want to see them either since I find their questions on all sorts of subjects unpleasant.

What a contrast to Cambridge where I had to divert the stream of people who wanted to talk to me. I feel myself completely useless, and the lack of work is making me lose faith in my strength. I thought that I was a good scientist and public spirited and even in Cambridge I left a mark. Take, for instance, the club which is usually connected with my name, and of which, like old Pickwick, I was a permanent president – I think it will continue for a long time . . . I do everything to keep my nerves in order. I must admit that I have a bottle of Valerian, but nowadays I do not touch it very often. I make a special effort to walk every day. The theatre also helps and I try not to think of unpleasant subjects. But now and again it hurts me to tears when I think that I did everything I could for our people here and I can't be blamed for anything. Not only am I sincerely loyal, but I have deep faith in the success of the [plans for] new construction [in the Soviet Union]. I thought they would have been proud of my scientific success and of my behaviour, but the result is that they treat me like dog's excrement, which they try to mould in their own way.

This is only stupidity, stupidity and nothing else, otherwise I cannot explain why Maisky is silent, and other things. For instance, preliminary work on the transfer of the Institute started two months ago, yet nothing has been done in these two months, except that I visited factories and made a list of equipment (other than that in Cambridge) and drew up plans for the laboratory. All this took me only a fortnight. It is funny to watch O. struggling with papers and bureaucrats. What is done by a telephone call in England, requires hundreds of papers here. You are trusted in nothing. They don't even believe you when you give your own address, but demand a paper from the house committee. People do not trust each other at all here. They only trust paper – that is why paper is so scarce! Bureaucracy is strangling everyone, even Mezhlauk, whose orders are diluted and destroyed in the paper stream.

Yesterday, to my great joy, your parcel arrived. All my clothes are completely worn out, so it is very welcome. But don't think it is an easy job to get it. The duty on it is more than 1000 rubles, so two of these parcels will take away my monthly earnings. It is clear that I cannot pay the duty and O. said he will try to get it for me duty-free. The question is clear and simple, my trousers and shoes are worn out, and these articles are not available at the shop to which I am attached, so they should be only too glad you can send them to me. At the same time I am a useful man (or so I was told) and I cannot go about without trousers, nor can I run around the shops if I am to work on science. Well, I have to produce endless papers . . . and no one knows when I shall get the parcel, though I am sure I shall get it in the end, for this bureaucratic apparatus does have the merit that it does really work in the end. I mean that if you have enough patience and skill, you can succeed in obtaining what you want. It is on this that men like O. make their careers.

If it were not for this bureaucratic apparatus, three quarters of the population of Moscow would be unemployed – what an awful thing for the people of Moscow. To destroy this bureaucracy will not be an easier task than it was to reform the countryside . . . But if they mastered the countryside, why can't they master the bureaucracy? The bureaucrat is the parasite of our government, but he can't be touched. There are some organisations in the Soviet Union which on the whole do get over the difficulties of the bureaucratic apparatus and one can only wish that their economic experience could be spread all over the country. Why this isn't done, I don't know. As far as I can see, this is more a question of education than of organisation, and to educate takes years. But even in spite of my cursing, I do believe that the country will come out of all these difficulties victorious. I believe it will prove that the socialist economy is not only the most rational one, but will create a State answering to the world's spiritual and ethical demands. But, for me as a scientist, it is difficult to find a place during the birth pangs, and as I wrote in my last letter, the time is not yet ripe and that is the tragedy of my position. The only way out is to be like a hot-house plant under the special care of the government.

But is this right? . . . There are lots of things not clear to me. But life will show . . . Your love gives me the power of life. I wish I could see the children, but everything takes its turn . . .

(A) From Anna Kapitza   Cambridge, 12 March 1935

. . . What you were promised has been done – an official from M[aisky] came and had a long talk with the Crocodile. It was a very friendly talk and the Crocodile made it very clear that he thinks they should treat you better to get a useful result. He expressed his opinion that their present treatment of you can only lead to your complete collapse. As regards the purchase of the lab, he explained that the lab was a gift to the University from the Royal Society and he doubts if the University would agree to sell it. However, if you need assistance, Rutherford would do everything he could to help, provided the help was for *you personally*. I think the position is that the ice is broken and negotiations can now begin. The Crocodile emphasised that you must be personally involved in any negotiations, so I think our people will soon recognize that they can't go on treating you as they have been doing and that they must take note of what the Crocodile says – after all, he is not just anybody, but someone to whom the whole world listens. He will now write a letter to Maisky and all should go well – our people will see that no one is trying either to deceive or to cheat them and that when you say something it can be believed. Now that the Crocodile has taken this up it shouldn't be long before everything is settled to everyone's satisfaction.

At the start of the conversation the official said that you are quite happy and all is well with you, but the Crocodile answered that there was no point in saying this because he knew you too well to believe anything of the kind. He had read your letters and talked to people who had met you and he knew very well that you couldn't be happy and content far away from your lab. So this question was immediately liquidated.

You know what a genius he is, so you can well imagine what he said and how he said it. He pointed out that it would be the ruin of you if you were put in charge of a big Institute and that you need a quiet and peaceful environment without administrative worries in order to work effectively but in such an environment you could produce brilliant results. Above all, to work well a scientist has to feel free and trusted and if he, Rutherford, were to be forced to work in the same way as you are, he too would react as you are doing . . . I should like to believe that our people will at last realise that we are doing the best we can and that no one is trying to deceive them.

(A) From Anna Kapitza   Cambridge, 28 March 1935

. . . It is very quiet here – the undergraduates have gone but the labs are still working . . . There have been some small difficulties in the Mond but nothing so serious that they can't deal with it. Pearson [the head of the workshop] complains that some of the researchers ask him things beyond his knowledge – particularly the Pole and the American [H. Niewodniczanski and H.A. Boorse] who are novices as regards the lab techniques. But they have already learnt a lot and are finding their feet. Milner and David [Shoenberg] are quite content.

Jimmy [Chadwick] is taking Feather to Liverpool so the Cavendish will change a lot. I saw the Crocodile, who doesn't feel quite himself in the spring and is very sympathetic with your troubles. I think he loves you very dearly and feels very well towards you – just what's needed. He seems to have got used to me too and often starts talking about more general topics. I like him very much – he has a remarkable personality . . . I had a letter from Mrs Millar [see pp. 89 and 124], who, poor soul, seems to be at death's door. She has a bad heart and not only can she not get up but she can't see anything. She has to have a nurse day and night and I don't think she has much longer to live. As always, she writes very warmly about you.

(a)   Moscow, 5 April 1935

. . . Leipunsky came yesterday evening and we talked for an hour and a half. He had seen Mezhlauk and tried to persuade me of all sorts of impossible things but he got tangled up in his own arguments, struggled helplessly, and was ashamed of his position. Of course, my arguments are so logical and so complete that it is senseless to speak against them. Without trust it is impossible to work. Why should it be only after a year or a year and a half that they can trust me, when during all these fourteen years I never did anything which could destroy their confidence in me? Nothing can be brought up against me? And not only did I not destroy their confidence in me, but I did everything possible for the Soviet Union and its reconstruction – even though it is more difficult to be true to the Soviet Union abroad than it is here. Well, why then in spite of all my good conduct is it that I am not only not thanked, but abused, accused of ridiculous things and am being completely ground down? Leipunsky couldn't answer at all. "That is what they think and that is what should be done", was his only argument. He came to me not like a friend but rather as an emissary and we parted very drily.

. . . Yes, Mezhlauk does not want to see me, but now it is no longer necessary, everything has been said and all is clear. Perhaps I shall try to talk to his senior colleagues, or write to them. This I haven't yet decided – I don't see any point in it. It is so difficult to say about oneself: "I am an honest man, a good scientist, I am used to being treated like one, I think I have rendered service to humanity and this gives me certain rights in life" . . .

(a)   Moscow, 8 April 1935

. . . Everything is still the same, but my mood is a little better, and I am proud that I don't let myself go, and am not losing my courage. My position reminds me of my moral condition of sixteen years ago, when I lost my wife and two children. At that time I was very troubled with the same apathy and the lack of wanting to live, but I was saved by deliberately not letting myself think about the past. I put away all the letters from my wife, I did not go to the cemetery, I hid all the portraits of her, and everything that could remind me of the past – even now I haven't the courage to read our old letters. I am doing the same now. I avoid everything that reminds me of my laboratory, and the interrupted thread of my thoughts. That is why it is easy for me to study physiology, organic chemistry and biochemistry. But until my laboratory is here I must draw the line at physics – I never thought I should be put in such an amazing situation.

As for the Institute, everything is still at the planning stage . . . It is difficult for O. – they don't give funds and supply only bad helpers. He himself says they are no longer interested in it. This is because the ridiculous colours in which it was all painted, and because of which the whole of this business started, have now disappeared. And no one is interested in pure science. I am almost certain that even if it will all be built and the equipment purchased from Cambridge, the Institute will not be able to live in such an atmosphere and I shall be accused of not wanting to make it work or something of the sort. The most sensible thing for me is to seek refuge in biophysics. First of all, my gifts in that field are unknown and there is no scale of comparison with my previous scientific career, and, second, there is no need for a complicated technical base, such as is necessary for my work and such as exists in Cambridge, and which can be created here only if I am given exceptional conditions. It is most amazing how their attitude to me has changed. For instance, Mezhlauk, who wanted to receive me twice a month, every 2nd and 14th, now receives me only once in a month and a half and even then I have to telephone for a whole week and am told, perhaps tomorrow or he wanted me to come today but something came up which prevented it, etc. No one shows any interest in my paper on pure science except you – and only you are important to me.

Of course, I did curse a bit, but who would take any notice of it if they really thought I was an important scientist and my work was needed by the country. The incident of the parcel is very characteristic. Of course, everyone understands that when I came I didn't intend to stay here, so I did not take any spare clothes with me, so it is quite natural for you to send me underwear to replace what is already almost worn out. It is understandable if they charge duty, but if the duty amounts to one and a half times my monthly pay*, then of course it affects my standard of living very seriously, and I can only be glad that it hurts me rather than you. I am still paying for the duty in instalments. I asked that I should be allowed to get the *New Statesman* but this was refused. In a word, it is difficult to imagine an attitude of less consideration. After seven months my suit is beginning to wear out, my shoes are completely torn (I use the ones you sent for wear at the theatre), here they don't make them in suede. Why not allow mine in from England? "Not allowed and won't be let in". A

---

* The apparent inconsistency with what appears on p. 226 is probably no more than an example of Kapitza's frequent carelessness in quoting figures. The 1000 rubles of p. 226 is more likely to be correct.

most "delicate" attitude to a Royal Society Professor who was once a prominent scientist! But you know, funnily enough, all this really doesn't touch me very deeply and I don't feel all these privations much. Either they are compensated by the pleasures of the theatre, or perhaps love of comfort is only an acquired habit and by nature I am very simple. If I write about it, . . . it is only because it is a good test of their attitude to me . . .

It is not nice to boast, but I think that I was not without some clairvoyance when at the start I decided to go over to physiology. Then it was looked on as a kind of demonstration, but now the logic of life sends me to it – logic which in a month or two will be understood and clear to all. So you see that on the whole I have rather a happy nature. I can take a problem in any domain, provided only that it interests me, and I am not afraid to go over from physics to biophysics, because in present conditions I prefer a corner in Pavlov's Institute, to an Institute of my own. Very likely I am the only scientist in the Soviet Union with such a feeling, so my psychology is completely incomprehensible here. But the cause of it is very simple: the only things which could make me wish to have my own Institute are (1) the real development of my old work, and I am almost certain this will not be possible, and (2) personal ambition, which of course I have like everybody else, but it was already fully satisfied when the laboratory in Cambridge was built for me . . . I think that I left a mark in physics and I can now change over to another domain and it would be a great pleasure to work with Pavlov, who is such a great man . . .

Well, dear, what a long letter I have written – it is such a pleasure to talk to you – there is no one except you. Now I must go to the Academy of Sciences, where I spend an hour and a half a day, though even that is too much. I am involved in everything and directing it all [i.e., the planning of his new Institute]. I am beginning to think I am not a bad administrator since I don't spend much time and all goes well, without any fuss. I know all the main things and look at the details only when they show the character of this or that worker, since I must know everyone's capacity. Give my regards to Rutherford. He is very good, and how nice it would be if he could come here, as I can't go to him . . .

67. D.L. Talmud, Cambridge, 1934.

(a)   Moscow, 9 April 1935

. . . Yes, it is very painful without work, painful not to be believed,
even when I tried to tell them: "I willingly forgive all the indignities
which you made me suffer, all the injustice, etc., and you can see I
am trying to do everything to help you in the creation of science –
why then this rude and senseless attitude and mistrust?" If I am as
I am and not good enough, why keep me? There is a place where I
am appreciated as a scientist. Very distressing and incomprehensible.
I am still waiting for a talk with Mezhlauk and then will ask for an
audience with M[olotov] . . . Today Talmud came from Leningrad.
He is a very pleasant companion and even if you do call him a hypnotist he is a very pleasant and kind hypnotist . . .

11 April. . . . Interest in my work has vanished but the scientist
comrades were so alarmed, even though the conditions laid down for
my work could only be called normal at best, that they started to be

indignant without any restraint. "If we could have the same, we should do better than Kapitza." Of course, it is all nonsense, but with envy, suspicion and so on, the atmosphere is unbearable and fit for the theatre of "Grand Guignol". I am afraid that all this is being confused by Mezhlauk, who hasn't understood either my mentality or that of others. In any case, the problem is more difficult than at the start. I personally see only one way out – to retire quietly into biophysics. I am still waiting for the long-promised audience but unfortunately it has been put off for a long time because Mezhlauk is ill again and away from Moscow. And I have to get permission for returning to Leningrad.

Well, you see my dear wife, how sad things are, but there is nothing to be done. I know that time will solve all, but it is difficult to wait. The present position is useless for everyone. The scientists are definitely opposed to my transfer here. "There he enjoyed himself, while here we experienced all sorts of privation, and now he comes and wants to be a boss!" That is why the Academy has such a cold attitude and is unwilling to cooperate. Of course, once they have understood my work, and discovered that it is all purely scientific, they do not see any use in me, but even if they now regret that they started all the trouble, they have to consider the question of prestige, and for this reason they support it all . . .

12 April. Someone came in yesterday and I had to interrupt my letter, but today is the free day* so I shall finish it . . . Well, to continue: the situation is such that I feel myself a useless and foreign element and this certainly does not help the development of my scientific work here. I am in an impasse from which I can see only one issue – to leave the scene, and go into a new domain where I am a new man and have to build everything from scratch. I know it is difficult to start a new path in the 40th year of my life [actually his 41st], but it is not impossible and it seems 1000 times easier than to recreate my old work here. Then perhaps the scientist comrades would be convinced that I don't want to be a boss and that all I want is to build science in the Soviet Union. If they all have such ill-feelings towards me it can only make me happy to retreat into such a modest corner. At the same time my refusal to take all the worldly goods can only underline the sincerity

---

* For several years the usual seven-day week was replaced in the Soviet Union by a six-day week, with a "free day" every sixth day instead of on Sunday.

of my decision. When I tried to go into physiology before going to Moscow, it was interpreted as a demonstration (all nonsense, but it was difficult to prove the contrary); now after four months in Moscow, when almost nothing has been done, the weakness of my position is apparent. I shall not only be unable to continue my work in physics, but cannot even join in the social life of the scientific community . . .

I wrote a paper about pure science but no notice was taken of it – they only said why does Kapitza tackle such a problem? He is a newcomer in our country, and assumes the role of a reformer. This is, of course, a statement characteristic of the present time! If the country really needed pure science, it would be discussed everywhere and the only thing for me to do would be to join in the discussion. But there is no discussion and the scientists are frightened of it and alarmed because their conscience is not clear. You probably understand the mentality of scientists. An honest scientist needs solitude in his work and individuality, but he also wants acknowledgement and for this he needs organisations like the Academy of Sciences, with its titles, periodicals and meetings. A dishonest scientist is scared of too much order and planning in science, because his bankruptcy would become too apparent. Hence, the individualism and charlatanism (of which we have plenty) – they resent the organisation of science.

The rather condescending attitude of the Government towards science is quite approved by the scientists. They are better looked after than others, though no one understands a scrap about their business, and it is so pleasant for the scientist to paddle in the disorder. Of course, this is only possible because during the imitative period, the country does not need science. The country is developing on account of Western science and so they need interpreter-scientists, not original creators. No one in the Government is able to appreciate whether a man has invented something himself or read it in a foreign paper. And it doesn't matter how the man got his knowledge – the important thing is how he will apply it for the growth and development of the country. It is this that is valued, and for this there is, of course, no need for an organised science . . . That is why there is no wish to create the necessary conditions for my work here . . .

. . . Talmud brought me coloured reproductions of the Shchukin collection, so the room is brighter with Degas, Cézanne, Picasso, Gauguin, and Manet hanging on the walls. Today is a free day and I am going for a walk. I now have some friends who enjoy my company. One of them is a writer and critic, rather well known. He is rather a muddle-headed intellectual, but a nice fellow. His wife is beautiful but as someone said about her, she is charmingly stupid. My mood is better after my quarrel with Mezhlauk, I feel more energetic, and even if the senselessness of my position is the same, I somehow feel not completely without hope in the world . . .

(a)   Moscow, 13 April 1935

. . . My life is so empty now. Sometimes I rage and want to tear my hair and scream. With my ideas, with my apparatus, in my laboratory, others live and work. And here I sit all alone. What for? I don't understand. I want to scream and break the furniture and sometimes I think I am beginning to go mad. Only Valerian drops (my mother's prescription) save me . . . My dear, try to feel how I suffer – except for you, no one cares, so soothe me, I implore you! You must excuse the hysterics in my letters . . . You must know, even if only slightly, in what mood I am . . .

14 April. My dear . . . yesterday I wrote in a very depressed state. Today I feel better, but I am sending yesterday's letter so that you will know how I feel in the present conditions . . . I shall go to Leningrad on the 19th and will work on physiology there. It will be pleasant to live with Mother. I also love your father very dearly. Looking forward to the journey gives me courage and happiness (a very little bit). My head is very empty – at other times I like to let my imagination roam and think about impossible experiments. Just to exercise my brain, but now it is quite empty. I feel I am stupid or at least begin to be more stupid. Science is so far away and so foreign to me. It begins to seem that I am quite alone and separate from life, but it doesn't frighten me to disappear from life's arena. Well, as soon as I let myself go, I start on sad things again . . .

(a, A)   Moscow, 3 May 1935

. . . Here I am in Moscow again. I arrived yesterday, and the weather here, just as in Leningrad, is not like May, but cold and unpleasant. Yesterday I went walking with Talmud and today O. came to report what has happened here. As usual I have to be careful with what he says since he does not care for accuracy. But my darling I feel I can't live like this any more. If they go on torturing me like this I shall land in a home for nervous complaints. At the moment I am waiting for a telephone call to tell me when Mezhlauk is going to see me. This doesn't mean I shall see him, but I shall continue the letter as soon as everything is cleared up.

4 May. Yesterday Mezhlauk received me towards the evening and it was too late to write to you after that. Your letter, in which you say that everything has appeared in the Press, came only after I saw Mezhlauk, so when he put the issue of *The Times* with Rutherford's letter in front of me, I was amazed. He also gave me a Press review of all the newspapers. I felt that something like this must have happened since various foreign journalists had telephoned O. while I was away in Leningrad. I am glad that Mezhlauk regards it all as of little consequence, but all the same our talk was not very pleasant. How can I say that I am happy here now, when I am completely isolated? How can I say that it is not dreadfully painful without my work? But of course the question is very difficult, and I really don't know what to do.

. . . Why should I be involved in all this? I asked Mezhlauk to bring me together with Molotov. There was a time when people counted it a pleasure to meet me, but here it is the opposite. Mezhlauk says Molotov would be hardly likely to want to speak to me and he pointed out that I should take it as a great honour that he himself talks to me, an honour not granted to either Karpinsky or Volgin. Altogether he says I must obey them if the people gave them the right to rule the country. I said I obeyed all the orders I have been given. But some of the orders sound as if Beethoven was told to write the Fourth Symphony to order. Of course, Beethoven could conduct an orchestra to order but he would scarcely have been willing to write a symphony to order, in any case not a good one. So I obey everything they ask me, but I can't begin any creative work and of course Rutherford is right in saying that Kapitza "requires an atmosphere of complete mental tranquillity". After the talk with Mezhlauk I had to lie down for two hours. Even with all the worldly goods, I can't feel happy when one part of the population does not talk to me because it is afraid and the other part because it feels too important.

But of course all this does not make me shut my eyes to the colossal work which is being done; I do agree with the socialist ideas and agree that the right to govern must be in the hands of the people who work. I don't have any disagreement with Mezhlauk and his colleagues, except on the question of science (internationalism and freedom), and the attitude to me. I said to Mezhlauk that in that respect they must obey me as they understand nothing in science – either they must drive me out or obey. This he did not like. Well, Anna, probably it will stop at that, Rutherford will complain and all will finish.

By the way, the attitude to me, except for such things as parcels and the ban on getting foreign papers, etc., is above criticism. I am well served with cars and soon it seems I shall have the right to drive the car to be assigned to me. This I haven't done for eight months. I am well provided with theatre tickets and steps are being taken to arrange a flat. The "guardian angels" [i.e., secret police shadows] went back to heaven so I do not see them any more and I can go to Leningrad or whichever place I want to in the USSR quite freely, though it seems they prefer me to stay in Moscow. Well, now about your visit. Mezhlauk says he will arrange with pleasure for you to come for two or three weeks to help me and promises they will let you out again. I am worried about the children being ill such a lot this year and how you will be able to leave them. I know that you do look after them well . . .

I spent the 1st of May [in Leningrad] with Semenov. He came the previous evening and stayed overnight and we had a good talk. In the morning we went to find our colleagues in the demonstration and wandered about for two hours before finding them. The people on the demonstration were very jolly, with lots of music and dancing during breaks in the marching. The people were gay, full of energy and the wish to live and it made a tremendous impression. Among all these happy people I felt even more alone and unhappy. Why don't they want to trust me? Am I too much of an individualist to mix with the people?

I went along with the demonstration; Talmud was there and some of Semenov's and Joffé's students and I talked a bit with them. Then I went home and after dinner went to see Talmud who is very kind to me. Later Semenov came and we talked about Soviet science. The Academy is lifeless and rigid – the scientific institutes are full of second-rate or even hundredth-rate scientists and all this doesn't make a very bright picture, and calls for great reforms. Of course, there is only one way out and that is that the younger people (say up to 40 years old) must get together and offer constructive criticism. We could think of six or seven people who are young and want to create science. But all this is only plans.

On the evening of the 1st I returned to Moscow. Of course, the car from the garage of the Academy was late again and I almost missed the train. The train was nearly empty, with only three people in the carriage, since no one travels in the Soviet Union on the night of the 1st to 2nd May. It was strange to read the letter of Rutherford in *The Times*. It was as if it was about a different Kapitza, who was a scientist, and whose fate interests scientists throughout the world while the actual Kapitza is a tiny, silly and unhappy thing dangling on a string like the little spiders with wire legs, hanging from a Christmas tree. They hold me by the string and I dangle my feet helplessly and without avail so that I don't move from the spot. Well, what did I do in these eight months? Nothing! . . . You are a good little wife, and I live only because of your letters . . . I was very happy with your sketches of the children – so send me more. I know that they take you only a few minutes but they give so much pleasure. The ones you sent me are hanging on the wall and I do admire them often.

5 May. I was sitting in my bath and rushed naked to the telephone. It was the Observer correspondent and I asked him to ring me again in 15 minutes. I hate correspondents and just now more than ever. I have to go to the Academy to tell the Praesidium about my work in England. This is the first time I shall be speaking about my work. It will be very painful, but there is nothing to be done. I hope that Mezhlauk will see me on the 14th as arranged. The Observer correspondent rang again. I refused to give him any information or an interview. To the devil with them, if one gets mixed up with them one might land in a very unpleasant mess. Well, I must go to the lecture now and will continue the letter when I come back and then send it off. You see, I write every day and I think it should be interesting for you to have something like a diary.

I have just returned from the Academy where I was fighting for a bit and had to put the issue explicitly and I threatened to resign, but all ended well and we parted as friends . . . I was very interested in your description of how it all got into the Press. Our Maisky can't be very bright if the English journalists could provoke him so easily. But sooner or later it had to get known. Rutherford's letter is very well written and he puts the question broadly. I can recognise the style of my beloved old man in expressing his thoughts. The best thing for our authorities is to keep quiet; abroad they will storm but eventually they will stop. I personally will get into a corner and stay silent – to the devil with them all . . .

(a)   Moscow, 8 May 1935

. . . I have been busy writing a letter to Comrade Molotov, about the present polemics, and the impossibility of my taking part in them [see p. 319] . . . Yesterday the Institute received a very good car, big and comfortable, which I can use as long as I am Director. It is rather fun to have such a car. When I came to the door of the Metropole the doorman helped me out as if I were an old man. Yesterday I also bought shares in the Government loan, so now I am a shareholder. When I lived in capitalist England, I never bought any shares or loan stock, and always drove the car myself (it is true it was my own car), while here I own loan stocks and have a car with a chauffeur. But all this is laughter through tears. Perhaps the letter I have written [to Molotov] will make them change their opinion of me at last.

Yesterday Talmud visited me; he is staying not far away, and we drink coffee together in the morning. He is a nice fellow and urges me to come to Leningrad. You know it was amusing that when I wanted to potter a bit in Semenov's institute because Pavlov was ill, Semenov got frightened and said I had better ask permission etc. He couldn't refuse me but felt safer that way. Yesterday O. again tried to intimidate me. They will be agitated for a while and then it will pass (but they are definitely impressed). Thank Andrew for the greeting and tell him to write to me, my dear; tell him Daddy wants him to write a letter like Serezha [Sergei] . . .

P.S.   Do not send the cuttings – they will only worry me, since there is probably all sorts of nonsense in them and I hate the Press.

(a)   Moscow, 9 May 1935

. . . Unpleasant news today – they are giving me almost no tickets for

68. V.I. Mezhlauk (l.) and V.M. Molotov (r.), 1935.

the theatre. I don't know why, but apparently this is a punishment. I feel like a schoolboy – if I do something the teacher doesn't like, they put me in the corner and my teacher is Mezhlauk. But now that I have a wireless I don't mind . . . Everyone is discussing Stalin's speech which is really quite remarkable and energetic . . . I feel great apathy – if I talk to someone I have to lie down afterwards. But my mood is not too bad, it is as if I was under an anaesthetic – nothing gives me pleasure, but nothing upsets me either. I think this is natural adaptation – by this means I protect my nervous system and automatically make it less impressionable. Perhaps that is why I don't even want to do any science, since it always agitates me . . .

11 May. I haven't much strength just now or to put it more accurately all my strength goes into thinking. Today I wrote a letter to Mezhlauk commenting on Rutherford's letter. I said the main thing in it is the complaint of a representative of the scientific world that important work is being interrupted and that the mentality of a scientist is not being understood. Also, that the letter is a friendly piece of advice and that it would be ridiculous for me to intervene. I even detect the influence I had on Rutherford, in the way he speaks very well and

kindly of the Soviet Union, but I'm afraid they haven't sufficiently appreciated it.

I am sending you the photo of Mezhlauk and Molotov [fig. 68] for you to show to Rutherford. He is a good judge of people's souls by their faces, so ask him what he thinks in this case. Mezhlauk has a nice little face, and even though I am cross with him, I sometimes like him quite a lot, though he shouldn't put on such airs. I suppose it is just a bad habit and difficult to get rid of. You know, I was told in the Academy that I don't make myself important enough. Someone invited me to come, and I went, and then I was told that I shouldn't have done it, since this person should have gone to me! Well, I don't understand this business of self-importance and ignore it, so I felt no difficulty in going . . .

I had your letter and as usual was very pleased. It is a great comfort to me, especially when they want to punish me. But unfortunately for them it is difficult to punish me, if not quite impossible, since nothing can be worse than my position with no work, no you, no peace, no Rutherford, no comfort, what else can they do to me? . . . Mezhlauk has just telephoned to say that Molotov will see me the day after tomorrow, and he hopes all will end well. This quite upset me – well, we shall see what happens . . .

(a)   Moscow, 31 May 1935

. . . Let me tell you how I see it all. I have had a great moral change; not everything is clear to me, but here are the main features. You know that I always wanted to participate in the building of our science and I always told you I couldn't imagine a socialist state without science playing a leading role. And I still believe this, and believe that science in the Soviet Union will reach undreamt-of heights. That is why I always wanted to take part in the organization of this process, which would happen even without me, but I wanted to do my share. So as soon as I was transferred from Leningrad to Moscow in January, I took part, or more exactly, expressed my wish to take part in the reconstruction of the Academy. During all this time I took initiatives in this direction, wrote reports, discussed questions with Mezhlauk and so on. Now these efforts have lasted six months but have ended in nothing. Whose fault is it? Perhaps mine, in as far as

I did not know how to approach the matter, or perhaps the time is not ripe. One could write a whole treatise on the subject, but one thing is clear – further efforts are useless.

So I definitely decided that I will do only science in the Soviet Union and I will definitely refuse to do any organizational work . . . Of course, this is a serious decision, and I took it only after hard thinking. Perhaps you will think it foolish, since I am absolutely on the side of our socialist reconstruction, although not a member of the Party, and of course I do not differ from them in my own convictions, so with the scarcity of such people among our scientists it might at first seem quite a foolish decision. But this is not so, and in spite of everything I shall have to take the same position as Pavlov. I mean the position of a man who sits in his den, follows everything and tries to make his voice heard from there, without taking any active part in life. This will be more peaceful for me and more pleasant for others, and it simplifies the position.

Only one question is left and that is my science. I am trying to recreate my Cambridge work here, but I have no confidence that I shall succeed. As I wrote to Comrade Molotov [see p. 321], I look at this venture as being about as possible (in present conditions) as to make a hole in a stone wall with a pen-knife. Monte Christo did it, but it took him half his life to do it. But while I am Director of the Institute, I must carry out the definite programme I was given, as effectively as possible. Of course, I write letters "upwards" pointing out all the difficulties, etc., and I shall write to you about it in more detail later, but it's a very long and dull business.

At one point I had to resign because they wanted to make me do something which I didn't consider possible. Later I withdrew my resignation, and returned to the "status quo". But I fear, and indeed am sure, that in another two to three months I shall have to resign again and leave it all. I cannot see any other way out. Of course, if everything comes right with the laboratory apparatus, then perhaps it may be possible to find another solution. But my instinct tells me that if our authorities will go on behaving like idiots, as they have till now, then nothing will come out of it all. I am not listened to and no one asks my advice – they only give orders – when I know so much better what should be done.

Mezhlauk thinks that the buying of the laboratory is only a question of money and it is difficult to imagine a more foolish attitude. He told me: "They will ask £200 000, well we shall offer £10 000 and we shall eventually come to an agreement". I warned him from the start that he was mistaken. Maisky seems to be not much cleverer. Well, it is their business, and so in two to three months, I shall resign. All the noise in the Press will have stopped by then, and our people will be quieter. But the building which will have been built will not be wasted in any case, since the country will get a first rate Institute. I put all my English experience into it and that should be useful here. Only the long [magnet] hall with a special foundation and the balcony will be useless, but everything else will be useful for any Institute. The organisation, the position and the living accommodation are all exemplary. I put enough energy into it so that whoever gets it will not need to reproach me.

Of course, they will be cross with me for my resignation, but in two to three months I should be a free man, with only the one difference that there will be no further point in exploiting me, and then I think they will leave me in peace. In one or two years everyone will have forgotten me completely. Meanwhile, I shall quietly be born again little by little as a harmless physiologist, without need of an Institute, expensive machines and so on. I shall be 43, and I shall have enough energy, if my head doesn't give up, to conquer a position for myself, such as is necessary for a scientist to be able to join the circle of his colleagues. This is the picture, or more correctly the plan, which I am following. Why not take up physiology straight away as I did six months ago? The explanation is simple. First, our people still see it in a bad light as a kind of demonstration and so they are cross. And since I have already made them sufficiently cross, and shall make them even crosser yet, I must not overdo it. After all, I like them, however stupid or strange it may seem. Second, my head is in such a state that I cannot concentrate quietly on scientific problems. Even without physiology, my head and nerves are so stretched that I shall be lucky if they don't break down altogether.

I want to know your opinion. I think you will have to come and the children will have to stay another year in England. With all the uncertainty and possibility of trouble, and also "punishment" for "obstinacy", why should the children be involved? Well, if physiology doesn't work out, then for material reasons, I shall have to take up one

of many other possible professions, and I do know how to do all sorts
of things. I don't remember a single piece of work I did in which I
didn't find something interesting and new . . . I can't send you a copy
of my letter to Molotov . . . since people might get upset . . .

(a)   Bolshevo*, 5 June 1935

. . . I have been here for four days already and feel much better – you
know how quickly I get better in the country . . . My heart is better
too, and I was able to row for almost an hour. Evidently everything
is a question of nerves and the injections and so on are all nonsense.
My mood is quieter, I think because of having reached the decision
I wrote about in my previous letter. I intend to do only pure science
in the Soviet Union and not to take any part in the organisation of
science. So let science in the Soviet Union develop as it will; sooner
or later it will find the right way – all I could have done was to make
this "later" be a bit "sooner". This will be no tragedy and I will no
longer fight and try to make our people cleverer; this would only drive
me insane.

It means that in future I picture very secluded work for myself,
purely scientific, and very possibly in physiology . . . If there was any
sense in transferring my work here it would have been because I could
have influenced the rise of science in the Soviet Union, but since the
question of my participation in the life of the country is excluded, all
this is very stupid. But crude violence is always stupid. A clever man
can always find a way to make the other do what he wants without any
apparent violence and in such a way that the other also wants to do
it. To put it explicitly, it is always better to change the method of
violence into one of mutual agreement. But our authorities consider
it harmful for the prerogatives of Government to come to an agreement with a citizen. What extraordinary stupidity! It is evident that
in any business it is better that the man who is doing it should think,
or even deceive himself, that he is doing it by his own wish, rather
than by commands . . . Commands are needed only for soldiers, since

---

* An Academy rest home near Moscow.

what they are doing is clearly senseless, and can't be justified in any rational way.

For a revolutionary army full of enthusiasm, of course, commands are no longer needed, though discipline is needed to coordinate the movements of units. For scientists all this is a thousand times more applicable. Until we create the moral conditions that make it possible for every scientist to work in his country, so that there is no fear that he will want to escape, it is difficult to count on the development and flourishing of science. All this is quite elementary, but it is extremely difficult to make them understand it. It seems the time is not yet ripe, even though what I say must be obvious to anyone who understands these questions even slightly. I have to bow my head as a victim of the times and the only thing to do is to retire into my den, as Pavlov did, and sit there doing my own thing. It should be possible to arrange this, not as well as in Cambridge, but well enough not to consider this time in my life as being lost.

It seems something can be done, but I am worried by one question – that is you, my dear . . . Well, see what may happen: I could work, one, two or three years in physiology, at the end of which they will get tired of keeping and torturing me. For me, the whole world is open for work, and they can keep me by force for only, say, five years. People have been in prison, in lunatic asylums, or in exile for similar periods and in much worse conditions than mine. I have all the material conditions of life – only the spiritual conditions are bad. In the end, they will have to give me back my freedom. Their wives waited like the famous Penelope, so you too must wait till everything is clear, be courageous, don't get upset, and obey your husband who is still clever . . .

There are several things I want to say. First, if someone is to be sent, Cockcroft *is a very bad choice** – anyone but him. The best would be F.E. Smith, Hardy, Jeans or Hopkins – that sort of man. Winstanley, Simpson, Dirac, etc., are all suitable but not John

---

* The reason for this remark is not clear. Possibly Kapitza resented the fact that Cockcroft *de facto* in charge of his old lab and so able to exploit the facilities that Kapitza had created. Thus he may have felt that Cockcroft was not a completely disinterested party. However, once the transfer of the lab equipment had been agreed, it was Cockcroft who bore the main burden of the work involved, and they remained good friends.

[Cockcroft]. Second, one must not put it off too long; I am sure our people will have nothing against this visit. But I don't want to see the physiologists who are coming to the congress, and I should prefer to go away during that time. I want to know your opinion about this. Third, please send me the most important of the newspaper articles, especially of the foreign Press, together with your own opinion of them. If you didn't read them, then ask Rutherford or someone else. I may be asked to discuss this question, so I mustn't be in the dark if they accuse me. So do give me this information at once. I thought of not taking any interest in it, but now it is difficult to avoid it . . . Well, my dear, don't worry, I understand everything very well and perhaps see it very much better than you think – I know people pretty well.

(a)   Moscow, 7 June 1935

. . . I am very glad you are sending my things with the physiologists. The custom officials won't be able to say anything since they will be distributed among 10 to 15 people. How foolish and stupid that I can get my own things only by means of such tricks – one of the many proofs of the idiocy of our people. This charging of duty on my own things after I was promised full assistance – well how can one trust them? How can one trust them in big things, when they can't keep their promises about details? But with your help we shall be cleverer and, if you are very energetic, you can send me a supply of things for another year . . .

Today I finished my affairs in town and I am going back to Bolshevo this evening and will continue to do so for several weeks. However much our idiots would like to ruin my health, I will find a way of keeping my nerves in order, so don't worry about me . . . Well, if eventually our idiots get more reasonable and let me out of the harem (I am like a white slave), I don't see why all this fuss couldn't end to everyone's satisfaction . . . Of course, they are whimpering a bit about prestige, but that is all nonsense. They are not such complete idiots as they are in Western Europe, not to see what is useful because of prestige. Perhaps one day they will see – after all Dostoevsky called his greatest hero "The Idiot".

Once again, please send me everything that was in the newspapers, so that I should know what is happening. At first I didn't want to take any interest in it but now I see that it would be rather difficult without it – our people think more about it than I should have thought necessary. One thing that strikes me is that they thought such an attitude of the Press would be impossible and even now they do not appreciate all the delicacy of Rutherford's tone. They don't know how Rutherford can swear . . . Well, he is an old farmer and he is aroused a bit at a time, and always keeps some curses in reserve. I know him well and there is little you can do – he is like me and never listens to anybody but himself.

Why is Webster not coming? But the best thing would be that if Cockcroft must come he should come with others – Hopkins, Keynes, Laski or others of that kind. Everything must be arranged strictly officially through Maisky, or else I shan't talk to them – but one thing is quite definite: Cockcroft must not come alone. I have my own reasons for this, about which I don't want to write. You have lived for 30 years and still don't know how to judge people . . . Try to send the *New Statesman* and *Manchester Guardian* again – perhaps they may be allowed now – devil only knows . . .

(a)   Leningrad, 19 June 1935

. . . Webster is a very good choice*. I think too that he must come as soon as possible and must have with him all the details of conversations between Maisky and Rutherford, since they don't tell me anything about it. I am sympathetic with all my heart to anything that will bring a peaceful solution, but I have not the slightest idea of what and where anything is being done. That is why I haven't taken up any kind of diplomatic attitude until now, but said everything just as I felt it. This is the only policy possible in my position. I want peace and good for the Soviet Union, and especially for science. I don't know if they have lost faith in our scientists or whether it is just habit, but our authorities here don't trust me at all . . .

You must realise that I am completely in the dark, and understand nothing of what has happened, and so as not to get submerged, it is

---

* Although Webster had contemplated coming to assist the negotiations of 1935, he did not, in fact, visit Kapitza until much later (in 1937, see fig. 16).

best for me to sit in the dark and keep silent, but I don't want things to be spread about me which are definitely untrue and harmful to the Soviet Union. Of course, our authorities behaved stupidly and meanly to me, but one mustn't accuse them of more than they have done, and of course it is never too late to put it right.

Now about Rutherford's doubts. He mustn't be afraid for me – tell him that I had already got all that was best in life a long time ago and now I am like a dead volcano, or rather a volcano which is temporarily dead (that's for you, so you shouldn't worry). And I never met anyone in the world who is better than my wife, so you can completely rely on me, just as I trust that you are a good and faithful wife. Altogether, if there were no you and no children, there would be no me now. The whole present position could have been easily and simply liquidated. Perhaps if Rutherford writes to Maisky, Maisky will promise him that you will have the possibility of coming here for three or four weeks to arrange things and if everything could be peacefully ended that would be very nice. Tell me, by the way, did Rutherford write to *The Times* again? What made him write? I understand nothing. When will our people get a little more reasonable? What a mess it is at present and who wants what? But I have to suffer for all this . . . Please send a wire to say when Webster is coming – the sooner the better and best if he comes to Moscow, where there will be more opportunity to talk to people.

(a)   Moscow, 24 July 1935

. . . All this time I haven't designed or built a single bit of apparatus and except for reading books and articles about physiology, I haven't done any scientific or technical work. I haven't given any lectures, nor published any scientific work, and I have definitely stated to everyone here that I would only do purely scientific work and nothing else except for consultations, but they were never asked for . . . I have not seen Mezhlauk but I'll write to you as soon as I have seen him. I shall accept whatever decision satisfies everyone with pleasure, but for this they must basically change their attitude to me. If this isn't done, I see no other way out but the one I wrote to you about – to go away as far as possible from the scene. This is not my fault. I am sincerely fond of our idiots, they do wonderful things and it will all make history. I was prepared to do all I could to help them and, even now, I shall do all that is in my power to help them. But what can be done if they understand nothing in science or rather they don't know how to create science? For this evidently one must wait till they get wiser. To take up the policy of Semenov or Joffé, to compromise and to wriggle, is something I don't know how to do and don't want to do.

They (the idiots), of course, may get clever tomorrow or perhaps only in five or ten years. But they will get clever, there is no doubt; life will make them do it, but the question is when? I have tried to hasten it, but up to now without results. But you can be certain that from the first day until now, I have never compromised my conscience and I am sure I shall never do it. All the time I say what I think, even if I am alone and none of the pleasures of life will seduce me. Nothing will intimidate me and nothing will seduce me. I feel myself very strong, my conscience is quite clear and there is not a single action of which I should be ashamed before our people, country, Government and even the Communist Party. On the contrary, I think I did a lot for them and for Soviet science, even though they don't want to acknowledge it. I don't want any thanks, but my inner feeling that I can hold my head high gives me happiness. This is the minimum of happiness possible without you, the children and my work, but it is essential for retaining the wish to live . . .

(a)   Bolshevo, 26 July 1935

. . . I had a wire from Dirac from Irkutsk to say he is coming on the 28th. Yesterday I had a talk with O[lbert], who complains all the time that I don't love him, and that he is so kind to me, and I'm unfair to him, etc., although he is so kind to me. My position is difficult . . . in normal conditions I would just put him in his place, saying "just attend to your own business and don't touch these questions", but at present that is impossible. If, as seems likely, I shall have to resign, he will be able to finish the job and adapt the Institute for some other work. My leaving will not affect anyone at present while there is no scientific work. During the last three months I not only haven't read anything, but haven't signed a single document. As I wrote to Comrade Molotov, I am as much a Director as I am the Shah of Persia and I could easily leave it all at present. Our authorities can't even grumble because I warned them several times that without people and apparatus, it isn't worthwhile building the Institute (I have written four times on this subject, once to the Government, once to Mezhlauk, once to Molotov and once in a statement to the Praesidium of the Academy of Sciences). Mezhlauk doesn't know when he can see me, but there is no hurry – better let them get used to the idea of my resignation, then it can happen more smoothly . . .

The weather is very bad – rain, rain and rain – and the whole summer is like this, though they say it is not always so bad. I was offered a holiday in the South, but I don't want it. Everything must be cleared up at last – all this nonsense has been dragging on for a whole year. Perhaps it is my fault and it might have been better not to agree with them over anything, but to be as stubborn as a mule. However, I wanted to behave well. Now I think I shall liquidate it all and try to arrange things as peacefully as possible, without fuss or cursing. I will restrain myself.

28 July. Yesterday I went to see Mezhlauk and this is the outcome of our discussion. I don't yet know if it can be called positive, but it can't be called negative. On the surface Mezhlauk was very unpleasant, as I had expected, but he did not say no to the questions in my letter to Molotov, so there is no direct motive for my resignation. He wanted me to telephone Rutherford. I explained my plan for a peaceful termination of the conflict and if it is accepted I think everything can be normal. My offer was not blankly refused by Mezhlauk and if his comrades agree to it, then it will be accepted. So I wrote my plan down on paper and have already sent it to Mezhlauk and am now waiting for the reply. A lot depends on Rutherford and the best means of communicating would be through Webster, whose visit here it seems could be arranged.

But what really touched me, and gave me pleasure, was our conversation about you. I pointed out that without you all the conflict would have taken quite a different aspect, and it was only because of you that there was no outburst of an anti-Soviet character. Mezhlauk said that he often heard about you as a very clever woman (blush!) and they all think that you must come as soon as possible and be at my side. He said that no pressure will be brought against you, that they promise you complete freedom of going and coming between England and here, as often as you like, and that you could bring the children only if you like it here. For this remark I gave Mezhlauk full marks – fine fellow, many sins were forgiven him for it. That is how they should treat not only you but everybody. Then Mezhlauk said that you will be given a ticket for the journey by ship, that I mustn't worry, that Maisky will let you know about it, and that nothing can happen to you since there is nothing better than his and Molotov's word, and nothing more is needed.

If my plan is accepted, then I see nothing against your coming for a month or a month and a half. When you get this letter, start quietly arranging things for the journey. Ask for a return visa to England, arrange family affairs, and let Granny [Anna's mother] look after the housekeeping. It looks as if you will have to go backwards and forwards. I am very pleased with everything now that there is some possibility of a peaceful solution to the whole conflict. I am now waiting to hear from them again. But if they let the opportunity pass, then I don't know if it will be possible to find another opportunity for a peaceful solution . . . I was told that our authorities have nothing against my talking to the physiologists. But I am a bit afraid that the same thing will happen with you as happened with the parcel [see p. 226]. Once you have been cheated it is difficult not to feel you may be cheated again. Well, we shall see – but I do want to see you . . .

(A)   Bolshevo, 30 July 1935

Yesterday Dirac [see fig. 69] arrived and we came here with Tamm and have been walking, boating and talking ever since. I haven't had such a pleasant time with anyone up to now. Dirac treats me so simply and so well that I can feel what a good and loyal friend he is. We talk about all sorts of things and this has been very refreshing. You see, my little Rat, that I'm almost completely deprived of any real contact with people. Many are afraid of me and many say quite openly – until the conflict with Kapitza is resolved it's better not to see him, much as we should like to. I've never felt so intrinsically lonely, and Dirac's arrival has revived my memories of the respect and reputation I enjoyed in Cambridge, in contrast to my present status which is like that of a hunted animal . . .

69. P.A.M. Dirac, 1925.

(A)   Bolshevo, 11 August 1935

. . . Dirac and I get on very pleasantly together, chatting and discussing only when we feel like it . . . I am waiting to talk to Adrian who is due on the 17th and then I'll write in detail [for Adrian's report see p. 267]. In the meantime, get ready to leave – I think you'll have to go back and forth at least once. As always, Dirac is somewhat eccentric but I find him easy to get along with. I tried to get Dirac into a flirtation with a good looking girl of 18, who is a language student and speaks English – but I had no success and she goes back to Moscow tomorrow . . .

(A)   Moscow, 11 September 1935

. . . I miss Dirac, who is now in Budapest, He will tell you all about me when he gets to Cambridge in about ten days time.

12 September. Yesterday I had a pleasant time with the English people. At midnight Patrick Blackett called in and we chatted till 1.30 a.m. I always have the feeling that he doesn't care for me and his wife even less . . .

After spending two months in Moscow, Anna returned to Cambridge at the end of November to wind up things there and finally she and the boys returned permanently to Moscow. The letter below was written just after Anna's departure for Cambridge.

(I)   Moscow, 25 November 1935

My treasured little Rat,

It is already the second day that you are no longer with me . . . You should just about be getting to the Hook of Holland and tomorrow you'll be in Cambridge. I am following your journey the whole time. Yesterday morning I went to Bolshevo and I felt sleepy all day . . . this morning I went to look at the building work, and found everything in rather a poor state – for instance, the paint on the walls was full of bubbles. In comparison with the building of my Cambridge lab, it is a sorry sight . . . The builders seem to have gone to sleep and there is hardly anyone working on the site – I just can't understand what's going on . . .

Shalnikov is a great help – he doesn't get upset and evidently he is used to such conditions. But every now and again I begin to have doubts – am I right in asking Rutherford for all the equipment? For in such conditions it is doomed to destruction or at the best to a wretched existence. There in Cambridge they can somehow or other make some use of it, while here if we can't get sinks, pipes or steel what shall we do with it? Perhaps nerves of steel are needed, but I don't have them. Perhaps you have to know how to swear, how to thump the table with your fist and shout, but I can only rarely manage to do this and then I feel completely broken.

How shall I be able to work quietly? I am waiting impatiently for Olga Stetskaya [who soon became his Deputy Director]. Perhaps she can help – Olbert thinks only of prettiness, neat pathways, curtains and portraits. And I don't like it – the laboratory is beginning to look like an official government office. He tells me, "This is a Soviet institution, Pyotr Leonidovich" and he has his own ideas on scale and style . . . from which he can't be budged . . .

27 November. I didn't write yesterday because I had a cold and hardly managed to prepare my lecture . . . But I did give it at 8 p.m. . . . Some of our professors came but they all seemed to be sleepy and inert and sat like statues. They haven't any enthusiasm for science . . . they seem so oppressed and hungry, so exhausted by tedious routine. I have never seen such an inert audience before. This is just impossible!

I have lectured in nearly all the chief universities of France, Belgium, Holland and Germany and I have distorted the German and French languages, so I am sure I must have lectured worse than I did yesterday, but over there people did react. Here, however, there wasn't a single question. I can't continue in this way. They must be stirred up and captivated, which I shall try to do. My best hope lies in our youth, where I should hope to find enthusiasm more easily. For it does exist, as I found when I visited some of the new factories at the beginning of the year – I recall how enthusiastically they showed me everything and discussed it all. This means we can indeed work with enthusiasm, but why doesn't it exist among our scientists? Worst of all, of course, is the Academy [of Sciences]. I recently estimated that the average age of the Academicians is 65 and the best chance of being elected is at 58 – and of dying is about 72 . . .

But I put all my hopes on the younger generation. Sh[alnikov] is shaping well – at least he is enthusiastic and I am happy to have him. I should be very surprised if I couldn't find eight such people to work amicably with me. As for the rest – to the devil with them. And I shall bite and swear and do anything to keep them away from me . . .

(I)   Moscow, 13 December 1935

. . . When I talk to various scientists I am surprised that many of them should declare, "Of course you can easily do everything because so much is given to you, etc., etc.". As if they and I didn't have the same opportunities when we started work. As if all that I have achieved had come down as a gift from heaven and I hadn't exerted the devil

knows how much effort and nervous energy in the process. It is bad
of them to reckon that somehow life has been unfair to them and that
everyone is guilty except themselves. After all, what is struggle for, except to adapt to our environment, so as to develop our abilities and
create the right conditions for work? . . .

70. Rutherford 1923.

5

Correspondence with Rutherford (1921 – 1937)

This chapter starts with a very respectful letter informing Rutherford of Kapitza's intention to start work in Cambridge on 21 July 1921 and continues with letters both to and from Rutherford right up to the time of Rutherford's death in 1937. The chapter concludes with letters to Bohr, Cockcroft and Dirac expressing his great sense of loss in the death of his teacher. Most of the letters are from the time following Kapitza's detention in the Soviet Union. Kapitza's original English has been left mostly unedited, except for correction of spelling and of the more glaring grammatical errors.

(a')

12a Langham Mansions
Earls Court Square
London SW5
18 July 1921

Dear Sir Ernest,

Please let me tell you once more how very grateful I am to you of having admitted me among your pupils. I hope to do my best to profit my staying in Cambridge and the honour to be under your rule.

If you don't object I will come to Cambridge this Thursday and take up my work at once.

Yours truly,
Pierre Kapitza

(a) From Rutherford   New Plymouth, New Zealand
                                               9 October 1925

My dear Kapitza,

I was glad to get your letter of the 26th August with photographs –
one of your machine and the other of Chadwick and yourself [see fig.
58]. I had a real good laugh when I saw the spick and span appearance
of yourself and Chadwick at the wedding. I hope that you did your
duties as best man to the satisfaction of all concerned. I understand
from Fowler that his top hat completed your toilet. If you went to work
in your laboratory in the same attire you would create a great sensation . . .

I hope your research has gone on well and that your machine will
be in full working order on my arrival. All the scientific men are very
much interested in your experiments, of which I have given a brief account at the various University Societies [in Australia and New
Zealand].

With kind regards,

                         Yours sincerely,
                          E. Rutherford

(a)   Cambridge, 11 October 1925

My dear Professor Rutherford,

They elected me. But poor, poor Skinner. He is so distressed. When
he came and told me the results of the election I felt that it may be better if he would be elected and not I. I never thought at the last moment
that they elect me and I was so sure that it will be Skinner that it would
not so much disappoint me as him.

The first thing which I am doing, coming home from the lab it is
this letter to you. I write to thank you as I know perfectly well that
in spite of the fact that you are far from us it is your influence which
has done the trick.

The experiment is going satisfactory and [there is] nothing to report
to you of special [interest] at present. We are quite close to test the

switch. (Indeed very, very cautiously, as you were saying me several times!) I shall write you at once as soon as we have made the test, probably next week, if nothing happens wrong.

Everything else is going well too (DSIR, Laurmann, Pearson, Cockcroft). The numerous newly married couples are spending their time more in making their nests as on scientific investigations. But this is indeed temporary. We are missing you very much. I feel myself very uncomfortable as nobody is scolding me sometimes a little. But we hope you will be soon back. Thank you once more very much. Best wishes for your return journey. Kindly remember me to Lady Rutherford.

I am very sincerely yours,
Peter Kapitza

(a') From Rutherford   Christchurch, New Zealand,
3 November 1925

My dear Kapitza,

I was very pleased to get your telegram announcing that you had been elected a Fellow of Trinity College, and am exceedingly pleased to hear it. I offer you my warmest congratulations on this great distinction and I hope you will enjoy the privileges of your new position. I am glad to hear that your experiments are going well and hope that by the time I return you will have succeeded in getting your breaking apparatus* in good working order . . .

With kind regards,

Yours very sincerely,
E. Rutherford

---

* This refers to the circuit breaker of the impulsive high-field equipment.

(a)   Cambridge, 17 December 1925

Dear Professor Rutherford,

I am writing you this letter to Cairo to tell you that we already have the short circuit machine and the coil, and we managed to obtain fields [of] 270 000 [gauss] in a cylindrical volume of a diameter of 1 cm, and 4.5 cm high. We could not go further as the coil bursted with a great bang, which, no doubt, would amuse [you] very much, if you could hear it. The power in the circuit was about 13.5 thousand kilowatts (13 500 amperes at a pressure of 1000 volts), approximately three Cambridge Supply Stations connected together, but the result of the explosion was only the noise, as no apparatus has been damaged, except the coil. The coil has not been strengthened by an outside band, which we are going to do. Up to 200 000 gaus we have gone quite safely and we are not in very special difficulties.

At present all these experiments have been done with a higher speed machine, and I am very happy that everything went well, and now you may be quite sure (98 percent) of the money is not wasted, and everything is working. The accident was the most interesting of all the experiment, and gives the final touch of certainty, as we now know exactly what has happened when the coil bursted. We know just what an arc of 13 000 amperes is like. Apparently it is not at all harmful for the apparatus and for the machine, and even for the experimenter if he is sufficiently far away. I am very impatient to see you again in the laboratory and to tell you all the little details, some of which are amusing, about the fight with the machines . . .

Kapitza's idiosyncratic spelling can be seen in the facsimile of the following letter (see fig. 61). Here the spelling has been corrected.

(a)   Paris, 27 April 1927

My dear Professor Rutherford,

I am going to be married tomorrow and Saturday. I mean to say that tomorrow at the consulate and Saturday in the Russian Church in

Paris. When I shall be back in Cambridge I do not know. What do you think about it??!! I fear you are rather angry. This is why I propose to have no honeymoon and bring my wife in few days time after my wedding to Cambridge. I hope you shall be so kind and help to get a British visa for my wife. Kindly ask the Foreign Office to give the visa at once as having no visa I shall not be able to get back to the lab. I got engaged about a week ago and was very busy all this time in arranging the formal part of my wedding. It was very difficult because my future wife is a Russian emigrant and the Soviet Consulate were making great difficulties in registering the marriage. But now after much talking they agreed.

I hope you understand that I am a victim of my own 300 000 gauss and I have to confess that the dose which I received is rather a strong one. I presume you are interested to know who is the lady. I shall refer [you] to Robertson or Simpson, who met my future wife during her visit to Cambridge recently.

I hope you are not more cross with me for my lost time, but you see that even in more important questions I have a quick decision and great speed of action.

Yours *very* sincerely,
Peter Kapitza

(a') From Rutherford   Cambridge, 29 April 1927

My dear Kapitza,

I received your letter at breakfast this morning, and I read it with much interest and amusement. I wish you and your wife all happiness in your new state. I told Chadwick this morning and I think he is communicating with you.

You ask me to get a visa for Mrs Kapitza but your state of mind is shown by the fact that you did not mention any of the necessary details, maiden name, age, etc., for that document. If the young lady has already been in England she must have a visa of her own which should be available for the purpose of her re-admission, unless her marriage invalidates it. In any case send me all the necessary details.

I will expect to see you back in the Laboratory before long. I will, of course, attend to the payment of cheques, etc.

My wife and I unite in sending our warmest congratulations on the event and our best wishes for the future. They say it is a bad wife that does not help a little, so I shall expect your work to make even faster progress. I am not very surprised at the news as I had heard rumours of your magnetic susceptibility under intense attracting fields.

> Yours very sincerely,
> E. Rutherford

(a)   Normandy Hotel, Deauville, 3 May 1927

My dear Professor Rutherford,

Many thanks for your and Lady Rutherford's congratulations and best wishes. Gradually I am recovering the presence of mind. I hope to be in Cambridge at work next Monday to share my love between my wife and the Laboratory. This will be possible only if the visa will arrive in Paris during this week. I am a little afraid to introduce to you my wife, as she is rather timid.

At present we are on the seaside and shall return to Paris at once after the visa arrives there. I am sorry to give you so many troubles and appreciate very much your kindness . . .

(a)   Hotel Normandy, Paris, 11 May 1927

My dear Professor Rutherford,

Good heavens! Today is the 11th and no visa. The work and the bismuth are waiting. My wife is also very keen to get to Cambridge. She is archaeologist and has to write a thesis on Semitic temples and she is very keen to start her work. And owing to some silly asses in the passport office we have to stay idle in Paris wasting time and money, which are so necessary for starting life. I can't understand why this delay happens? I thought that after six years of faithful work in England I could count Cambridge as my second home and my wife should be welcomed.

Can you not explain all this to some of the "big bugs" and make them feel sorry? I sent you a wire yesterday. But I think you are also keen to see me at work. I promised you to hand over my paper on the "dynamo" on the 15th, now it will be impossible. I hope that soon everything will be arranged and you will approve of the new step which I have made in my life as you do it usually.

Kind regards to you and Lady Rutherford.

> I am your most sincerely and impatient,
> Peter Kaptiza

(a)   173 Huntingdon Road, Cambridge, 29 August 1931

My dear Professor,

Accept my and Anna's congratulations for your 60th birthday. Reaching this age a man may be sure that he passed his adolescence, but still it does not mean any more! It is a great mistake to attach too much importance to the age of a human being – it is only a chronological data which is meaningless without the history to illuminate it. I often thought that periods of men's life have to be determined on a more rational and scientific grounds, as time scale gives only a very approximate average estimate, the fluctuation from which sometimes may be enormous. Not only that but different aspects of a human being may in the same specimen develop to a different extent. Thus a man may have a soul of a child, the appearance of a grandfather, the interest for the surroundings of a youth, etc.

In taking the average of all these separate data a more correct estimate of the age of a man is obtained. I do not like to trouble you with detail calculations, and feel certain timidity (my natural quality as you know very well) of presenting it to you, but as a matter of curiosity I get that you personally will only reach the age of 60 in about 20 years.

Our kind regards to you and Lady Rutherford.

> Yours most sincerely,
> P. Kapitza

P.S.   I hope you will kindly accept the small birthday present enclos-
       ed herewith. With the greatest respect to the high honours
       which you now hold and which makes us all feel proud for hav-
       ing such a "boss", but I have to admit that sometimes you
       challenge your high title. On numerous occasions I watched
       you extracting from your pocket a most disgracefully looking
       pencil or even more often a microscopic piece of such a pencil.
       I hope you will agree that since your head has to bear all the
       weight of a golden crown [Rutherford had recently been
       elevated to the peerage] it is unworthy to your hand to touch
       any more such wooden pencils.

(a)   Near Moscow, 4 September 1933

My dear Professor,

We arrived safely in Leningrad, and after a week there we went to
Moscow. At the moment I am writing from the rest house near
Moscow where we are waiting [for] the arrival of Bukharin from the
Pamirs. In Leningrad I went to the most famous Russian heart
specialist and he requested that I should take a five-week cure of
mineral baths at one of the Caucasian health resorts. I do not like this
way of spending my time, so to check the correctness of his diagnosis
I went to another famous specialist in Moscow and after two hours of
careful examination, by X-rays and other modern methods of examination he came to the same conclusion as the first doctor.

Actually, it appears that nothing is organically wrong with my
heart, it is only badly overworked, probably through nervous strain,
and the doctor thinks that to avoid more serious trouble it is time to
take a complete rest. This is all very well but the trouble is that if I
go to the Caucasus I will miss the Leningrad Conference to which I
am invited and for which I am paid! I spoke to Joffé and he took a
very sympathetic attitude and not only encouraged me to go to the
Caucasus but was prepared to arrange all the formalities to get a place
in a sanatorium. If Bukharin also releases me from my moral obliga-
tions, I shall go to the Caucasus about 10 September and return back
to Leningrad at the beginning of October. But still I hope to be back
in Cambridge towards the end of October . . .

71. Annual Cavendish group, 1933. This is the last in which Kapitza appears.

W.J. Henderson, W.E. Duncanson, P. Wright, G.E. Pringe, H. Miller,
C.B.O. Mohr, N. Feather, C.W. Gilbert, D. Shoenberg, D.E. Lea, R. Witty, – Halliday, H.S.W. Massey, E.S. Shire,
B.B. Kinsey, F.W. Nicoll, G. Occhialini, E.C. Allberry, B.M. Crowther, B.V. Bowden, W.B. Lewis, P.C. Ho, E.T.S. Walton, P.W. Burbidge, F. Bitter,
J.K. Roberts, P. Harteck, R.C. Evans, E.C. Childs, R.A. Smith, G.T.P. Tarrant, L.H. Gray, J.P. Gott, M.L. Oliphant, P.I. Dee, J.L. Pawsey, C.E. Wynn-Williams
Miss. Sparshott, J.A. Ratcliffe, G. Stead, J. Chadwick, G.F.C. Searle, Prof.Sir.J.J. Thomson, Prof Lord Rutherford, Prof.C.T.R. Wilson, C.D. Ellis, Prof.Kapitza, P.M.S. Blackett, Miss. Davies.

(a)   Leningrad, 23 October 1934

Dear Professor,

I am gradually recovering from the shock*. You know probably all about it from Anna, this is why I have not written you before. Thank you very much indeed for all your kindness and also for the help to keep an eye on my boys in the lab. There are only two things necessary. First, to keep Milner from making too many gadgets (like Wynn-Williams). And the second to tell Shoenberg that the experiment is more important than the theory (he is inclined to think like Skinner). And this is all. I will try to do remainder by post.

I hope you and your family are well. Most kind regards, [best] wishes from your most sincerely,

Peter

The following is a draft written at V.M. Molotov's request (see p. 323). However, it was *not* approved and was not sent at the time. Eventually it was brought to Rutherford by Anna in November 1935.

(a, c)   Moscow, about 14 May 1935

My dear Professor,

I was informed that the case of my retention got into the papers. Whatever is the point and the object of this discussion, I only would love to keep myself completely out of it. But as a point of fairness I would only like that everyone concerned should know that first of all, I am and always was in sympathy with the work of the Soviet Government on the reconstruction of Russia, on the principle of socialism and I am prepared to do scientific work here. Secondly the Soviet Government does its best to build me a laboratory.

There are indeed points about which we disagree, especially as regards the treatment of the question of pure science, as they are far greater valuing the solutions of applied problems and also we disagree

---

* Of not being allowed to return to Cambridge.

about the most efficient way of dealing with scientists, but no doubt there is the best intention to develop science in the Soviet Union.

Personally I am very miserable indeed that all this happened. I miss you, my laboratory and specially my work and it is not to be expected that I soon will be able to resume it, and all this makes me very unhappy. The stupidity of the created position is that it is based on complete misunderstanding as everyone concerned really acts with the best intentions. I have no objections for you to make part or all of this letter public.

Below follows the report to Rutherford by E.D. Adrian on his visit to Kapitza

(a)   Cambridge, August 1935

Kapitza considers that there were three reasons for his detention:

(a) unfounded reports from England that he was doing war work: these reports must have come from Cambridge and from a well informed source.

(b) Gamow: when Gamow was out of Russia he wrote to Molotov asking for the same standing as Kapitza used to have, and he made this a condition of his return to Russia.

(c) the reason that his abilities would be valuable during a war.

He resented these arguments and asked for leave out of Russia. Therefore his detention was not voluntary. His attitude has been that he has never refused to do any work here (in Russia) – e.g. consulting work, etc. – except that he could not repeat the construction of the Cambridge apparatus because it would take so long and be so distasteful and demoralising. It would therefore be much easier to start a new line of research, namely physiology.

There were three courses open to him:

(a) to base his hopes on the result of international protests by scientists – as to this his position is that, though it would be pleasant to regain his freedom, it would be at the cost of strained relations between the outside world and the U.S.S.R., and he does not feel justified in regaining freedom at such a cost.

(b) to resume the Cambridge work here as soon as possible by getting the apparatus from Cambridge and getting restricted freedom by retaining some official connection with English scientists.

(c) failing (a) or (b), he might leave physics altogether and start physiology, by making the Soviet Government realise that he cannot go on with physics without his English apparatus and assistants.

Therefore he thinks the right course is this:

(a) His moral obligation to English scientists who helped him so much will be mitigated if the Government gives him *a sum to transmit to England* to meet the expenditure involved in his work during the last 14 years. He reckons that the sum should range between £30 000 and £50 000. If Lord Rutherford agrees, he would like this sum to be divided between the Royal Society and Cambridge University according to the decision of the Mond Laboratory Committee or Lord Rutherford, but not less than 12% of the sum to be offered to Trinity College.

(b) *To resume his work in Russia he requires the following conditions:*

(1) *All the apparatus* which is in his laboratory in Cambridge must be sent here except the accumulators, the liquid air machine and the old 200 kw. dynamo from the power station. The standard apparatus (ammeters, microscopes, spectrometers, lathes, clocks, etc.) can remain provided that duplicates are sent here – preferably ordered by Cockcroft, the cost to be defrayed out of the above mentioned lump sum. The same applies to the pictures and fittings on the walls and to the switchboards everywhere. All the stores, or duplicates of them, should be sent here too and all the old apparatus whether in use at the moment or not. All the wiring of the big machine must be sent. The other wiring can be left. All the documents (e.g., drawings, papers, letters, catalogues) should be sent. The packing and transportation would be arranged by the U.S.S.R.

(2) *Assistants.* To induce Laurmann and Pearson to come and help him for the first three to four years he asks Lord Rutherford to guarantee that their present jobs and pension rights in Cambridge should be given back to them if they decide to return there after three to four years in Russia. He hopes to be able to arrange favourable conditions for them in Russia and does not anticipate that they will be unwilling to come.

(3) It is essential that he should retain the title of *Royal Society Professor* or obtain that of Professor at Cambridge or obtain some

equivalent professorship, and it is essential also that this post should have a salary of about £400 a year attached to it. In return for this he hopes to give a course of lectures in England every year. He hopes that the tenure of such a post will enable him to arrange with the Government to visit England for two to three months every year. The salary is needed to cover his expenses in England . . . The matter is very urgent and must be settled within the next two months.

Rutherford's reaction to this report is contained in his letter of 28 August, which follows. In a somewhat more formal version dated 25 September and taken by Anna when she went to Moscow, Rutherford replied also to Kapitza's suggestion that he should retain the title of Professor: "The suggestion of a Professorship seems to me not feasible. I am sure neither the University nor the Royal Society would look at such a proposal."

(a) From Rutherford   Chantry Cottage, Upper Chute, Andover,
28 August 1935

My dear Kapitza,

Anna is coming to see me today before she leaves for Russia. Adrian came to see me and brought me your statement. I was glad to hear about you and your health and I hope to see Dirac on his return. First I must tell you the present position of the Mond Laboratory. We have had to act on the assumption that you will not be available. I have taken charge as acting Director with Cockcroft as second in command . . . We have a good programme of work in view in the Mond next year and hope to keep things going quietly till the new people have gained experience. We shall not try to continue your experiments with high magnetic fields but leave these for a time to see whether you are able to continue them in the USSR. It is your own line of work and I hope you will be able to take it up again before long.

The lab will be a good concern running with about the same staff as before and a full complement of research workers. The services of Pearson are of course essential to keep the lab going. Laurmann is prepared to help in any way he is required and I shall of course look after him unless you are in a position to persuade him to join you. I have of course discussed such matters with him. The new secretary Miss Pantin is shaping well under the guidance of Miss Stebbing who is a tower of strength.

In your statement, you discuss the question of reimbursement of the University, Royal Society, etc., for the expenses incurred in helping you with your experiments. You probably know that the Embassy asked to purchase the main apparatus of the Mond Lab six months ago on terms to be discussed. I found neither the University nor the Royal Society were prepared to consider such an offer which I declined with thanks. At the same time, I offered to help you in any way I could to obtain duplicates of any of the apparatus you wished but on condition there was a free and direct consultation with you as to your wishes. Naturally I was not prepared to help anyone but yourself as my sense of gratitude to the USSR is not particularly strong.

I am myself very doubtful if the USSR could be persuaded to pay the full expenses of your work in England. I do not think either the Royal Society or the University or Trinity College want compensation. Under normal conditions, there was nothing to prevent you taking another post elsewhere if you wished to do so, apart from any personal sense of obligation for help rendered. After all, the help was given unconditionally on broad lines to promote scientific activity in a promising field and apart from the enforced interruption of your own work we have no grievance of any kind. You did the best you could with the help given and there is no obligation on you or the USSR for compensation to any institution concerned. You have added a new department of work in the Cavendish and this will go on but no doubt with much less distinction than if you were present to guide it.

Another point I must mention. If you decide to take up work along magnetic–cryogenic lines, I should be only too pleased to help you by arranging for duplicates of the apparatus in the lab to be constructed and tested for you and sent to you, provided of course the cost is financed by your government. For example the helium and hydrogen liquefiers could be duplicated under the charge of Pearson while I have no doubt I could arrange for the big generator to be built along the old lines . . . We could also send over any old apparatus which is of especial interest to you. I am sure Cockcroft would be very pleased to get special things for you so as to stock the new lab.

Well I think I have said enough for the time being. We all miss you in Cambridge and you will be glad to hear your "K" Club continues. I miss our Sunday night discussions and the escort home. You will have heard that Chadwick and Feather have gone to Liverpool – we are sorry to lose them but [it is] in a good cause. Searle has retired and I have arranged for Ellis and Oliphant to help me in the direction of research. It has been a very busy year for me but fortunately I am in good physical form. The X-ray crystallography (Bernal & Co.) are now a part of the Cavendish and I keep a fatherly eye on them. They have not yet tried to convert me to Communism! The experiments in transmutation go ahead and we have in view a new high potential D.C. plant for a million volts. The site of the Cambridge Philosophical Society Library has been handed over to us and will be reconstructed soon for the new purpose. Oliphant will be largely responsible for the design of the new apparatus for the purpose with the general advice of Cockcroft. It is another adventure of the "old man" and I hope will prove a success – at any rate I believe in following our own methods and not merely imitating others – a sentiment with which I know you will be in whole-hearted agreement. Well, I hope you will be able to surmount your difficulties and get down to active experimental work again. I shall be only too glad to hear from you. With best wishes for your work and health.

Yours ever,
Rutherford

The following letter was intended to serve as a formal acceptance by Kapitza of the proposed arrangement between the University and the Soviet Government. Kapitza's appreciation of Rutherford's efforts on his behalf is expressed in the more personal letter which followed a day later.

(a)   Hotel Metropole 485, Moscow, 19 October 1935

Dear Lord Rutherford,

Mr Rabinovich has informed me and shown me your letter of 8 October about the outline on which an arrangement could be made, between the University and the USSR Government to transfer the equipment and apparatus of the Mond Laboratory to enable me to proceed with my research work in the Institute of Physical Problems of the Academy of Sciences now under construction in Moscow, which is under my direction. I am sympathetic with this scheme.

Mr Rabinovich has also passed the list of apparatus for my approval. In all I regard it as quite complete and sufficient for proceeding with my work, except [for several extra items included in the enclosed detailed list] . . . I do not mind if you supply copies or originals of the apparatus, but in case of compressors, if new, and similar apparatus, I should very much like that a responsible person like Cockcroft [together] with Pearson, should submit them to laboratory or factory tests similar to those undergone by the originals.

I understand from the authorities here that the sum of £30 000 is being transferred to the trade delegation in London to meet the payment as soon as the authority of the University for the transaction will be granted. I have been promised that the trade delegation will inform you officially as soon as they will have the money in hand. Meanwhile in case you find it desirable, an advance up to £5000 could be made at once to meet the cost of ordering the duplicates. Some agreement has to be arranged to cover this advance. I suggest that you make a proposal yourself, but I am only very anxious that not a single minute should be lost owing to bureaucratic formalities, as I am really quite sick without my work and only dream how to resume it as soon as possible . . .

I should only like to mention once more how anxious I am to get all the stuff as soon as possible and should very much appreciate if you will find some way of starting sending the apparatus which you will find possible before the formalities are settled. Any scheme to speed up the matter I shall appreciate very much, as I was promised that there will be no delay on the part of the USSR authorities . . .

(a, c)   Hotel Metropole, Moscow, 20 October 1935

My dear Professor,

Life is an incomprehensible thing. We have difficulties in clearing up single physical phenomena, so I suppose humanity will never disentangle the fate of a human being, especially as complicated as my own.

It is such a complexity of all sorts of phenomena that it is better not to question its logical coherence. After all we are only small particles of floating matter in a stream which we call fate. All that we can manage is to deflect slightly our track and keep afloat – the stream governs us.

The stream carrying a Russian is fresh, vigorous, even fascinating and consequently rough. It is wonderfully suited for a constructor or an economist, but is it suited for a scientist like me? The future will show. In any case, the country earnestly looks forward to see science develop and take a prominent part in the social organisation. But all is new here and the position of science has to be newly determined. In such a condition mistakes are inevitable. We must not be too hard judges and never forget that the objective is a pioneer one. I have no ill feeling, only I am not confident in my personal strength and abilities. Indeed I will do my best to resume here the scientific work along the line in which Nature endowed me and also help to develop science in Russia. I am even surprised to find that I have force to withstand more than I ever could expect, like this year. I never dreamt that to be deprived of doing my scientific work should be for me such a trial, but now it is over, in any case the worst of it.

I write this letter mainly to tell you how enormously I appreciate the support and help which you render me by arranging the transfer of the laboratory equipment and helping me to get Laurmann and Pearson's help. In general, it will be for me very difficult to work without them as you remember one of them was with me for 17 and the other for 10 years. Without their help, at least at the beginning, I could not see how to reconstruct my installations.

I miss you very much, more than anybody else and now I realise what a great part in my life played the personal and scientific inter-course with you, which I enjoyed for 14 [actually 13] years of my stay in Cambridge. Now left on my own I am sure this experience will greatly help me. I would very much like you to realise how grateful I feel and will always remain for all what you have done to me and what you are doing now. I will always have the best recollection about my Cambridge years and also of all the kindness and aid which I had from my fellow scientists. Indeed, I will acknowledge in writing to the College, Royal Society and University as soon as everything is made known.

I am very happy to have Anna with me now. She is arranging for the transfer of the family and as soon as it is finished is going to Cambridge to arrange matters there.

With most kind regards to Lady Rutherford,

Yours most affectionately,
Peter Kapitza

(a) From Rutherford   Cambridge, 21 November 1935

My dear Kapitza,

. . . I am inclined to give you a little advice, even though it may not be necessary. I think it will be important for you to get down to work on the installation of the laboratory as soon as possible, and try to train your assistants to be useful. I think you will find many of your troubles will fall from you when you are hard at work again, and I am confident that your relations with the authorities will improve at once when they see that you are working whole-heartedly to get your show going. I would not worry too much about the attitude or opinions of individuals, provided they do not interfere with your work. I daresay you will think I do not understand the situation, but I am sure that the chances of your happiness in the future depend on your keeping your nose down to the grindstone in the laboratory. Too much introspection is bad for anybody. I am hoping to see Anna before long, and she no doubt will be able to tell me how you are. I am glad to say that notwithstanding all these additional troubles, I am feeling very well and quite competent to drive the Laboratory along in the normal way!

You will be interested to hear that C.T.R. Wilson has been awarded the Copley Medal, Darwin the Royal Medal, and the Hughes Medal went to Davisson of the Bell Telephone Company. You will probably have seen that Chadwick has got the Nobel Prize in Physics, and we are all very pleased; also the Chemical Prize went to the Joliot-Curies, and this is a happy solution of a difficulty, and also along historical lines.

. . . Buck up my dear fellow, and get to work,

With best wishes, Yours ever
Rutherford

(A) To J.J. Thomson   Hotel Metropole, Room 485, Moscow
Master of Trinity College   23 November, 1935

My dear Master,

I am very sorry that I shall be unable to enjoy any more the College and the intercourse with all my friends there. I miss it already very much and probably will miss it in the future no less. During my life in Cambridge the privilege which I enjoyed and valued the most was no doubt my connection with the College. Coming as a lonely stranger to Cambridge, I am indebted to the College not only for facilities for work, a home and friends, but also for intercourse with people which stimulated my work. Through the College I learned to like English people, and will now always feel towards England like my second home country. I should very much like you to transfer to the Fellows of the College "good bye" and tell them how deeply I appreciated my relations with them and the College, as I do not think that it will be soon when I will be able to come to Cambridge.
    Accept my most sincere regards.

I am yours sincerely,
Peter Kapitza.

The minutes of a meeting of the College Council on 6 December 1935 record:
    "Dr. Kapitza The Master read a farewell letter from Dr. Kapitza. It was agreed that copies be sent to all the Fellows."

The following letter is typical of many to Cockcroft, in which Kapitza goes into great detail and complains of slow progress. This time he asked for various items not previously agreed and this provoked Rutherford to reply very bluntly (see p. 277).

(a) To J.D. Cockcroft   Moscow, 20 January 1936

My dear John,

. . . I am very grateful [for what you have sent, which] will be very useful.
  (1) I am very anxious to have the foundation bolts as soon as possible as I understand they have been lost during transport.

(2) The safe lamps from the Kodak have duly arrived.

(3) In alteration to my previous letter I should like to have the set of synchronous clocks. About 25 small ones and one big one similar to the arrangement in the lab. As from the summer we shall have the frequency normalised.

(4) Please send me the fittings for the ceiling wiring, no need to bother about the tubes, mainly the angle pieces, and the connection insulators, etc., are important.

(5) As regards the panels for the research rooms it is desirable that they should be fitted with the set of terminals on the top to be used for the overhead wiring. The two galvanometer terminals must be fitted with two small fuses and a switch.

(6) I would like to confirm, what I have transmitted through Anna, that we should require 5 or 6 sets of thermostatic regulators for the water heating system, similar to the one installed in the Lab.

(7) Though I have not mentioned it before, but it appears that we shall have great difficulties here in installing the automatic telephone system, similar to the one in the Lab. I should be grateful if Professor will find it possible to send one here with a dozen of telephones and a 20 – 25 numbers automatic switch connector.

(8) Answering the question of Pearson, we should like to have the dies and gauges in metric as well as Whitworth. We have particular difficulties here with small drills and tools in sizes 2 mm and downwards. Kindly arrange a very good stock sent.

(9) I am very particular to have all my stock of small wires, specially in beryllium-copper and cadmium-copper sent to me. Kindly supply the mutual inductors used in my experiments.

(10) I should very much like that you send me the 4 fultograph sets complete; also send me according to Laurmann's instructions some wireless junk, hand condensers, grid leaks, telephone relays that we have from Wynn-Williams.

I hope that we shall be able to instal the liquefier in the middle of March and I regard this as the most suitable time for Pearson to come here. I am very sorry to hear from Pearson that he is able to work only in the evenings on duplicating my apparatus. I hoped that Professor and you appreciated my point in my original letter that I must start work very soon again, and the principal condition is that I should get my apparatus as soon as possible. I hoped I could count on your and

Professor's sympathy in my present position and also that you appreciate that it is very painful for me to be deprived from working with apparatus which I created, while other people are enjoying the benefit of my ideas, which I am unable to do myself because of circumstances beyond my control. The same applies to my Mendeleev cabinet, which as you know is mostly made up of elements which were sent to me as private presents and which I was surprised not to receive with the first lot of goods.

I think this is all of urgency for the moment. We are getting on gradually with the wiring of the lab. Things are going on not too badly and I hope by the beginning of March the wiring will be complete and awaiting the switch boards. The absence of the foundation bolts is a great nuisance. I am glad to have Anna again with me, she arrived safely with the children. I hope to ring you up soon and I shall write you a more detailed letter about our affairs, at the moment I am very short of time. Once more I am so thankful to you for all the trouble you are taking in helping me to resume my work here, and I hope one day to be useful to you in return. Most kind regards to you and Elizabeth.

(a) From Rutherford   Cambridge, 27 January 1936

My dear Kapitza,

Cockcroft has shown me your last letter asking us to send a complete set of automatic telephones as well as a good many other things. In addition to the apparatus mentioned in our agreement with the USSR we have already either promised you or provided you with apparatus which is costing us a large sum. While I am anxious to help you as far as possible, I feel that these drains on the laboratory should come to an end. After this date, if you want anything of importance, you will have to arrange either to buy it yourself or to pay us for getting it for you. While I quite understand that you are very anxious to get your work started, I feel that you are a little ungrateful and inconsiderate in the way you write to Cockcroft about your wishes. You have to remember that Cockcroft has had to work like a horse to get the apparatus ordered and shipped to Russia, and, in fact, his own work for the first six months will be largely stopped for that reason.

I notice also that you complain about the slowness of the work on the new liquefier. You should bear in mind that a considerable part of the time of Pearson and others has been taken up in preparing apparatus for shipment to you, and in trying to keep the work of the laboratory going. As you know, the essential condition for the transfer of apparatus to you made by the University was that the work of the Department should not be seriously interfered with. While we will try and hurry matters as far as we reasonably can, you must remember that you are not the only person concerned, and that we on our side have to waste a great deal of time in attending to you, so I hope in future that you will discontinue complaining letters of this type, otherwise we shall begin to consider you a nuisance of the first order. You know from old experience that I am not slow to express my frank views to you, and I trust you will, as of old, give them serious consideration . . .

As you no doubt have guessed, the papers in this country have been completely occupied this week with the death of the King and the accession of the Prince of Wales. Tomorrow is the funeral, and the Laboratory will be closed, and Memorial Services will be held by the University and in many of the College Chapels. I hope Mrs Kapitza and the children are well and happy and that you are getting down to useful work again which will keep you from worrying about other matters.

(a)   Institute for Physical Problems, Moscow,
                                                          18 February 1936

My dear Professor,

I thank you for your letter very much. I will soon write to you at great length. At the moment we have a spell of bad luck with the children. Peter [Sergei] had inflammation of the middle ear (both sides) and is just recovering after three weeks in bed. The doctors had to pierce his drum. Andrew had a very strong flu. He is also recovering. Mother has pleurisy – at her age (70) it is a bad thing. Myself, I had a week of bad colds. If you add all the routine work and having no secretary besides Anna, you will see easily why I am delaying my letter to you which must be a long one and I propose to describe to you many things. But in spite of all this trouble I feel much happier having Anna and the boys with me.

I think very often of you, and often feel the desire to have a good talk with you to get some further fatherly advice.

Most kind regards and best wishes from us all.

Yours most affectionately,
Peter Kapitza

(a, c)   Piatnitzkaya ul., 12, kv. 4, Moscow,
                                         26 February to 2 March 1936

My dear Professor,

We still have some difficulties in getting out of illnesses. After I have sent you a small note about the boys being ill, I myself followed them to bed, and after the flu had an inflammation of the middle ear. The ear was so bad that the doctor almost pierced the drum. I asked for this myself as I thought this will relieve the pain, which during the day was very nasty. Today is the first day that I am normal, but still have to spend a few days indoors. Anna was the only hero in the family who was not in bed. Mother is nearly alright too.

Your last letter I liked very much. Indeed, you were not too kind in it, but I felt you so well, and it reminded me of all the innumerable cases when you called me a nuisance, etc. . . . I feel myself very miserable here, better than last year, indeed, but not so happy as I was in Cambridge. Anna's return brought me much comfort and happiness. In any case, my family life is resumed and this is very important as I was here very lonely, quite alone, so the family is very much to me. Your letter reminded me of happy years in Cambridge and then I felt you as you are, rough in words and manners and good in your heart such as I like you, and this makes me feel rather happier. The lost Paradise!

Some of my friends here call me Pickwick. I take the people better than they are and they think of me worse than I am. Probably this is right and this was the source of my misfortune. Nobody here appreciated that I tried to be useful and good to my country. They only saw in it all sorts of rotten things, and probably see some even now.

How can I help? But relations with the authorities recently improved slightly. I do not know what they have at the back of their minds but in any case they seem to do everything possible to help me resume my work. And in official relations you could not reasonably expect any more. The relations are official and normal.

But my scientist colleagues are very scared of me and behave like pigs. You see, my Institute is attached to the Academy of Sciences, of which I am not a member [he was then only a Corresponding Member], but they govern my Institute. Fortunately, I do not need to attend their meetings and functions, but they have the say in the running of the Institute. I cannot appoint a research student without their sanction, all my finances must be approved by the council, etc. This would not be bad in general; someone must look after the science, but what a Council in this wonderful institution! The President is Karpinsky, he is 90! In his young days he was not a bad geologist, nothing extraordinary, but now he keeps going only by continuous sleep. During the meeting of the Council he sleeps with a happy and kind smile on his rather attractive face, probably dreaming of his young days. He is a kind harmless man, an excellent President, never in anyone's way. The two Vice-Presidents are young people for the Academy, as they are only 65. The first one, Komarov, is a botanist. He knows what a plant is, and could tell a daisy from a poppy, and probably he knows more names of plants than any other man in Russia, and for this he got into the Academy. Otherwise he is absolutely stupid. I have seldom seen such a stupid face, looking at him you really get sick. Listening to him is even worse than looking. Our friend Lowry is a genius compared to him. The second vice-president is better. His name also starts with a K[rzhizhanovski], but is such a complicated one that I dare not spell it in English. He is an electrical engineer, and was responsible for the scheme of the electrification of the Soviet Union, a great achievement as I understand. But the man has got no scientific experience at all, he is a great dreamer and very romantic. He has terrific schemes but is lost in the details and everyday things. He is, like the President, a very kind man he is very popular in the Academy. He promises, while you speak to him a lot of things, but never does a single one. May be this is better than to promise nothing and to do nothing. But an inexperienced man like myself, has for few days some pleasant hopes, and this I think is the reason for his popularity. Then comes the Secretary – Gorbunov. He was elected this year to the Academy specially to fill this post. Between the 90 members of the Academy, the average age of whom is 65 years, it was impossible to find an active enough member to do the secretarial work.

He is not much of a scientist, in the recent years he made some expeditions in the South-east of Russia, very daring ones. So he is rather an explorer. Probably he is the only person on the Council who has some personality. In any case, when you talk to him, he does express views and opinions, what others scarcely dare. Then comes our friend B[ukharin]. Do you remember a short man with a small beard that I once brought to College? He is a peculiar mixture of journalist, economist and philosopher. He is alright, but terribly frightened of me now, and avoids seeing me. The next one is a chemist B [probably Baikov]. A technical chemist, supposed to be shrewd, but no more. It is difficult to say much about him; he is kind and non committal in conversation, probably a very suitable member of the Council, fills the space and does not disturb anyone.

Finally, one comes to a physicist, Vavilov, who is young, only 45. I doubt if you know him by name, his work was in the fluorescence of liquids. You know the sort of work when you pass a beam of light through a vessel filled with liquid and observe the light perpendicularly. Once installed, you can play with the apparatus for all your life, changing the liquids, the number of which is immense, and you can also vary the spectra of the primary beam. And thus you have such a number of combinations that it can keep a research student busy all his life and give him the feeling of satisfaction that he is doing scientific work. He never did anything else. I was always surprised why Vavilov got into the Academy when even with our poor stock of physicists we have such people as Skobeltsyn, Fok and others, who are miles better than Vavilov. I think you will find the secret in that Vavilov is a very polished man, who knows what to say and when to say it so as to please everybody.

In general I regret so much that I am not a polished man, as this would make my life so much easier. But I know a great scientist who, without any well polished manners, got as far as only you can get. But this is in England, where there are too many people with good manners and their value is not too great, it appears that here they value good manners much more as they are not so common.

Now the last member of the Council is Frumkin, the physical

chemist. He is the only man on the Council with scientific standing. If he is not too brilliant, he is shrewd and honest and devoted to science. He is a melancholic looking man, never gets excited, is cynical in his attitude. He was kind to me and was never frightened of keeping in touch with me during the period of my detention. So I have a feeling of deep appreciation of his personality. Well this is the Council of the Academy of Sciences, as you see a not very attractive picture. I think the Royal Society at its worst never had such a collection of specimens in its Council. They never take any interest, whatsoever, in my Institute; not a single member of the Council has visited me and never a word of sympathy or interest. I myself feel pretty indifferent towards them too. At present I had only two collisions with them and both times I managed to get what I wanted, and they know by now that I am not a lamb.

The other scientists are also quite indifferent to me. My former teacher Joffé ignored me all this time, and only now suddenly burst out in kindness. He is the head of the physical group of the Academy, and is the leader in physics in general. As I have no desire to take part in his doings, I keep away. So you see how lonely I am. All my hopes are in the young people, whom I want to pick out from the University undergraduates. I give a course of lectures there, to make friends with them and get them interested in my work.

You see how much I appreciate any signs of friendship from Cambridge scientists in my loneliness, and I hope to start corresponding with them. But still more I shall be glad if my friends will come to visit "the man behind the bars". Now all my desire is to resume work as soon as possible and to do the experiments with the magnetic field and helium which have been interrupted. I hope this will take three or four years, and what will happen next I cannot anticipate. But my actions are determined and I do not scatter my energy outside my scientific work. I agree with you 100% that I must "keep my nose to the grindstone". Your excellent advice. But now you can see why I am so anxious to get my apparatus as soon as possible, the sooner I get it, the sooner I could start my work. And then I am sure my state of mind will get more balanced, and I am sure that then all these old idiots of the Academy of Sciences will stop irritating me as much as they do now.

Things with the lab do not go so badly, even if it is impossible to do as many things as well as in the Mond, but considering the comfort of having work to do these are all small things. I have now a new assistant director [O.A. Stetskaya], a lady engineer, very experienced and good. She is a most excellent worker and with her help (she is helping me very well), by the end of March the lab will be ready for the installation of my Cambridge apparatus. The greatest difficulty which I encounter is the supply of small articles, even if they are manufactured here. And now I shall explain to you why this is so.

You see, Soviet industry is growing at a terrific rate and everything is done to make its growth organised and well planned, so that all the system of supply and production is also well planned and organised. But the supply of a factory which is operating according to a definite plan is anticipated in detail at the beginning of the year, and evidently its demands are in very large figures. Such a system is indeed quite unsuitable for the supply of laboratories. I have written and spoken to the authorities that the labs must be supplied in a different way, and I think that the people here begin to recognise that the system of supply must be altered for the scientific institutions. But at present, if for instance, we require four bars of phosphor bronze, the need of which we had not anticipated in the beginning of the year, we must get it as an exception and the permission of the sub-secretary of the Heavy Industries [Commissariat] is required and this means a lot of correspondence which is equally extensive whether we require 10 kilograms, 10 tons or 10 car loads of stuff. And so, even if the Institute is given sufficient money, and even if the industries are quite decent, we are actually very badly supplied.

Indeed, all this is temporary and in two or three years time all will change, but before it is changed we must live and work. And you know how slowly a bureaucratic government machine moves and changes, especially now when all the interest is to see industry growing and interest in science is very academic. As for getting the various supplies from abroad, the bother is about the same. We are given quite sufficient foreign exchange, but to get anything, again you have to make a plan at the beginning of the year. And then all that you ask must pass through a certain controlling organisation which must satisfy itself that the given article is not procurable in the Soviet Union, competitive tenders are asked, etc. And all this applies to small articles as well as to big ones. You may well imagine what an amount of writing we must do and what a terrific stock of people we must keep to carry through all this bureaucratic work. I dare say all this will be eventually changed but at the moment you see how immensely helpful you are in sending me all sorts of small supplies of general articles. And this is the reason for the difficulties about which you express surprise in your letter.

I know that you run your laboratory with great consideration to economy and to financial welfare. And I discern in your letters a certain conflict between the fatherly attitude towards me of desiring to help, and the Director of the Cavendish Laboratory trying to keep the business end as high as possible. Let the father win! After all you must not complain, as after I left, the Cambridge Physics Department is left with a laboratory worth 25 kilopounds equipped with the best cryogenic equipment and I reckon you will be left with a five figure reserve in capital. You will have several excellently trained people like Pearson, Laurmann, Miss Stebbing. Indeed, you may say that you had spent a lot of energy in all this, and you are right, but even if you pulled the ropes I provided them!

I even hoped that the Varsity and the Royal Society would acknowledge my contribution and my letter of resignation, but the man behind bars is not considerately treated. And you must not grumble if you are left with a few hundred pounds less than you anticipated. Just keep in mind my position. Absolutely alone, half chained and very miserable. All my hopes of regaining some happiness is in the anticipation of starting my work, and without your help and sympathy this is impossible, so if you take a formal attitude I am done in.

I take it from Rabinovich that you have consented to transferring any apparatus and equipment in the Mond in original or duplicated. You certainly could not have thought that making the original list by memory I have not omitted a few items which I hope may be included eventually. That I am not greedy you may see in that I omitted all that I did not need. Like the big liquid air plant (value at least £1000), the accumulators (£400), the 33 kW generator set (£200), the wiring (£1500). The synchronous clock you proposed to send, but in the first instance I omitted it as I did not think that the frequency would be standardised, but now that we introduced the standardised frequency I asked to re-include them, besides the pendulum seconds clock. And now from all that I said you may appreciate why I am so keen to get all sorts of materials and supplies even more extensively than we usually had them in the Mond. There we had the possibility at any moment to get all we needed from the shops, and here you are generally

72. With John and Elizabeth Cockcroft at the Bolshoi Theatre, Moscow, 1936.

speaking deprived of this possibility. Indeed, I shall correspond about it all with Cockcroft and he will bring the matter before you.

Now the next point very important to me is the help of Pearson and Laurmann. Their help is indispensable to me to start and to run the lab for the beginning, and this is why I am so keen to get them. You must not be afraid that I shall keep them, much as I desire it, for the simple reason that I cannot provide the means which will induce them to stay here permanently, since both assistants and scientists are very badly paid here. For instance, I myself get even at par, only half of what I used to get in Cambridge, and in reality actually only 1/6 of my Cambridge means. But to live modestly I do not need more and never have complained of my pay. Provided I get the facilities to work, this is all I need at the moment.

To get out of this difficulty my colleagues the scientists take a number of jobs, five or six, and in this way they manage to get quite a lot of money. I resent this way of living as it will dissipate my energy and I will be unable to do research work. I am glad Anna shares my views and we are prepared to have a modest existence. Anna was wise to bring plenty of clothes from England as all our salary goes into food. I hope that perhaps in the future things may change and I shall not be forced to dissipate my energy.

But you will easily see that it will be impossible for me to expect Laurmann and Pearson to stay here longer than it is possible for me to compensate them by their normal salary to keep their families in Cambridge. As regards the times of their travel, I think they must follow the liquefier as soon as it is ready and sent off. I understand 1 April is the most probable date, so about this time I suggest they should leave Cambridge. Should you like me to write officially to the Mond Committee I will do it. Let them better come together, then Laurmann can help Pearson with the language.

Now about John [Cockcroft], I am very touched by the way he acted and highly appreciate his help. Unfortunately, I cannot see how I could compensate him for his work in England, but I am sure that you will see that he gets adequately compensated. But when John comes here I will do everything possible to see that he gets a good time [see fig. 72] and something like a free trip to Caucasus or Crimea. But you know that I have plenty of sense of gratitude in my heart.

Now I have written you a long letter, and I hope this will draw a picture of my life and position which are far from enviable. I also hope that I can reckon on your sympathy and support, and I know I shall get it, as you always were good to me, and especially now that I so badly require it. In my impatience to get to work I try to hurry you, and you must not get angry as it is only natural. You would not expect me to ask you to delay the work on the liquefier? Would you? Then everything is O.K.

Now I wrote a long letter to you, and should you like it I shall continue to do so. To talk to you is a great pleasure to me and I hope you will not mind. Meanwhile, my most kind regards to you and Lady Rutherford. I am sending you an "ex libris" which was done for me – you can see that the tenderness for the crocodile has not changed.*

Affectionately yours,
Peter Kapitza

P.S.   Anna has retyped my letter to make the reading possible, and to correct to some extent my spelling.

---

* The ex libris was illustrated with a crocodile.

(a) From Rutherford   Cambridge, 23 March 1936

My dear Kapitza,

I was pleased to receive your long letter, but sorry to hear you have had so much trouble in your household with influenza and some of its consequences. I hope, however, you are now quite recovered and hard at work . . . I was interested to hear your remarks on the scientific position and about the seniority of many of the men in power. I was very sorry to see the other day about the death of Pavlov. He has had a stirring and interesting life and was certainly a man to be proud of.

. . . I read with interest, and of course with some amusement, your special pleas why all should be done for you. I think I quite understand your position and I think I have been exceedingly generous already. At one time I expected you to ask whether we could remove the paint from the walls to decorate your new quarters! Actually, of course, I quite understand how difficult it is to get at short notice laboratory necessaries with a newly developing industry. We will do the best we can to help you over your difficulties with regard to the smaller matters.

You would have thoroughly enjoyed seeing last week the display of crocuses in King's and Trinity College avenues. They are, I think better than usual, and brought home to me that this rather trying winter is coming to an end. I hope to get away to my Cottage for a holiday in a fortnight's time . . .

With best wishes to you and Anna.

> Yours very sincerely,
> Rutherford

(a)   Moscow, 26 April 1936

My dear Professor,

. . . The apparatus is arriving alright, except a Wolf lathe (packed in Germany), where the stand got broken . . . You need not worry – I am arranging the matter with John. The delay in receiving the switchboards is very sad. The wiring for the big machine is already done two months ago, the big machine stands firmly on its foundation and the director with a sad face and bad temper is wandering round awaiting the remainder. All the people here feel rather disappointed, I put all my energy in the wiring with two shifts working day and night, and it seems in vain.

I must acknowledge that you are very generous in sending all sorts of materials and fitting. They are priceless here and will help immensely in doing work. But in spite of everything I do not feel happy. Actually I got the desire to work but no desire to live. The other day I quarrelled again with the authorities. A nice chap [Mezhlauk] to work with and rather kind to me, so I was rather sorry to say nasty things to him. But I could not keep myself from saying what I think. Maybe this is all because I am still idle. I did a good bit of reading to pick up the lost time; I have had some new ideas, but all this without my lab makes me even more nervous. You scarcely can imagine how impatient I am to see Pearson and Laurmann here, for at least six months. I hope it will not be long till they come. It was great of you to help me so much.

The boys are alright, not at school yet, but playing in the yard with some boy friends. But soon they will go to the country; we rented a miniature bungalow in a lonely spot near Moscow. It is not far, so I shall be able to spend almost every night with them. Anna is running the house in her usual efficient way. Now the food conditions in Russia are quite normal and you can get almost anything, including oysters, lovely caviar and plenty of most delicious smoked fish, such as sturgeon, which would make even our "gourmands" at the High Table dribble. With wine it is worse, no good old aged wines, but cheap wine is plentiful.

We are living very unsociably, having only very few friends. Most of the people are still scared of us. This I do not regret, such a seclusion has its attractions and leaves more time for meditation. The people I see the most are the laboratory staff. I have only one research student at present and I have also got another one, but he has not yet come to Moscow, since it is useless to have him before the lab starts work. These people are nice. Especially efficient is my assistant director.

. . . By the way, Pohl from Göttingen visited Moscow to give a lecture. He came to see me and the future building of the lab. He remembers you with enthusiasm. But he was frightened to accept my

invitation to lunch. I was thoroughly amused. The dangerous Kapitza
– the leprous man. How funnily circumstances can change in life.
Well, my old Professor, kind regards and love from yours ever,

Peter Kapitza

The letter below follows another letter of the same date dealing with business
matters.

(a) From Rutherford   Cambridge, 15 May 1936

My dear Kapitza,

Now that I have dealt with official matters, I will return to some items
of news. You will have heard that Sir Herbert Austin, through the
Chancellor, Stanley Baldwin, made a gift of about £250 000 to the
Cavendish Laboratory. This came as a great surprise, for we were in
the initial stages of getting ready to issue an appeal for funds for
rebuilding, etc., but personally I did not expect we should raise any
substantial sum for a number of years, and I hoped to hand the baby
over to my successors! Actually, part of this gift will be used to defray
the cost of the new High Tension Laboratory which is now rising
rapidly . . . and as you know, we shall probably build a new research
block on the site of the old Zoological building, but this will probably
not be started for a couple of years. [The Austin Wing was completed
in 1939.]
   Immediately after the gift, Austin motored over to Cambridge on a
Sunday and we showed him the Laboratory and our general schemes
of rebuilding, and also some experiments. He was entertained to tea
by the Vice-Chancellor and I took him to dinner in Trinity when J.J.
was very affable to him. He motored back to Birmingham that night
and must have had rather a hectic day – especially as he told me that
on the way down he had the narrowest squeak of a motor accident he
has had in his life just outside Birmingham. I may mention that he had
the securities in his pocket at that time, so one could make a good story
out of it!
   You may have heard also that Housman died suddenly of a heart
attack about a fortnight ago. I think I told you that he was obviously
going down hill and it was merely a question of time when his trouble

would carry him off. I had to be in London on the day of his funeral, but I understand one of his poems was set to music as a hymn. I believe some of the sentiments contained in it were not exactly orthodox! I think you knew him pretty well, and there is no doubt that he was a man of extraordinary ability in many directions and one of our greatest figures in Trinity . . . Ellis is leaving to take Appleton's place in King's College on October 1st, and it has been arranged that Feather will return to us from Liverpool and will be made Lecturer in Trinity and given a Fellowship, and also be made a Lecturer in the Cavendish. It is quite possible that Oliphant may be leaving us in a year's time for the Chair in Birmingham. You know that Cockcroft was definitely elected F.R.S. a short time ago.

I am sorry you feel a little disappointed about the progress of the transfer of apparatus, but I can assure you that but for Cockcroft it would have taken twice as long. He has been indefatigable in putting things through as rapidly as possible, but of course we are in turn dependent on the makers, and the sudden burst of armament [building] does not make things easier. The whole world is in a mess, but we hope things will clear up before long.

We have had Appleton here giving the Scott Lectures on the upper atmosphere. He has so far given two and does them very well. I shall be very glad to have him as a colleague in the Laboratory next year as he is ready to accept responsibility and pull his weight in the Laboratory proper. Shoenberg and Milner are both shaping very well and have got some interesting results recently. This term I have been busier than I have ever been, but as you know my temper has improved during recent years, and I am not aware that anyone has suffered from it for the last few weeks!

My best wishes to Anna, and I hope you are all flourishing. Get down to some research even though it may not be of an epoch-making kind as soon as you can and you will feel happier. The harder the work the less time you will have for other troubles. As you know, "a reasonable number of fleas is good for a dog" – but I expect you feel you have more than the average number!

(a, c)   Moscow, 18 July 1936

My dear Professor,

I intended to write to you a long time ago but partly due to some over-work, partly to the badly organised life I could not find the necessary energy for the pleasure of having a talk with you. And during this time there happened a lot of things: In the family life, Andrew was bitten by a dog and had to be inoculated. We are living in a small bungalow [in the country], rather a crowd of us, besides ourselves and the maid, an English girl is staying with us to help Anna look after the children and the household and to keep up the freshness of the English language in the family. There is also my nephew and for a few days we had Dirac and Joan Thomson. The density of the population in the three rooms was so high that some of us had to sleep on the veranda, that we enjoy immensely as the summer is particularly hot and the nights are warm. But although to work in such conditions is hard, I am managing a letter. Otherwise it is very pleasant as we are near the forest and except for the dust of the road we enjoy the country im-mensely. To stay in town is very trying.

Now about the lab. It is still not quite ready. The builders are very slow and continuously break their promises, now we have their final promise that all will be ready for 15 August. Pearson and Laurmann, thanks to you, are here and both hard at work. Laurmann has done the wiring of the big machine and we hope to have a trial run next week. Pearson brought both liquefiers (helium and hydrogen) in parts and half done and it will take a long time to finish, assemble, test and run them. I am sure that he will be unable to finish it by October and I am afraid not before December or January. Both the liquefiers are only half done! And testing takes most of the time as only then a number of leaks, burst tubes, etc., appear. Unfortunately, we have also a holiday in the lab from 15 August to 15 September, and this cuts into P.'s and L.'s time. We shall shut up the lab completely as we are all very tired and had no rest for almost a year. Most of the apparatus arrived safely and of the big things only the Weiss magnet, the lab magnet, the solenoid and few other little things are left. But I will not trouble you about this and adjust it all with Cockcroft. I shall write to you again about all this, as things clear up and I see how

P. and L. get into their work. How is it about the financial settling – is it all alright? If you have any difficulties let me know at once.

Now regarding my personal work. I started to do a little research and in that respect I feel much happier. But all sorts of administrative matters still prevent me from devoting all my energy to my work. I propose to start with the Zeeman effect in the strong fields which I never did again after originally doing it with Skinner, when we only went up to 120 kG. Now with the generator we can triple our field, and until I get my helium it is a suitable thing to do . . . I now have got four research people, young ones, and I do not propose to increase the number for a year, not until the lab starts work. Now that's all about the lab.

Two main events have happened in our academic life. The old Karpinsky, president of the Academy died. He was 90, but still in working order. A kind soul and they say a good geologist, who did a lot for the study of the earth's crust. But indeed at 90, a man is a slow working engine and it is no surprise when it stops. He had an excellent state funeral, with Stalin and Molotov carrying the urn to the Kremlin wall where it was entombed. And then a huge public meeting in the Red Square. This was to show the care of the government for science. On the whole, I approve of such demonstrations, but I think it is much better to give attention to living people. This, however, is indeed risky, for when a man is still alive he can always do something wrong one day, while it is certain that a dead man will do no more wrong.

Now here is a case when a living man did something wrong. This is a mathematician by the name of Luzin [see p. 331], quite a good man – Hardy knows his work. You know that our school of mathematics used to be always a strong one, probably partly because Russians are keen on useless mental exercise, like chess, vint [a Russian card game], etc. and partly because mathematics is one of the happy subjects in science which do not require apparatus, laboratories etc., but just pen and paper. I think it is right to say that present day mathematics is very far ahead of other natural sciences, at least 100 years. The mathematics used in present-day engineering, physics and astronomy was developed in the beginning of the last century. Very probably some of the modern mathematics will be useful to experimental science, but a lot of work has to be done before this happens and probably a genius of some sort will be required who will possess a sense of real life and also be able to master abstract mathematical ideas (some kind of Poincaré). At present, all the mathematicians are detached from life and from other scientists and are boiling in their own juice. They are a crowd by themselves and, as it happens, a quarrelling crowd.

I don't know what you find from your experience, but I never yet met one of these mathematicians who was completely sane. Luzin is a good representative of this crowd. He got into a quarrel with the younger generation who are probably not less able than him. And these young people feel their superiority and forget that they themselves will get old one day and their brains will lose flexibility so that they will have to keep going mostly by experience and live on the capital earned in their young days – these young people started an attack on Luzin. Apparently, this old man did not behave faultlessly; he did not sufficiently recognise the ability of the young generation and was very forgetful in referring to their achievements, so it looked sometimes as if he took credit for other people's work, and probably due to jealousy gave bad testimonials to able people and praised mediocrities and even gave favourable references to obviously useless people. So the young people started to campaign against the old L. trying to give a political meaning to all his conduct, suggesting that he did it all with a political purpose. L. is rather a timid person who keeps floating mostly by very crude flattery, which he overdoes to such an extent as to annoy people. I doubt if such a man as L. has any strong political feelings at all – his whole ambition concerns his reputation and success in mathematics, but the absence of strong personality sometimes made him unscrupulous in achieving his reputation. You know that very few scientists can keep absolutely impartial and control their ambition and jealousy.

Our bosses, who manage masses of people perfectly well, but I think are not accustomed to dealing with scientists, took the case of L. very seriously at first, but now I think they begin to realise that the case is not worth much and L. will only be severely reprimanded and then left alone. I tried to interfere and explain that in the history of mankind all sorts of people appeared among scientists having all sorts of mentality, and various moral levels. And if we start to select our scientists by valuing them apart from their abilities in their direct work, but also as regards their moral and other qualities, we will deprive humanity of 3/4 of its geniuses. The main thing from my point of view is to create a healthy atmosphere and develop good traditions in the scientific community so as to keep all the "evil" instincts of scientists down, but no good will be done by crude interference from outside. However, this point of view did not meet with approval and I got kicked, so I had to remember your good advice "to keep my nose to the grindstone". I must confess that your advice and the experience I gained during the happy days of my intercourse with you are very good and help me now.

There is a third thing happening. I started a row about the bad supply of our scientific institutions – a piece of brass or copper which I was able to get in Cambridge in one day, I cannot get here in four months! By writing endless letters I finally managed to persuade our comrades that it is impossible to do serious research work if you are so badly supplied even with the most common materials. Finally, this seemed to make an impression and some signs of a desire for improvement was noticeable, but now the symptoms of improvement have disappeared again . . .

P.S.   Anna retyped my letter to make your reading it easier, but this deprives you from laughing at my spelling.

(a) From Rutherford   Cambridge, 5 August 1936

My dear Kapitza,

I have received your last letter a week or so ago, and am interested to hear that you are on holiday in the country while I have to stay in Cambridge to help things along! I hope you find Laurmann and Pearson helpful in getting things going. I am sure it will be a very interesting experience for them in Russia, but I hope you are not going to spoil them for hard work when they come back! I am interested to hear that you are thinking of trying the Zeeman effect with your machine, but I do not suppose you are likely to find anything very unexpected; however, it will amuse you to verify the theory in the highest fields obtainable.

You tell me that you get all the news of the laboratory, but I am sure you will be glad to know that the Mond is pulling its weight, and a good deal of interesting work is going on. Shoenberg is making good progress in his experiments on supraconductors. It is clear that the effects depend greatly on the shape of the specimen and on its purity. Milner has got some interesting results on the effects of a magnetic field on the resistance of cadmium down to liquid helium temperatures. Peierls is much interested in theories of supraconductors, and is very helpful in discussing matters with the men in the Laboratory. We have made preliminary experiments to obtain very low temperatures and results seem promising . . .

We are at present considering the plans for the new extension which will probably be started about two years from now. I shall probably send some of the boys to have a look at some European laboratories so as to be in a position to advise about the internal arrangements. Which would you think would be the best laboratories for this purpose? – and from this point of view, would it be worth while looking at one of the Russian laboratories, say at Kharkov? I imagine that Cockcroft, Ratcliffe and Dee, and possibly others, might go on a holiday of this kind next year. There is a good deal of discussion at the moment of what re-arrangement should be made in the old laboratory so as to make it as useful as possible in the future. At the moment we have as many ideas as there are people on the Committee, but I hope that by Christmas we shall have reached a general agreement . . .

I was interested to hear what you had to say about Karpinsky, etc. I saw him when he was over here for the Centenary of the Geological Survey, and I gathered indirectly that he was very angry with Sir John Flett, the retiring Director, because he was not asked to speak when I presided at the opening of the Congress. I am sorry myself that I did not hear him as he seemed to me very active for his age.

I think I have given you most of the items of news. I am glad to say we are all very well, and like everyone else I am rather alarmed by the potentialities of the situation in Spain. It looks a very bad business and is not likely to be settled for some time. This country as a whole is very prosperous, and the re-armament is giving a lot of fresh work. We have got great arrears to make up, and I hope they are going ahead about it as efficiently as possible.

I hope that when you have proved yourself a good Soviet citizen, you will be allowed to come and see us again, but I suppose that will not be for a year or two. Give my kind regards to Anna, who I hope is enjoying life in her own country. I am expecting to hear by the next letter that you are beginning a definite piece of research. You will find that this will help to restore your mental equilibrium quicker than anything else.

(a) Moscow, 19 October 1936

My dear Professor,

It is a long time since I have written to you. As a matter of fact, I have been away on my holiday, part of which we spent with John and Elizabeth Cockcroft. We had intended to go to the Caucasus, but unfortunately John got bronchitis and we had to stay in Crimea. John, no doubt, will tell you in detail about our life here and the Institute. Myself I was very happy to have learned from him about you, the Cavendish, its population and work. I felt that quite a bit of my soul is still with you and it gives me a lot of happiness to hear all about Cambridge.

I doubt that I shall be allowed to go about for quite a time yet, and if only I could have the smallest hope that one day you will come to see me here (it could be arranged to be quite a private visit), it would be great. I still feel myself half a prisoner, to be deprived of travelling abroad, of seeing the world around me, and visiting labs; it is a great privation and no doubt will eventually result in narrowing my knowledge and abilities. Now we are gradually returning to work again. Things in the lab are taking shape. The staff which we engaged is quite good and, if not sufficiently experienced, is very enthusiastic, and willing to work hard. I hope in a month's time to have the Zeeman effect spectrum taken. This will be the first start – real work will begin when we shall get the helium plant working. Pearson is hard at work but he does not think this will be before the New Year . . .

Besides the work in the lab, I am trying to improve scientific life here. Just imagine that we have no Science Museum here and I am trying to stimulate the Academy to start one similar to the Kensington one [see p. 343]. Don't worry, I am only suggesting it but not actually taking part in the organisation. Secondly, I am trying to improve the supply of scientific apparatus and materials used by the labs. Now it is shockingly bad. I have already written to you about it. This is most important and I am taking it most seriously, but it is not easily done. Now we are able to start work owing to your kindness in supplying me so generously with all sorts of stuff, but this supply will not last forever. Thirdly, I am engaged in reorganising the life in the Institute more reasonably. As an example of the stupidity of the system in use – we have had five bookkeepers. This was due to a very detailed system of accounts. Imagine that each particular research work had to have its own particular account – so much for rubber, so much for cardboard, paper, etc. This is good for a factory but silly in a lab. We are simplifying all this. Now we have only two bookkeepers and hope to reduce them to one, or even to one half. People are sympathetic to all these reforms.

Fourthly, I begin to think about organising the social scientific life. Imagine that there is no physical society in Moscow, where you could give a paper. All scientists have colloquia with their students, but no common life exists. This is rather a difficult problem until our Institute will have started real scientific work. The second reason for it is that scientists in Russia are, in general, very badly paid, and they have to have a great number of different jobs, mainly teaching, to get sufficient living means, and so no time and energy are left for meetings and preparing papers. But I hope that this will change.

Anna and the kids are well; in about a month's time we hope to move into our new house, which is situated quite near the Institute. There we shall be much more comfortable and I will be able to supervise the lab more closely without interrupting my reading work . . .

As regards Laurmann, he is very much needed, as we must train the assistants how to operate the machinery; I already have an able young man whom he will teach. So I would ask you to spare me Laurmann, as long as he wishes to stay, and no less than six months after the New Year, when he is going on leave to Cambridge. Pearson, as I had already mentioned, will have to stay at least three months to finish the liquefiers and then I should like him to stay at least six months more to train the chap to run the apparatus and to do inevitable repairs in the new apparatus. To do it in less time is impossible, as the chap, even if clever, has no experience in cryogenic work. If the liquefiers had come completed, indeed, the time would have been less, but all the testing of the gear takes such a long time, since this must be done very carefully to avert unpleasantness in future. Actually I do not regret that the liquefiers came unfinished as this gave me the opportunity to teach the people here the technique of making them.

I hope that you will help me in this case as you did before. If people in the lab grumble you may tell them that it is only on account of me that they enjoy using the best helium liquefier and elementary human gratitude requires a little privation on their part to help me in my difficulties and troubles. After all, I now have in mind a much better liquefier, which if successful, may be useful to them too. An exchange of experience will make both the labs stronger and will benefit science and humanity.

Well, my dear "old" Professor, you see what long letters I am writing to you. I love to have letters from you too, and how much more I should like to talk to you. Most kind regards to you and Lady Rutherford from us both.

(a) From Rutherford   Cambridge, 29 October 1936

My dear Kapitza,

I was very glad to get your good letter, and have had the opportunity of having a chat with Dirac and Cockcroft about their visit to you. I am glad to hear you are feeling in much better spirits and prepared to get down to a job of research straight away. I hope you will have good luck with your experiments on the Zeeman effect in strong fields. I am very interested to hear of your efforts to improve the conditions of research in Russia, and I wish you all success with regard to them. They all seem to me very reasonable and proper . . .

You will be interested to know that J.J. T[homson] celebrates his 80th birthday on December 18th. The Trinity people are giving him a special dinner in Hall and a special address and a bit of silver plate. We as a Laboratory will probably transmit to him on the day our good wishes signed by a number of the scientific staff and students. J.J. as a whole looks very well, but has undoubtedly aged markedly the last year. He has been a good deal worried by the illness of George [his son], but I am glad to say the latter is now well on the way to convalescence, but will not be able to get down to work for another six months . . .

(a) From Rutherford   Cambridge, 20 January 1937

My dear Kapitza,

. . . I was interested to hear in Anna's last letter to me that you were leaving Moscow for a few days to see some scientific laboratories and

73. Rutherford and J.J. Thomson outside Mond entrance, 1933.

to have a change from your work. I am glad to hear from Pearson that things are going well with you and that your laboratory is now getting in good working order. You no doubt have heard that we have done our best to keep work going in the Mond and things are going very well there. You may also have heard that Simon has obtained a Readership in Oxford and has thus better conditions for carrying on his work. He was over here yesterday and gave an account of the latest work in fixing the temperature scale for very low temperatures and of the permanent magnetisation which appears in certain substances at low temperatures. He is an energetic worker and is making good progress.

Incidentally you may know that Lindemann is very anxious to

become an M.P. and there will probably be a contest between three of them for the seat to be vacated by Lord Cecil. He is very keen to take part in public life under the aegis of his friend Churchill. I imagine however that he has not much chance of success. In a letter to *The Times* he made it clear that he alone could save the country in these difficult times of air warfare . . .

Our latest news is that Dirac has succumbed to the charms of a Hungarian widow with two children, I believe aged 11 and 13. They were married in the Christmas vacation and are looking for a suitable house in Cambridge. I understand she is a sister of Professor Wigner who is Professor of Mathematics at Princeton, and he met her there. I think it will require the ability of an experienced widow to look after him adequately! I hope it will turn out well: I have not yet seen her . . .

You asked me some time ago to send a recent photograph of myself. I have had one taken by Ramsey & Muspratt and am sending you a copy under separate cover. Miss Stebbing [Rutherford's secretary, formerly Kapitza's secretary] and the family consider it a reasonable one, but I myself am not sure that it does justice to my natural good looks – I forget that I am becoming an ancient veteran! You will notice that my hair is disordered in the characteristic fashion!

With best wishes to you all and good luck to your work.

(a)   Moscow, 16 February 1937

My dear Professor,

It is a long time since I intended to write, but a letter to you is a long business. I have to write it and Anna has to copy it and make it readable, and that I could not do for some time. As a matter of fact, I am very tired and somewhat overworked. To make the Institute work is, in general, not so easy; you know this from your own experience, but in my case it is even more difficult, as to get the smallest thing here is a long job. Now it gets slightly better and I am all the time pressing the responsible people and try to impress on them the necessity of proper shops, factories, etc., for organising the supplies of materials and apparatus for scientific work. You know my temperament, I want everything to be done quickly and I simply cannot stand it when things are going slowly; the result is that I am very tired.

I also had some family troubles. Mother (she is now over 71) had a very severe heart attack, which very nearly ended fatally. Anna spent a fortnight in Leningrad helping the nurse, and I too was going there periodically. Now Mother is out of immediate danger, but the doctor says that she will have to spend at least a month or even two in bed, before a recovery can be expected. The rest of the family, including myself, are well.

Now about the lab – Pearson and Laurmann are hard at work. Pearson is just finishing the helium liquefier and by the end of the month we hope to have liquid helium on the tap. Meanwhile, he is starting a new [helium] liquefier, which I have just designed. This one will be much simpler than the old one and give nine litres of liquid helium per hour, using the same compressor and less liquid nitrogen. I hope I did not make a blunder in my calculations!

Laurmann, a Russian [P.G. Strelkov, see fig. 34] and myself are doing the Zeeman effect, and the work is more or less done. We could not find anything wrong in the theory. Up to 300 kilogauss the splitting is proportional to the field. The excessive splitting at 100 kilogauss which we observed with Skinner in 1924, for two particular lines is easily explained by the appearance of "prohibited" lines [close to those in question], which become quite apparent [i.e., resolved] only in the strong fields which we are now able to use. I do not propose to continue this research much longer. Meanwhile, we prepare for the galvanomagnetic researchers and as soon as liquid helium is available, we shall start them.* I hope that by the summer I will be more free from administrative work, and will have the research work in full swing. After two years of fasting – a good feast!

Last summer when Shoenberg was here he told me that he would like to come here to work for a year, starting with the summer term, provided of course that you approve of the arrangement. I shall be very glad to have Shoenberg here for a year; it will do a lot of good to the research people here, especially as he speaks Russian. I shall be very grateful to you if you could help him get a year's leave from Cambridge, since coming to work here for less than a year is quite useless. The news of Dirac's marriage was quite unexpected for us. As you know we like Dirac very much and hope that he made the right choice

---

* In fact, Kapitza himself never did return to this research but became completely absorbed in research on liquid helium.

of his life companion. But very often people are good at theory but bad at experiments . . .

P.S.   Your ex-king gave a wonderful entertainment to the whole of Europe. Even here, people were thoroughly thrilled with the story of little Simpson.

(a) From Rutherford   Cambridge, 22 March 1937

My dear Kapitza,

. . . I was sorry to hear of the illness of your mother, and am glad to hear she is out of immediate danger. Give her my best wishes. As for myself and family, we are all very well, and I still can do a day's work without serious effort, and I may tell you in confidence that my temper improves with age! – at any rate my wife says so! This is very largely the result of getting my knee in order. I now walk more like a young man – at least I think so – and have had no trouble with it for the last three years. This is the result of the ministrations of a local pork butcher, Mr Waller, who is an unlicensed expert in such matters and a most amusing and interesting fellow.

We have now installed the 1 250 000 volt D.C. generator in the High Tension Laboratory and expect in the next few days to have at least a million volts applied for acceleration purposes. The present appearance of the High Tension Laboratory is very impressive and many people say it reminds them of the illustration of Wells's film "Things to Come". It is naturally rather a big job getting this new installation going, and the Austin money comes in very useful for the purpose. We are also installing within the next few months a Lawrence cyclotron, and Cockcroft has designed an electromagnet of about 40 tons for the purpose. The plans for a big magnet to take the place of your generator are now also well in hand. Your old Laboratory has been very full up and is doing good work.

The Trinity Commemoration [Feast] went off very well. I thought J.J. seemed tired and made rather a rambling speech, but shorter than usual. Austin was my guest. He is a pleasant man but very limited in his reading. Aston is still troubled at times either with his tummy or heart, and is rather worried about himself. Either the climate of Cambridge or the Trinity dinners do not entirely agree with him!

However, he manages to keep on with his work and with his golf, but did not feel well enough to turn out yesterday . . .

I hope that you are now hard at work and getting things going in good style, and that you are getting good work done by Laurmann and Pearson. You mentioned about Shoenberg and a stay in your Laboratory. I have no objection to this arrangement, but it is a little dependent on whether he can get permission from the 1851 Commission to hold over his studentship for a year. He has been working very well and I think has developed into a very competent researcher.

(a)   Moscow, 7 April 1937

My dear Professor,

It is always a great pleasure to have a letter from you and hear about Cambridge and the Cavendish. As regards us, I had a very sad time, as on 18 March Mother died. It was somewhat unexpected as she seemed to improve, and quite suddenly the heart collapsed. She was 71, and in full working strength – some of her books are just appearing. I have been great friends with Mother and I feel her death very acutely. The rest of the family is quite well, and I am very hard at work all the time.

The Government Commission has examined the lab, and gave a favourable report. Now we are waiting for the Government to approve it. The helium machine is in working order, and if not for all the family troubles I should be already working with the magnetic fields and low temperatures . . . In May we are expecting Bohr, he is coming here on his way from America and Japan and I have already arranged all the necessary visas for him and I am looking forward to seeing him. At the moment Pearson is busy making the new helium liquefier, which according to the calculations will have a capacity of 8 – 9 litres per hour and will have its starting time reduced to half an hour. The expansion engine for this machine is very much simplified – it is already made and tested. If we find it O.K. we shall proceed with the heat exchangers and the rest. We hope the machine will be working in the summer.

As regards Laurmann and Pearson's stay in Russia, Pearson wants to return to England in the summer, but he hopes that you will be kind enough to give him an allowance of five or three months, if necessary, to finish the new liquefier. Laurmann is quite prepared to stay here, of course if you will find it possible, another half of the year after the summer, till the beginning of the year 1938. He is not planning to visit England this summer, but wants his family to come and stay with him here. I hope you will kindly help me in persuading the University authorities to approve of these arrangements. You know how very important it is for me to have my two best assistants working with me. I was very glad to hear that you are very fit and in full working power. It seems ages since I last saw you, and I cannot imagine how I shall be able to go on like this, I love you very much and not to see you is the greatest privation for me.

(a) From Rutherford   Cambridge, 23 April 1937

My dear Kapitza,

. . . I was very sorry to hear the sad news of the death of your mother. In your previous letter you had told me of her illness, but I hoped that she would recover in due course. I can quite appreciate what her death means to you. The death of a parent is always a great break in a family, and I sympathise with your and Anna in your great loss. I had of course met her on several occasions in Cambridge and thought that it was very wonderful that she kept up her teaching and literary work so long. I am glad you were able to see her in her last days in Russia.

I am interested to hear about the progress of your laboratory and your experiment with the Zeeman effect. Even though the results in strong fields do not show anything very novel I am sure it must have been a great pleasure to you to get down to definite experimental work. I am interested also to hear about your new helium liquefier and hope that it will fulfil your predictions. We shall no doubt hear about its performance in due course. I did not know that Bohr was coming back through Russia and will see you in Moscow. I think he was very much bucked up by his new general theory of nuclei, but I am not sure whether he has been able to push it mathematically very far. Give him and Mrs Bohr my best wishes when you see them . . . The Mond Laboratory has just started up work again and is going strong.

(a)   Moscow, 10 May 1937

My dear Professor,

. . . Just at the moment the lab is having holiday, but lately things were not going well, the helium liquefier is leaking and Pearson is pretty fed up to take it to pieces an endless number of times and I am annoyed as I have no liquid helium for my experiments. Then the coil for the magnetic fields has burst, for nearly 12 years we never had a burst so we practically forgot how it looks, but we tried to raise the limit of field and this did it in. Everything else is going well, but I feel very tired. Unfortunately, I cannot leave the Institute, since life there is not sufficiently organised to be left to itself.

The family is feeling quite well, the children had a very good winter – Moscow is very good as regards climate. It was very kind of you to write so well about my mother; it is true that she was full of life and energy and you could never have told that she was old . . .

(a)   Moscow, 13 September 1937

My dear Professor,

I have not written to you for a long time. Probably it was due to my being very tired and overworked. We stopped work in the lab on 1 August, and went to stay in a small bungalow about 35 miles away from Moscow. This was a temporary place – very primitive. A little house is being built next to it, but it goes very slowly and I am afraid it will go on for another year or two. The spot itself is beautiful, we are on the river in a pine wood.

Even though the lab was closed I had to come to town, as we had some building going on in the Institute, or rather round the Institute, mostly plastering and repairs, but this moving in and out rather spoiled the holidays. Well, I hope this will be the last year of building. In the country I had plenty of physical exercise; we were clearing our plot of old stumps and pulled out quite a good amount of them. Dirac and his wife were staying with us for three weeks and he also got quite keen on this kind of work. It is great fun to see Dirac married, it makes him much more human and I am glad to see him settle down in his private life. We liked his Manci, I think she will make him quite a good wife.

Seeing Dirac and Bohr in the same summer was a great pleasure for us, as we are very lonely here. Bohr came in June on his way home from his world tour, he was with Mrs Bohr and his son [Hans]. They stayed for six days and he gave a big public lecture which was a great success, I thought he did it very well. We talked a good deal with Bohr and I got to know him much more, and I liked him very much indeed, he is a very good and fine man. He said that he will be coming more often to see us, and I think this is a very good decision. In a few days there will be a conference on nuclear physics – Blackett, Ellis and Peierls are all coming, but it is a great pity that John Cockcroft was not able to come. Webster is coming for a short time just to see us, it is very sad that he gave up science; there is no doubt that his work on crystals and ferromagnetism is now regarded as quite fundamental.

We have finished the Zeeman effect, it came out very nicely, all quite regular. We also studied the Paschen-Back phenomenon and it agreed with theory to 2%, which is our experimental accuracy. Now we are preparing the work for publication. As we expected, nothing exciting or out of the ordinary happened, but it had to be done one day. This month I shall start on my old problems on the change of resistance.* Now that we got the helium liquefier in the lab, there is nothing to stop me proceeding with my work, which was interrupted for three years.

I am very grateful to you that you arranged for Pearson to stay here for another term. I don't think I shall require him any more after that. We have now got two well trained men, who can proceed with the work on the liquefiers, and in the next three months Pearson will finish the new helium liquefier, which will be much more powerful than our old one. It will give about 6 – 9 litres per hour (I hope), and also will be much simpler to manipulate. All the parts are made and we only have to put it together. The new type of expansion engine also looks very promising. If we shall really get this large amount of liquid helium, then I have an idea how to make an adiabatic machine to work on the magnetic principle to get to much lower temperatures which would be difficult with only a small amount of helium.

Shoenberg arrived a few days ago and is starting to put all his gear

---

* However this programme was abandoned after a few preliminary experiments, as Kapitza became increasingly absorbed by his new researches on liquid helium.

together. The family is alright. The summer was very good; we had plenty of sunshine and both the boys got sunburned and were very well. Peter [Sergei] is going to school and he looks at it very seriously but it is a little bit difficult for him to start studies in a new language. I should love to have a little stroll in the world, to see the French exhibition, to look you up in Cambridge. It is absolutely idiotic the way we Russian scientists are pickled here. And the people seem not to realize how bad this is for the development of our science. But probably I shall be the last one to be permitted to go out. Dirac and Bohr were telling me about you and I miss you very much. Would you not one day acquire the old travelling spirit of your great nation and come to see us here? It is quite possible to arrange all very quietly without any lecturing, etc. I hope you had a good summer rest and have again plenty of energy for your work. Please do write to me, it is a great comfort to me to get news from you and from Cambridge. I hope to write more often to you now that it is much easier, back in town where I can find peace in my study. I imagine all your grandchildren are quite grown up, and soon you would be a great-grandfather . . .

(a) From Rutherford   Cambridge, 9 October 1937

My dear Kaptiza,

. . . I am glad to say we are all very well, including the grandchildren. Peter is now fifteen, and is at Winchester where his father was before him, and is doing reasonably well. Elizabeth is at the Perse School, and is academically very bright. Today we are having a visit from Niels Bohr, who is on his way back to Copenhagen. He is giving us a talk this afternoon. He showed me the nuclear model that you had made for him – a very nice experimental demonstration of some of his points. I do not know if you remember that I had a somewhat similar model, which was made by Andrade, to show the scattering of α-particles. This was carefully shaped, so as to give the scattering according to the law of inverse squares . . .

My wife and I are going out to India at the end of November, and are returning in February. We shall travel out by P & O via Suez, and I am to act as President of the combined British Association Meeting and the Indian Science Congress. The meetings are to be held in

Calcutta at the beginning of the year . . . An elaborate programme of sightseeing has been arranged for us, and we shall travel in a special train and sleep on it for ten days, doing tours by day and travelling by night. From Bombay, we go to Hyderabad, and then up to Delhi and Benares, etc., and on to Calcutta, probably returning via Madras and Bangalore. I am afraid it will be rather a hectic time, but fortunately the weather is quite cool in the North, they say, and not at all oppressive in Calcutta. We hope, if we have time, to make a flying visit to Darjeeling, to see the Himalaya Mountains . . .

Bohr told me about his trip to you, and I am very interested to hear of the work that you have been able to accomplish. No doubt Pearson, when he returns, will be able to give us the latest information about your big helium liquefier. The Mond Laboratory is very flourishing, and a large amount of work is in progress. For reaching very low temperatures, we have had built a special water-cooled magnet to give us a field of about 40 kilogauss with a current of 600 amps. We are getting in a machine to produce still higher currents. This will be water-cooled too, and we have installed a reservoir and pumps for using the water over and over again. Some interesting experiments are also in progress on the extraordinary heat conductivity of helium at low temperatures. The conductivity is very large for small differences of temperature, and falls rapidly with the quantity of heat transmitted. Cockcroft is, of course, very busy with a multitude of things on his hands. He is invaluable to us in all these problems of building and reconstruction. I am glad to say that I am feeling physically pretty fit, but I wish that life was not quite so strenous in term-time. All of your friends are, I think, very well. J.J.T. is ageing a little, but otherwise he is as lively as ever . . .

I hope it will not be too long before you are able to come over and see us all again. At the moment I cannot make any plans ahead, but some time I hope I may have the opportunity of seeing you . . .

Best wishes to you all.

Yours ever,
Rutherford

Rutherford died of a strangulated hernia on 19 October

(a) To J.D. Cockcroft   Moscow, 1 November 1937

My dear John,

It is difficult to believe that there is no more Rutherford. We all had the feeling that Rutherford is immortal not only by his work, but as a human being – he was so strong and so full of life. We both owe a lot to Rutherford. Please let me know if there will be any subscription or anything of the sort to commemorate him, either in the Cavendish, the Royal Society or Trinity, I would like to take part in it. I have written an obituary notice in *Izvestia,* then on the 14th of November there will be a memorial meeting arranged by the University and I am asked to take part in it, and also to write an article about Rutherford. It was very unfortunate that you could not hear me when I rang you up, on my part I could hear you quite well.

Things in the lab are not going badly at all. We just started the new liquefier and the first time it gave four litres per hour; I hope to increase the output considerably. Now it is quite certain that Pearson will be free before the New Year, I will not claim his services any more after that. But I should like Laurmann to stay here until next autumn and he is quite agreeable, since you told me over the telephone that there is no difficulty about it, I would be grateful to you if you could confirm it to Laurmann.

We are all quite well. Peter [Sergei] goes to school and a few days ago he came back very proudly wearing a red tie and told us that he was elected to the Pioneers. I think Andrew is rather jealous of him but he will have to wait till he is at least nine years old. Well, our best wishes to you and Elizabeth.

(A, I) To Niels Bohr   Moscow, 7 November 1937

My dear Bohr,

I am sure you feel very sad about the death of Rutherford. For me it is a great shock. All these years I lived with the hope that I shall see him again and now this hope is gone. It was not enough to correspond with him, you know when you speak to him you get much more from his eyes, from the expression of his face, from the intonation of his voice, than from his words. I loved Rutherford, and I am writing to you, because I know that you had a great feeling for him. From his words I always gathered that he liked you the most amongst all his pupils and to be sincere I was always a little jealous of you. But now this has gone.

I learned a great deal from Rutherford, not physics but how to do physics. Rutherford was not a critic; you, probably, like me never heard him arguing, neither in science nor in questions of life. But his influence on everybody around him was felt by his example and by his opinion, always brief and in the long run correct.

Once talking with him in the Trinity Combination Room, shortly after the Maxwell celebrations, he asked me how I liked it all. I answered that I did not like it because all the speakers tried to represent Maxwell as a super-human being. Indeed Maxwell is one of the greatest physicists ever existing, but still he was a human being and this means he had a human character. And for us, the generation which never saw Maxwell, it would be much more valuable and interesting to know about the real Maxwell and not a sugary extract from him. Rutherford, indeed laughed loudly and said: "Well Kapitza, I leave it to you, after my death to tell about me how I really was". I don't know if it was a joke or if he was half serious.

But now he is gone. I have to speak about him at a big meeting on 14th of November and I have to write about him. I must do it, since of all the Russian scientists I knew him the best. But all the little weaknesses I noticed when I was with him, now look to me so trivial and insignificant and a great unreproachable man stands before me in my memory. And I am afraid I will do the same as the pupils of Maxwell did when speaking to us on the Centenary celebrations in the Cavendish Lab.

One of the first features of Rutherford, which comes to me, is his great simplicity. He hated complicated apparatus, complicated experiments, complicated theories, complicated arguments involved in diplomacy. He stood for the simplest, which always proved to be right and the most powerful argument. He was simple himself and therefore very sincere. It was so easy to speak with him, the answer was written on his face before he spoke. How sad that I will not see him again. I am even glad that I was unable to come to his funeral. It would be so painful to see his face not living.*

---

* Presumably Kapitza is thinking of the Russian custom of displaying the face of the deceased at the funeral.

He was very good to me. Extraordinarily good. How much encouragement in my work I owe to him. He never was too severe on my stupidity or mistakes, even when I irritated him. I consider myself a very lucky man to have been close to him for 14 years.

Dirac writes to me that people begin to like the carving [by Eric Gill] of Rutherford in the Mond, for which I was partly responsible. I am very glad. I am working very hard at present, trying to recover the lost years. I have to write to you about a number of questions, but I will put it off for another time. At present, I have no one here, who will share my feelings after the death of my good old Crocodile. I remember with great pleasure your visit to Moscow. Do come again next year . . .

(A, I, g) To P.A.M. Dirac   Moscow, 7 November 1937

My dear Paul,

Thank you very much for your letter – you are quite right that the death of Rutherford is a great blow to me. I always lived in the hope of seeing him again and even in his last letter to me he hinted that a visit to see me was not improbable. Rutherford's reputation as a scientist is indeed recognized everywhere, but Rutherford as a great personality was known only by the people who worked with him. What I liked the most in him was his simplicity . . . and it was this that made life and work with him so pleasant.

Without Rutherford, C.T.R. Wilson and younger physicists who have left Cambridge, there is a feeling that the School of Experimental Physics in the Cavendish is in danger . . . And I think that you, who are now the leading personality in physics in Cambridge, should take some active interest in keeping up the great traditions of the Cavendish Laboratory, so important for all the world.

This year proved to be very strenuous and losing Mother and Rutherford in the same year is very painful. Work in the lab is getting on not at all badly . . . The family is quite well and Anna will write in more detail to Manci. . . . we all hope you will keep to your habit of coming to see us every year, now not alone but with Manci.

With very kind regards to Manci.

74. Panorama of the Red Square, St. Basil's Cathedral and the Kremlin, 1956.

6

Letters to the Kremlin (1929 – 1980)

This chapter contains extracts from some of the many letters Kapitza wrote to the top political leaders, about practical matters concerning the needs of his scientific work, about general questions of scientific and technological policy and, particularly courageously, appealing for colleagues who had been unjustly repressed. A few letters to high officials of the Academy of Sciences and Moscow State University and some to Bohr, Born and Langevin about Soviet and international scientific matters, are also included. The first two letters are from Cambridge (written in English): one is a reply to an invitation from L.B. Kamenev (an important political leader, then in charge of scientific matters) to visit the Soviet Union regularly as a consultant and enquiring when Kapitza would return permanently to the Soviet Union, and the other to Bohr about Gamow. The rest of the letters are all subsequent to Kapitza's permanent return to the Soviet Union in 1934.

(a)   Cambridge, 2 February 1929

Dear Mr Kamenev,

I beg to acknowledge receipt of your letter in which you kindly invite me to help in the organization and the working of a new Kharkov Institute. I greatly appreciate this offer as it will give me an opportunity of helping my own country in its scientific development which is now developing very rapidly.

   With regard to the conditions in which you offer me these consultations I am afraid I can only accept them with these alterations: first, it is quite impossible for me to undertake any obligation to come every year for a definite time, say two or three months, to Russia. The time which I could spare for my travelling to Russia will depend very much on the time required for my scientific researches here and as I am in charge of a Laboratory I can only abandon it at some times, according to the conditions of work; second, I highly appreciate that in the present economic conditions of Russia the scientist cannot be expected to be paid at the scale usually adopted in western countries and I think I will be quite justified in offering you the following conditions for my consultation work. I shall regard £200 yearly as a retaining fee and my expenses will be covered if you pay me £150 travelling expenses plus 70 rubles per day actually spent in Russia.* The time and the length of my visits can be adjusted by mutual agreement and in accordance with the actual necessity . . .

I have been following with great interest and satisfaction the tremendous development of scientific research work in Russia and am very glad that I now have an opportunity of helping. I am especially indebted to you for the passport facilities which you now kindly offer for my visits to my own country. It was rather strange that for me, a Russian citizen, the travelling facilities to Russia were so prohibitively difficult. As the passport regulations created so many complications it was easier for me to go for my holidays or scientific visits to foreign countries . . .

Now returning to the last point in your letter of the possibility of my returning to Russia. I appreciate immensely your kind remarks but unfortunately the next few years will be so full that it will be impossible for me to leave this laboratory which was specially built for me and for which I am responsible and before accomplishing a complete line of researches which are now in full progress. I hope you will appreciate the remarks in my letter as made by a person who sincerely likes his own country and is keen to see the scientific progress developed more efficiently in present Russia.

I am, yours sincerely,
P. Kapitza

---

* This is considerably more than Kamenev's offer of 2000 rubles for 2 or 3 months. In a letter from Kamenev of 15 April 1929, Kapitza's conditions were accepted in full. A strange feature of this correspondence is that Kapitza's letters were written in English, rather as if to demonstrate his independence.

75. N. Bohr.

(I) To Niels Bohr   Cambridge, 15 November 1933

Dear Bohr,

At your request, Dirac has just told me about the troubles with Gamow. I think it is better for every man to work in the country and under the conditions which he likes most, and this is why I think it would be much better for Gamow to work abroad provided he can find a job, especially as there is very little experimental or theoretical work on nuclei being done in Russia at the moment, and Gamow's abilities would therefore be much better exploited abroad. Gamow's particular character also seems to make him work to the best advantage when he has a wide circle of social contacts.

The main objection to Gamow's staying out of Russia appears to

76. J.D. Cockcroft and G. Gamow, Cambridge, 1932.

me to be that it will make it extremely difficult for young Russian physicists who wish to study abroad to get permission. At the moment there are under discussion in Russia about ten young physicists who wish to go abroad, and no doubt if Gamow stays in Europe without a permit from the Russian Government, their position will be very much prejudiced.

I can therefore only see one way out of the difficulty, and that is that Gamow's stay in Europe must be authorised in Russia. The best way of doing this would be for Gamow to get official leave for say one year, then for the second year the leave will be easier to get, and so on, till his absence is looked on like a chronic illness which people get used to! I think this would also be the best for Gamow's sake because of his changeable character – he may after one year or two change his mind – his wife may get homesick as it is her first visit abroad – and in this way he will not have burnt his bridges behind him.

I think the best hope of arranging this leave would be through Joffé. I am sure that Joffé is powerful enough to arrange this provided he is approached in the proper way. For instance if you try and speak to

Joffé I am sure he will do his utmost to meet your wishes . . . I understand from Dirac that you may be going to Russia yourself, and in this case, you could discuss the question frankly with Joffé.

It is all very sad indeed that the political conditions are such that the country which claims to be the most international in the world, actually places her citizens in such a position that it is very difficult for them to visit other countries, and I feel very sorry for the present conditions, but am glad to say they are now rapidly altering for the best. Accept my most kind regards to yourself and Mrs Bohr.

Yours sincerely,

In the event, Gamow took matters into his own hands and did not return to the Soviet Union. Soon afterwards he emigrated to the United States and was appointed Professor at Washington University in Seattle. Gamow's own account of the affair is given in his autobiography (Gamow 1978).

(I) To V.I. Mezhlauk
     Deputy Chairman of the Sovnarkom   Leningrad,
     (Council of People's Commissars)   2 November 1934

Comrade Mezhlauk,

In reply to your request of 26 October (received only in the evening of 31 October), I am writing to inform you about the scientific work I propose to carry out in the USSR. As you know my main work up to now has been in the field of cryogenic magnetic research and was performed in my Cambridge Institute. These are some of the most technically complex investigations of contemporary physics and demand an exceptionally well equipped technical base and highly qualified supporting personnel. In Cambridge I developed my work over 13 years and my colleagues developed their expertise in the course of constructing the unique and original apparatus with which my laboratory was equipped . . .

To begin this work afresh the whole laboratory must be recreated. Without carefully selected and specially trained assistants and mechanics, without the technical drawings, data, etc., and under my sole guidance it would take several years of intensive work in any country to do this and only if there was good support from industry. In the Soviet Union, where technical resources are extremely overloaded, many materials are unavailable *and above all, in the absence of trained assistants*, I do not see the possibility of taking responsibility for scientific research similar to that I worked on in Cambridge. As I have already told you, the only possibility of achieving this would be to send young scientists to work with me in my laboratory and so gradually transferring the technical experience from Cambridge to the Soviet Union. I should like to remind you that over the last two or three years I have more than once suggested that some of our young Soviet scientists should be sent to work with me and I offered to give them priority over several other foreigners wishing to join me . . . To my great regret this was not accepted. In the present circumstances I definitely do not consider it possible to undertake the creation of a new laboratory and I have therefore decided that my work in the Soviet Union should be in a different field.

In fact I have long been interested in so-called biophysical phenomena of living Nature which are amenable to study using physical laws. In particular, I am interested in the mechanism of muscular action. Since this field lies at the boundary of two fields of knowledge, it has always been somewhat neglected in spite of its great scientific interest. In recent years A.V. Hill and his school have made considerable advances in this field, as was recognized by the award of a Nobel Prize a few years ago. My close acquaintance with Hill, who often consulted me on questions of physics, gives me the opportunity of getting to know the direction in which his work is going and its methodology. No one is working on these problems in the Soviet Union and since they do not require any enormous or powerful apparatus, but only sensitive and accurate instruments, I propose to enter this field. Moreover, Hill, who is a specialist in mathematics, has somewhat overemphasized the thermodynamic aspects of muscular processes and left the physical aspects – just those that interest me – on one side. I have also consulted I.P. Pavlov and found that he approves of the general trend of my proposals, which he finds interesting, though he has never worked on these problems himself. Moreover, Ivan Petrovich [Pavlov] has amiably agreed to provide space for me and the necessary technical facilities in his laboratory. As soon as I have finished studying the essential literature, I shall start experimental work. If any of our scientific institutions should require my consultancy, it goes without saying that I shall be very willing to oblige them as I have done up to now.

(I)   To V.M. Molotov
       Chairman of the Sovnarkom   Moscow, 7 May 1935

Comrade Molotov

It should have liked to tell you in person all that I am writing, since I am not good at writing, but unfortunately you have not wished to see me. Comrade Mezhlauk has informed me that articles have appeared in the foreign Press discussing my detention here. Sooner or later, of course, the scientific world was bound to get to know about this, but I very much wanted it to be as late as possible, in the hope that a quiet solution might be found without my name being splashed all over the newspapers, thus leading to all sorts of misunderstandings. But apparently this problem was beyond my powers.

Comrade Mezhlauk asked me to announce that I prefer to do my scientific work in the Soviet Union, but I regret that I am unable to do this at the present moment and I should like to explain why. The main reason is that I have been placed in such conditions as to make me feel very bad and I have explained this to Comrade Mezhlauk several times. I am not speaking about material conditions – I have never been specially concerned with them and in the present circumstances I think very little about them. But I am referring exclusively to moral conditions and the conditions of my scientific work. Naturally in evaluating them I am thinking in terms of comparison with the conditions I had at Cambridge University.

While in Cambridge science develops freely and scientists can freely travel abroad, here in the Soviet Union everything is under the direct eye of the Government. Of course in principle this should be welcomed since this dependence should guarantee that science . . . should be a leading and basic factor in the cultural development of the country. But since science represents the highest level of intellectual work, demanding very particular care and attention, it can be distorted in the hands of dignitaries who appear to condescend when they talk to scientists. Such gracious and lofty condescension to the scientist has offended me many times. One of our dignitaries here made me wait 1 ½ hours in his reception room before seeing me and another who had agreed to see me twice a month has hardly ever kept his promise.* The number of telephone calls to arrange an appointment, standing in line for a pass and walking down long corridors, all these depress me so much – and I am not joking – that I have several times had nightmares about having to go to an appointment. And I am increasingly getting the impression that a conversation with a responsible government figure is beginning to be regarded not as a businesslike discussion useful to the country and the Government, but rather as some kind of reward or honour.

Letters are not only not answered, but not even acknowledged. I fail to see how in such conditions normal scientific work can develop or how a scientist can acquire self-respect. Since my detention everything has been done to make me lose my self-respect. During the first four months nobody took any notice of me and I wasn't even given a ration card for bread. Moreover, apparently merely to frighten me, two agents of the NKVD have followed me in the street for three months and they have sometimes amused themselves by tugging at my coat. Again certain responsible figures have tried to frighten me by a variety of means and continue to do so up to the present . . . I still can't understand what purpose all this serves, since the practical consequence has been to frighten most of the scientists and my friends away from me and as far as I am concerned the only effect has been to upset my nervous system.

Of course after 13 years absence abroad I may be wrongly interpreting some things and there is much I cannot understand, even though I have always tried hard to follow life in the Soviet Union . . . On my side, I have shown some initiative by writing two reports, which I gave to Comrade Mezhlauk – one on pure science in the Soviet Union and the other on our industry as a base for my scientific work. But neither of these reports has been discussed with me and my request to meet the comrades from the Department of Culture and Propaganda for a discussion has been in vain. Thus during these eight months I have been suspended in the air, completely alone and with

---

* Kapitza is referring here first to the Deputy Commissar for Heavy Industry, G.L. Pyatakov and secondly to Mezhlauk.

nothing to do – people have either been frightened away or else have not wished to talk to me. As a result my moral equilibrium has been completely upset and at present I am just no good for any serious creative scientific work . . .

When after four months the question of organizing my scientific work was brought up, I indicated two possibilities: (1) to start in a new scientific field or (2) to continue my Cambridge work, but for this I pointed out it would be quite essential for me to have my equipment, technical drawings and, at least at first, my former assistants. To achieve this I offered to act as an intermediary and to start discussions by telephone in the presence of someone authorized by the Government. This was refused and the matter was handed over to the Ambassador in London, while I was asked to prepare plans for the construction of accomodation for the equipment once it was received from England, if that proved possible.

At first I did indeed receive proper support for this work, but then everything changed. At first I was given very good architects with whom I could collaborate very well, but later they were taken off the work and replaced by almost hopelessly bad ones. It was only because I already had some experience of building work that I was able to design a satisfactory building. And for all these eight months the only serious work which I have had to do has been architectural, which of course is absurd. If the builders assigned to me should prove to be of the same poor quality as the architects there is reason to fear that they will produce a building far worse than my English laboratory. But as I have pointed out several times, the building will be completely superfluous as long as the question of getting my equipment and assistants from Cambridge remains unsettled. In spite of my requests for information about the course of the negotiations, I know nothing of what is going on. Of course I am not a delicate maiden who is used to having everything served up on a platter, and of course the position I hold in the international world of science did not descend on me like a gift from the gods . . . Without exaggeration I think that in the present conditions, my attempt to recreate my scientific activity here is like attempting to go through a stone wall with a penknife.

So of course I cannot say what I don't think. Not only do I not feel better here than in Cambridge, but on the contrary I feel very bad, and the best I can do is to stay silent. I don't want this letter to be wrongly interpreted, though I have noticed that whatever I do or say is taken badly at first. This is especially evident in the lack of a decent attitude towards me. During the 13 years of my stay abroad I have remained steadfastly loyal to the Soviet Union – not because fate determined that I should be born on Soviet territory, for I know languages and I have deeper roots abroad, where I am well known and I am properly appreciated as a scientist, so that spiritually I have long ago become a cosmopolitan. That I have remained loyal to the Soviet Union is solely because I am in complete sympathy with socialist reconstruction directed by the working class and with the politics of broad internationalism of the Soviet Government under the guidance of the Communist Party. I have never made a secret of my views and they are well known abroad.

During all these thirteen years I have unswervingly done everything possible to bring our Soviet scientists closer to those in the West and to support the prestige of our science. I have often travelled to the Soviet Union to give lectures and consultations and, in spite of everything, I have never once refused any work within my powers. And now, notwithstanding the unjust, cruel and offensive treatment I am receiving, I still respect the Soviet Government and my belief in the great work it is doing together with the Communist Party has not been in the least shaken. I believe in the ultimate victory of socialism, and that it is here in the Soviet Union that the foundation for the future of mankind is being laid down. No forces and no errors can now halt the development of socialism in the whole world. If I were a politician or an economist I should be proud to join the ranks of the Communist Party but fate has seen fit to make me a scientist and as such I must strive to establish the conditions in which I can best make use of my scientific gifts . . . Without my equipment, without my books, without my scientific colleagues and with my work rudely interrupted at a very interesting juncture, I feel at present sad, unhappy, shattered and worthless. If you knew how to treat a scientist, you would easily understand how I feel. But I believe strongly in the internationalism of science and that science should be outside any political passions or struggles, however much people try to involve it. And I believe that the scientific work I have been doing all my life is for the benefit of all mankind, wherever it has been done.

The draft of the letter to Rutherford (see p. 266) was not sent to him at the time, since it was *not* approved by Molotov, presumably because Kapitza would not say he was happy with his situation. However, the draft was eventually brought to Rutherford by Anna in November 1935, together with the letter to Molotov that follows.

(I)   To V.M. Molotov   Moscow, 14 May 1935
       Chairman of the Sovnarkom

Comrade Molotov,

Here is the letter to Rutherford and its translation [into Russian]. Tell me what you wish to change in it. My opinion is that it honestly portrays the present position and my "credo". I am not writing about the buying of the laboratory until there is a definite answer, and then of course I shall need to know all the details of the negotiations. I want to tell you once again that Rutherford is quite the most remarkable and unique man and that I respect and love him very much and will never do anything that can hurt him or that is unpleasant for him.

As regards my wife's visit, would it be possible for Comrade Maisky to let her know that it would be quite safe for her to come to the Soviet Union for a fortnight to arrange for the moving of the children? You must not be offended about this, but you know yourself that your ways with people sometimes make them nervous. If I am threatened now and again, I do not mind; it only makes me sad, but I do not get frightened. But there is no need for my wife to live through it all.

I do so much want everything to finish well, but at the moment it is all very bad. I am not clear what it is that you value in me and what you need me for. It is only recently that you learned about me and before that you managed very well without me. You thought that since I was useful to English science, I could be useful to you too. Is this not a mistake? English people drink Epsom salts and the Russians do very well without it. You said to me that you have plenty of Kapitzas among your youth and I am sure that you have got not only Kapitzas but even super Kapitzas. But with your methods you will never "fish" them out of the 160 000 000 population. At present you must ask for help from England through Rutherford.

I will never admit that the attitude shown to me is the right one. It is not that I am offended myself, but I am afraid for the other Kapitzas. This I see very clearly and for the sake of the Soviet Union I cannot pass over it in silence. It would be such a good thing if you could trust me at last and give me my freedom. I should be happy and work again and you would not be cross with me and lose time. And when the laboratory is built, I shall start work on the Vorobyovye Gory* for the glory of the USSR and for the benefit of all people.

In 1936, Max Born, who had left Germany after the Nazis came to power, but had only a temporary position in Cambridge, was looking around for a more permanent post. Kapitza wrote to Mezhlauk pointing out what a good opportunity this was for the Soviet Union to acquire a theoretical physicist of world class and was authorized to approach Born, which he did in the letter that follows. In the event Born was offered and accepted the Chair of Theoretical Physics in Edinburgh just as he and his wife were about to make an exploratory visit to Moscow. Born comments on this episode in his autobiography (Born 1978).

(A)   Moscow, 26 February 1936

My dear Born,

Your letter to Anna arrived here, and I had the indiscretion to read it with great interest. I was very sorry to learn that you still are undecided where to settle in the hemispheres. You are an unlucky man, everybody wants you and the choice is so great that you cannot make up your mind. Maybe I am somewhat luckier than you are, as I have no choice. After the question of the transfer of my apparatus was settled, and now that I have a hope that I shall resume my work in a few months time, work which has been interrupted for nearly two years, and now that Anna and the children are with me, I feel much happier.

---

* The Vorobyovye Gory (Sparrow Hills) were later renamed the Lenin Hills.

77. M. Born.

The conditions of work are far from excellent, but no doubt are improving rapidly and at present all is done to give me the facility for my work. Indeed I have grown up in Cambridge, where I had spent 14 [actually 13] years in that place of old cultural tradition! And you, you worked most of your life in Göttingen, a place which used to be of equal intellectual standing, and of course it is rather hard to judge the present condition here by this standard.

But it is impossible to deny that it is most fascinating to watch the growth of new culture and on new principles, and I do not regret that I shall now be able to take an active part in this game. Indeed I resent and will always resent the piggish way in which it was done to me, but my dear Born, governments in all times of history and in all parts of the world act with no delicacy, and the individual is a single particle, which when put into "the stream of history" will get knocked about. Well all that you can do is to keep your skin hard. After all the Bolshies are angels compared with your Nazis, and what is more they have a true case to fight for . . . I agree with you they are the only ones who keep the right line, and indeed a winning line.

Your letter gave me the idea of playing a nasty trick on you and make your state of mind a little more perturbed than it is at present, and to suggest to you that you could include our 1/6 of the world in a possible choice of a place where you could settle, and I think you can consider this suggestion quite earnestly. Here are a few advantages, besides the one based on the broad historical arguments: (1) you will be able to start here a new school of your own of theoretical physics; (2) you will be welcomed here, and never feel that you are a foreigner or in anybody's way; (3) our theoretical physics is weak, and your leadership will be welcomed.

The bourgeois habits of yours will indeed suffer a little, but the welfare of the country is growing very rapidly and even now I hope you could be satisfied with comforts of life. Maybe you will find the shopping difficult, the roads not even, but you will get some compensation in the way of theatres and concerts which are better than in many places. With books and literature you will be well provided. In general the comfort of your soul will be greater than the comfort of your body!

Think the matter over and let me know. Personally I should be very happy if you came here, and I shall be very glad to have you attached to our Institute to run the theoretical side. There you can get a small five-room cottage, actually on the grounds of the Institute with gas, hot water, electricity, heated garage, standing in a park on the top of a hill overlooking Moscow. You may have pupils of your own, according to your taste, lectures in the University, etc. And we may run together a club like the one in Cambridge. You will take part in the problems on which we work in the laboratory; at present I have not yet invited any theoretical physicist, and I shall leave this matter in your hands if you will come.

Well, think it all over and let me know what you think; if you think there is sense in my suggestion, then I shall see how the matter could be arranged with the authorities, but I have reasons to believe that my idea will be welcomed and supported. But of course till I hear from you I shall not take any steps. And even if you will be tempted by my scheme, I shall advise you to come here during the summer for a month or two, and see it all yourself. I could arrange an invitation, which will cover your expenses and you could give here a few lectures. It is most important that all what I am writing to you should be kept *strictly confidential.* Looking forward to hearing from you and hoping that my plans may tempt you, accept my kind regards and best wishes.

(I) To V.I. Mezhlauk   Moscow, 25 April 1936

Much respected* Valerii Ivanovich,

. . . The agreed building plan for the Institute and the living quarters is not being fulfilled in many respects. The two shifts promised have not materialised. Today for instance there are only five painters in action, work on the floors has not yet started, nothing is happening about replacing broken window panes, etc. etc. It is evident that the planned completion of the laboratory by 1 May and of the living quarters by 25 May will not be achieved. We are ready to start work, most of the equipment has been received [from Cambridge], the workshop is there, the essential people are there and the only hold-up is because of the builders. I don't know what I should do. If it was a question of the internal life of the Institute where I am in charge and responsible for any shortcomings, I am sure I could find some way to restore order and enable work to go on. And, of course, if the Institute should not function properly you would be fully justified in laying the blame on me as the responsible person. Tell me in general how do you picture the mechanism by which builders in the Soviet Union can be made to observe their schedules? In capitalist countries the method is well known: first of all, breach of contract, then a fine, legal

---

* Forms of address such as "Much respected", "Highly esteemed" and the like are literal translations of the Russian and are meant to convey slightly more warmth than the formal "Comrade Mezhlauk", although less intimacy than the literal Russian equivalent of "Dear". This is in contrast to the English use of "Dear" which may imply anything from intimacy to formality.

proceedings and finally bankruptcy. That would be the fate of
Borisenko and Co Ltd [this is an ironical title] . . .

So I ask myself the question of why you, the Government, can't do
anything and I have two possible answers:

(1) Essentially you don't consider my work as a scientist sufficiently
important and necessary for the country, to provide the necessary
tempo of construction to enable me to start work as quickly as possible. But in that case why did you detain me? For I have been cut off
from my scientific work for nearly two years. I should have thought
nothing was simpler than building a laboratory and living quarters for
the scientific staff. I don't suffer from any *folie de grandeur* and
everything is on a small scale – indeed my Institute must be almost
the smallest in the Soviet Union . . . – a mere crumb for the Government! Such an insignificant crumb that Comrades Borisenko and
Venetzyan say that they don't want to take it on. Apparently their at-
titudes and declarations reflect the views of the Party and the Government and this I find very offensive since I have been spoilt by being
used to having my work treated with respect.

(2) The second possibility is no better – this is that you, the
Government are *unable* to make Borisenko & Co Ltd obey you. But
then, what sort of a Government are you, that you can't get a little
two-storey house built on time and put its ten rooms in order after
assembly? In that case you are simply wets! . . . I can't understand
how you seem to approve of all this and pat the builders on the
head . . .

Well, that's the picture of what's happening. Imagine you saw a
violin at your neighbour's and you were able to take it from him. And
what do you do with it instead of playing on it? For two years you use
it to hammer nails into a stone wall. Perhaps you will justify yourself
by saying that you haven't got a hammer to hand and the violin is a
strong one, so it won't break. Well, perhaps it won't break, but its
tone will certainly deteriorate. Of course the violin is me and hammering nails into the stone wall is equivalent to making Borinsenko & Co
Ltd work to schedule. And as for the violin you have taken by force
you can't even play "Chizhik" [a well-known children's song] on it.
So if you really want the violin to play and not fall to pieces, you must
make Borisenko & Co Ltd get the laboratory into shape by 1 May as
they promised and the living quarters by 25 May and not an hour
later. If you can't manage this you either don't want me to work or

you are simply a wet. Don't get offended and don't get cross with me for telling the truth.

(I) To V.I. Mezhlauk   Moscow, 27 April 1936

Comrade Stetskaya told me that you were offended by my letter and that you want me to take back what I said. I am sorry you should react in this way since there was nothing wrong in what I wrote and I hadn't the least wish to offend you . . . You are the Government and I am a scientist, but we are both citizens of the Soviet Union and we both have the same aim in view – to make the country flourish. We both agree that socialism is the most powerful means of realizing this aim. You have more responsible and complex problems to deal with than mine as a scientist, but we must both be equally zealous in dealing with our problems. Do you agree? Therefore I must do all that depends on me for the success of my work. At present the builders with their disgusting inertia stand in the way and they must be made to keep to their schedule. I, personally, have exhausted all possibilities, to such an extent that Borisenko doesn't want to talk to me, but let him go to the the devil, I don't care.

Then I turned to you for help. I wrote to you in March that I'm at a dead end, then I spoke to you on 5 April, I wrote again on the 11th, spoke to you on the 17th, and wrote again on the 22nd, but the construction is at a standstill. On 25 April there were only five painters at work (though it's true there are 30 today). So of cource I lost my temper and wrote that either you didn't want to or you couldn't make the builders work. As for the form of what I wrote, that's a matter of temperament. But, let me describe my psychological condition. On 27 March I was having a discussion with the engineers of the Autogenic Board and this brought home to me the importance for the country of the problem of obtaining oxygen from air and also that this problem is scientifically interesting and worth taking up, since theoretical considerations suggest that the separation could be achieved with a tenfold economy of energy*. I have been working flat out on this problem for the last 12 days – till 4 or 5 in the morning each day, and at last

---

* This was the beginning of Kapitza's important and successful work on this problem, over the next 10 years or so.

I think I can see a possible solution which I am anxious to try out. Of course theoretical solutions are worthless and I may have made a mistake, so it is essential to start experiments as soon as possible. The technical drawings were ready in three days, the mechanics started work after another three days and the turbine should be almost ready to test by the end of the week . . . So if it weren't for these hundred-fold accursed builders I could start experiments by 11 May and find out if I am on the right road to solving the problem. After all, the builders could have easily finished the laboratory a month ago and then there would have been no hold-up . . . Am I not justified in saying that all this business with the builders is like hammering nails into a wall with a violin?

So to sum up, let me say that my only motive in writing to you was, as always, to promote the welfare of the Institute . . . You have helped us a great deal, which I greatly appreciate and I am ready to continue working as before. But if you think I am behaving improperly and that I should think only of being respectful and polite, that I mustn't say what I think or feel in the interests of politeness, then we shall not be able to work together. I shall not be myself and our affairs will go sluggishly. In that case, better let someone else help me – someone less important than yourself, but someone who won't be offended by what I say. You mentioned such a possibility to me yourself . . . In Cambridge, Rutherford also had a considerable temperament and when we couldn't agree on something I often heard such compliments as "silly ass" or "silly fool", but things quickly settled down because we were drawn together by our mutual interest in the scientific work and by mutual respect for each other. I think that my readiness to draw you into our scientific work and to work with you, indicates my relation to you better than anything else. If I didn't respect you I wouldn't behave in this way.

P.S.   I should be very grateful if you could find it possible to discuss such matters with me without using Olga Alekseyevna [Stetskaya] as an intermediary. She takes it all very much to heart and gets upset. She is extremely helpful to me as Assistant Director and I don't want her upset, since this undermines her enthusiasm for the work.

During the summer of 1936 a virulent defamatory campaign was mounted in the Soviet press against the distinguished mathematician Academician N.N. Luzin, and Kapitza (at the time on holiday in the country) wrote a vigorous letter of protest to V.M. Molotov, then Chairman of the Sovnarkom (Council of the People's Commissariat). Molotov returned the letter curtly annotated "Not needed; return to Citizen Kapitza".*

(G, I) To V.M. Molotov   Zhukovka, 6 July 1936

The articles in *Pravda* about Luzin have perplexed, astonished and angered me and as a Soviet scientist I feel I must tell you what I think.

Luzin is accused of many things and I do not know if these accusations are justified, but even assuming they are completely sound, my reaction to the articles is still negative. I shall start with a few of the milder accusations against Luzin. He published his best work outside the Soviet Union. Many of our scientists do this, for two main reasons: (1) printing here is of poor quality as regards both paper and print. (2) by international custom priority is accorded (if at all) only if the work is published in French, German or English. If Luzin has published bad papers in the Soviet Union the fault lies with the editors of the Journal which accepted them. The complaint that he envied his students and that he sometimes treated them unfairly applies, alas, even to some of our most distinguished scientists.

Thus, there remains only one accusation against Luzin, a very serious one – that he concealed his anti-Soviet sentiments by flattery, though no serious crimes of any kind have been quoted. This raises a very important question of principle: how to deal with a scientist if he fails to respond morally to the demands of the age we live in. Newton, who gave mankind the law of gravitation was a religious fanatic. Cardan, who showed how to obtain the roots of a cubic equation and made important discoveries in mechanics was a debauchee and a libertine. What would you do with them if they lived here in the Soviet Union?

Suppose that someone close to you fell ill. Would you send for the

---

* This form of address rather than "Comrade" or "Professor" is intended as a snub, with its implication that Kapitza is not "one of us".

best doctor, even if his moral and political opinions were abhorrent to you? . . . I do not wish to defend Luzin's moral qualities but there is no doubt that he is our most important mathematician . . .

People like Luzin, who differ from us ideologically must first be given conditions allowing them to work in their scientific speciality without having any wide influence on society and, second, everything possible should be done to re-educate them in the spirit of the new age and thus make them into good Soviet citizens.

As regards the first point, the fact that Luzin was not a socialist was known to everyone in the Academy, and there are quite a few like him there. It wasn't an unexpected discovery by the Director of the 16th school, based merely on Luzin's flattering comments.* Nevertheless, he has been chosen to carry out a whole series of socially useful tasks, he has been invited to act as a referee, he has been entrusted with the leadership of the mathematical section of the Academy.

On the second point, has everything possible been done to re-educate Luzin and people like him in the Academy and can this be achieved by methods like the publication of the *Pravda* article? I am convinced that it cannot, indeed just the contrary – such methods make the re-education not only of Luzin, but that of many other scientists more difficult. How have you gone about reorganizing the Academy? First you began by electing Party members to the Academy and this would be the best method if there were distinguished scientists in the Party. However, leaving the social sciences on one side, the party members in the Academy are far less able than the old Academicians, so they have little authority.

We have not so far succeeded in producing new scientists from among our young people. I explain this by a very wrong attitude on your part towards science – much too narrowly utilitarian and insufficiently supportive. Therefore our main scientific capital is still the older generation which we have inherited. It would seem to follow that we should do all we can to re-educate them, to tame them and so on, but what you are doing totally fails to achieve this. Lazarev was arrested, Speransky was dismissed from the Academy and now Luzin is under attack. It's no wonder that this kind of "tender" treatment

---

* This refers to Luzin's visit to the school, following which he was criticized by the Director in a letter to *Pravda* for having given too uncritical and flattering an assessment of the school.

makes scientists like Uspensky, Chichibabin, Ipatiev and others run away. I know from my own experience how callously you can treat people . . .

What have you done to re-educate Luzin? Nothing. And what does this article in *Pravda* achieve? Either he will begin to dispense even more flattery or else he will have a nervous breakdown and cease his scientific work. You will only frighten him, nothing more. Dangerous enemies have to be frightened. But do you think Luzin is a danger to the Soviet Union? The new constitution is better than any other and it shows that the Soviet Union is quite strong enough not to be afraid of such as Luzin. And so, with all the agricultural achievements and political successes that the Soviet Union can claim, I do not understand why it should not be possible to re-educate any Academician, whoever he might be, provided he is given careful and individual attention. Think of Pavlov for instance, and there are not so many scientists of distinction among us to make this method impracticable.

From all these considerations I cannot understand the tactical idea behind the *Pravda* article and I see in it only a damaging step for our scientific work and for the Academy, since it neither re-educates our scientists nor increases their prestige in the country.

And I may add that Luzin's name is sufficiently well known in the West that an article like the one in question will not pass unnoticed. Thanks to its weak and unconvincing tone it may provoke the most varied and absurd comments.

In view of the damage to science in the Soviet Union that all this has caused I feel it to be my duty to write to you about it.

(I) To V.I. Mezhlauk
      Strictly personal   Moscow, 25 December 1936

Much respected Valerii Ivanovich,

I want to give you something like a report for the year, since looking back we can better judge where we should go next. After all, there has been an element of collaboration between us and a striving for the same objectives, and no one in the Soviet Union knows my work more intimately than yourself.

First of all, the Institute and its living quarters are complete and we had to devote a great deal of energy to achieve quality and speed – I think everything possible was done in this respect. In spite of everything, the builders were made to rise above their usual level. This was very difficult and without your constant support we should of course have been helpless. I consider that the laboratory has been built satisfactorily, the living quarters a bit better, and my own house well, although the reverse order of merit would have been preferable. But even so, it is good that the builders have shown that they have learnt something and are not absolutely hopeless.

The wiring and plumbing has gone very smoothly. The delivery of the English equipment has all gone well. Both the English and Soviet equipment is 90% installed and operating (all except the helium liquefier which was held up by a misunderstanding about tubes). On the whole . . . I should like to hope that in time you and I will come to be proud of this Institute and our efforts will not prove to have been wasted.

Now as regards people. The most successful has been the replacement of Olbert by O.A. [Olga Alekseyevna Stetskaya] and I particularly appreciate your support in arranging this. I thought that O.A. would be a good assistant, but she has turned out far better than I could have anticipated. She is an exceptionally good comrade and a first-class worker. It is only because of her persistence and endurance with respect to the builders that the Institute has acquired a reasonable shape. The electrical installation was entirely her responsibility and brilliantly carried out. She understands people well, she has raised discipline and manages the administration successfully. She is now taking part in the turbine work and is quite at home with scientific and technical problems. I think that I shall gradually be able to hand over all the technical aspects of the Institute's work to her. It is only thanks to her help that I am able to find time to work peacefully on my own problems. I should be very pleased and it would be entirely justified if you could on some suitable occasion give some kind of official recognition of her work.

As you know we are only gradually taking on scientific staff. At present there are four people and it is not intended that there should be many more. Scientific work started only in October so they have not yet had much chance to show what they can do. I am surprised at the lack of patience and inner discipline they have shown so far, but they do show enthusiasm which is a good thing. . . . But my chief concern has been finding suitable technical staff – five electricians, eight mechanics, two carpenters and one glassblower. They will be the "human" foundation of the laboratory. The scientific workers will, and indeed should, pass through the lab but the technical staff should be permanent. They have to be taught how to handle equipment and how to make scientific apparatus with interest and understanding. On the whole I am pleased with them and the main thing is that they are enthusiastic. But they lack discipline, they waste a lot of time in discussion and are unable to take quick decisions and go straight to their objective by the shortest path. To organize and instruct so many people at the same time is not easy and takes a lot of time, for they have been taken from factories and elsewhere and are completely raw. Once the laboratory is growing normally and gradually, the education of the technical staff will look after itself, but here we have sixteen people all starting at the same time. However, as I had hoped, the help and example of my English assistants Laurmann and Pearson has proved very important. They have already been working for half a year and I hope it will be possible to prolong their stay here for at least another half year. They get on very well with the other staff . . .

I won't write in detail about our scientific work until next New Year – if something has been achieved by then. For the time being the most hopeful projects are the ones you have seen – the little turbine [for liquefying air], the new helium liquefier and the Zeeman effect [using the high magnetic fields]. But please bear in mind that none of these are finished yet and any of them may still run into a blind alley, either provided by Nature or by the limitations of our mental abilities, and so come to nothing.

As regards the administrative and economic life of the Institute, O.A. and I have been struggling to simplify the administration and reduce the number of staff. I am laying great hopes on the introduction of a new system of financial control and book-keeping: It has taken seven months for this to be born and it is only thanks to your support that it has been born at all, in the face of every effort by the People's Commissariat of Finance to produce an abortion. It gives me no pleasure to recall the considerable time I have had to spend in getting to the bottom of our financial procedures. I found it fiendishly boring and indeed it reminded me of my schooldays when I had to learn the catechism – another boring and formal subject of no scientific value. But without this effort the birth would not have happened . . .

But our system of supplies for science is scandalously bad. I first wrote to you about this on 15 November 1935 and since then I have written letters and memos to you, to Comrade Bauman and a particularly sharp letter to Pyatakov, all over two months or so and with absolutely no result. So far you have only promised, but essentially nothing has been done. This is very unsatisfactory and I feel rather like Don Quixote battling with windmills. For me as a scientist, the most important question at present is bringing order into the scientific economy. Apparatus and materials are just as necessary for a scientist as is a good instrument to play on for a musician. At present our scientists can be compared to a virtuoso pianist who is asked to play a broken down untuned piano with several strings missing. I must repeat that I am able to develop my scientific work, only thanks to Rutherford who sends everything I need. You cannot reproach me that I haven't achieved any significant improvements in the matter of scientific supplies, and it is you who should blush that we are still standing where we were 14 months ago . . .

So 1937 is almost upon us. Will you continue to help me as before? Without your help we shall get nowhere. And this too cannot be regarded as normal, since I know very well that your personal work for the whole country is exceptionally important and responsible and yet I have to bother you with all sorts of trifles due to the lack of organization of Soviet life. But if I do this, it is only because I imagine that you will be able to use the experience gained in our Institute to bring about a healthier organization of our scientific life. If in the course of a few years our science is indeed based on healthier foundations, and I can feel that I have contributed my mite, then all that I have lived through will have been worthwhile. This hope is the basic source of my energy. In the meantime I think I shall have sufficient strength and spiritual courage to see me through 1937 and I hope you will have sufficient patience. My greetings for the New Year.

In 1937 the mass arrests of the great purges were beginning to get into swing and Kapitza wrote to Mezhlauk, then Deputy Chairman of the Sovnarkom, and also to Stalin, to plead for the release of V.A. Fok, the distinguished theoretical physicist who had been unjustifiably arrested. Fok was in fact released quite soon afterwards, following a harangue from an official of the NKVD, who he later discovered was Ezhov, the notorious head of the NKVD.

(G, I) To V.I. Mezhlauk   Leningrad, 12 February 1937

While here in Leningrad, I was greatly disturbed by the news that the physicist V.A. Fok was arrested yesterday. I regard him as our most capable theoretical physicist; his work on approximate methods of integrating the wave equations of modern electrodynamics is considered classic. These methods are known everywhere today and are taught in courses for students. He is still young (38) and is quite cut off from ordinary life by almost total deafness. His whole life is a persistent struggle with scientific problems. I cannot conceive that such a person could commit a serious crime. There must have been some mistake.

It is said that he is indicted in connection with some serious sabotage which occurred as a result of careless evaluation of electrical exploration techniques in geological work. Fok provided certain theoretical formulae and it is, of course, possible that others used them improperly but it seems to me totally improbable that Fok deliberately gave an incorrect theory. Not because this could very easily be found out but because Fok is far too good a scientist to do such a thing. Don't you see, he is like a great musician who cannot strike a false note because it would grate on his own ears first of all. These considerations compel me to conclude with 99% certainty that the arrest of Fok is a mistake. And if this is so there will be a whole lot of very dismal consequences for Soviet science. It will distance our Soviet scientific circles still further from building socialism and may, moreover, undermine Fok's ability to work and so provoke a bad reaction from scientists here and in the West.

It is said that, besides Fok, very many other theoreticians were arrested a few months ago in connection with this same affair. In fact, so many were arrested that in the university faculty of mathematics and physics no one could be found to lecture to students. I hardly know any of these people and not one of them is of the same calibre as Fok so I cannot feel as sure about them as I do about him. It is to be hoped that the investigations by the NKVD will show that the majority of them were not implicated in any criminal activities and in that case all those who were wrongly suspected will harbour a strong resentment. This will prevent our scientists from being, as you have put it, "won over". And what if they are proved guilty? This would be still worse, for there are so many of them that they must be called "enemies" rather than just "criminals". Most of them are still young and this implies that after twenty years the Soviet authorities have not yet understood how to win scientists over to their side, not even understood how to persuade them to be neutral but actually turned them into enemies.

I am afraid I am becoming quite fanatical in support of the ideas about which I have spoken and written so much to you. In order to "win over" scientists you have to provide them with conditions which they clearly see to be better here in our Soviet Union than in capitalist countries . . .

So why not bravely and energetically set about providing the right conditions? Surely this is a much simpler problem than those the Bolsheviks have already succeeded in solving. Only three things are necessary. First, eliminate the "rubbish" from scientific circles (improve the key personnel), second, create better scientific management and, third, create a healthy scientific community.

I am very greatly upset by Fok's arrest and torn apart by the fear that this is a crude and insufficiently considered action. It could cause great harm to our science. I am so worried that I have also written, very briefly of course, to Comrade Stalin about Fok. Otherwise I would feel that I had not done all I could to put right what seems to me a great mistake. You can be as angry as you like with me but I couldn't do anything else . . . Naturally, you take a wider view of the question and moreover you have all the facts available and greater experience. But still, it seems to me you should not be indifferent to what a scientist thinks about such questions.

(A, G, I) To J.V. Stalin   Leningrad, 12 February 1937

Comrade Stalin,

I heard yesterday of the arrest of Professor V.A. Fok, a corresponding member of the Academy of Sciences and considered both in the West and here to be an exceptionally strong scientist . . . In my opinion he is the most outstanding theoretical physicist in the Soviet Union . . . and I am extremely disturbed by his arrest . . .

Some years ago there were several cases of scientists being arrested, admittedly only for a few months, where it turned out later that they had done nothing wrong. If this should be the case with Fok it would be extremely sad since:

(1) It would enlarge still further the gap between our scholars and the country – a gap which it would be so desirable to eliminate.

(2) Fok's arrest is crude treatment of a scientist, which just like rough treatment of a machine, is bound to damage performance. And spoiling Fok's ability to work would seriously damage world science.

(3) Such treatment of Fok is bound to provoke indignation both here and in the West, like that provoked when Einstein was driven out of Germany.

(4) We have not got many scientists of Fok's calibre and Soviet science can be proud of him before the whole world – but this becomes difficult when he is thrown into prison.

I think no one but a fellow scientist can tell you about this and that's why I have written this letter.

(I) To V.I. Mezhlauk   Moscow, 22 February 1937

Today we have made liquid helium, so the laboratory can be regarded as finished. We are all very happy, I in particular, since I can take up my research again. Greetings!

(I) To J.V. Stalin   Moscow, 10 July 1937

The condition of science here is unsatisfactory. All the usual public statements that our science is better than anywhere else in the world are just untrue. Such statements are bad not only because any lie is bad, but are even worse because they hamper the process of improving scientific life in our country. That the situation of science is bad, I can assert with some certainty, since I worked for a long time in England and both living and working there were much better than here. However, the purpose of this letter is not to praise the English or to tell you how good it was there, but to explain the basic causes for our weak position and how we should struggle to raise the level of science in the Soviet Union.

The most astonishing thing about the state of our science is of course that it is weaker than in the capitalist countries and its growth and development does not correspond at all to the tempo of our economic and cultural life. However, in spite of everything, I still believe that under socialism science should be at a higher level than anywhere under capitalism. It is impossible that it should be otherwise, since science is the basic motive power and index of progress. So what are the most prominent defects? The economic basis of science is bad; there is hardly any scientific industry; the economic organization of our scientific institutes is muddled and absurd; the multiplicity of jobs held by each scientist breaks up our scientific forces and makes poor use of them and there is absence of unity and integrity in scientific organization, etc., etc. So who is responsible for this situation and how can it be improved? There must of course be some basic cause which has to be discovered and eliminated.

The first and most natural explanation is that it is because of the absence of attention paid to science on the part of leading comrades in the Government. This is the opinion of many scientists but I do not share it. It's true that I often disagree with a whole lot of measures, especially those concerning individual scientists, and in spite of everything I then speak out. However, basically I think that the Party leaders and comrades in the Government have a sincere desire to put our scientific life in order and a genuine recognition of the significance of science.

The second explanation is that our scientists are no good and lack talent. This too is not so. Comparing our scientists with those abroad I consider that on the average ours are no worse. But what astonishes me most of all is that morale in our scientific community is bad and that there is no enthusiasm for work . . . Our scientists are not united and are separated from the life of the country. But worse than that, they are not only not ashamed of their isolation, but assume a priestly attitude, regarding themselves as something superior and independent. The general mood is unhealthy – "everything should be done for us and we should work only at our science". Naturally in these conditions the position of our leading comrades [in Government] becomes somewhat difficult. How should they provide for such scientists? However clever the comrades in the Government may be, they cannot understand scientific work in detail nor can they decide which scientific researches should be supported more strongly than others and it is absurd to expect them to do so.

For instance in the organization of scientific work it is important to assess scientific workers and in normal conditions this is of course part of the business of a healthy scientific community – to point out who are the serious scientists and expose the charlatans. But this communal side of our work is unpleasant and since our scientists have no enthusiasm for raising the level of Soviet science, each of them selfishly tries to avoid such work and bothers only about his own work and life . . . In order to get recognition he looks for evaluation of his work not from his own comrades in the Soviet Union, but abroad, where there is indeed a real scientific community. And our scientists cannot be blamed for this since they have no other way. But this absurd and intolerable situation is only one more proof of the weakness of scientific life in the Soviet Union. It is in the absence of enthusiasm of our scientists for raising the level of Soviet science that we must seek the reason for the present sad state of science. If our scientists were to get together in a friendly way and try to achieve a healthier organization of science, things would change rapidly. If our scientists knew how to explain clearly what was essential for their work, I am sure their requests would be met better than in a capitalist country.

So why don't our scientists have the enthusiasm which there is for instance among the English scientists? Our scientists are cut off from contemporary life and that is the root of the evil. How to change this situation? The view I want to develop here may seem paradoxical to the majority of our scientists but I am more and more convinced that it is the only correct one. Scientific work is creative and modern scientific work is moreover collective. Every artist if he is to create with enthusiasm must feel that his work is recognized and understood.

Let us consider for instance the theatre in the Soviet Union which, without doubt, is the best in the world. Who created it and why has it become the best? I think the reason is the innate love of our people for theatrical art. By appreciating good actors and directors they raise the level of the art and give it the enthusiasm of creative work. Our theatre is, of course, created by the audience rather than by the actors and this is the case with any creative work. Its level is established not by the theatrical workers but by those for whom they work. The main difference between our system and capitalism is that with us the audience is the whole mass of the people rather than the select class of wealthy people who are the judges of art under capitalism. That is why our theatre is developing so successfully – because it is an example of art for the wide masses.

In other areas of artistic creativity where the masses have not yet been drawn in, the situation with us is worse. Consider for instance painting – we are evidently weak in this respect. We cannot have wealthy patrons and meanwhile other means have not been discovered to make a connection between the artist and the masses, such that the artist could feel his work is appreciated. That is why pictorial art is so weak with us. The situation is the same for any area of creative work – it cannot develop unless it is consciously accepted by the masses.

The situation is just the same for scientific creativity. The masses are far removed from it and the scientists work on their own, the most important of them being mainly interested in the recognition they get in the West. They are not proud of their Soviet science because they don't appreciate for whom it is needed. And until at least the most cultured of the working and peasant classes recognize and welcome every achievement of our science, our scientists will remain an isolated little group, which may provide fertile soil for every kind of wrecking activity, and who will desert the Soviet Union when a convenient opportunity offers itself. What should be done? There are two possibilities – the first is to let things take their course and wait until the industrial hunger of the country has been satisfied and culture can rise so that interest in science and scientific creativity will appear.

The alternative solution – and in my opinion the only sound one – is to begin now to instil the masses with an interest for science. But in order to arouse this interest an energetic propaganda campaign must be initiated. I think that if an appreciable part of the money given to science were to be spent on such propaganda, it would be repaid in ten or fifteen years by a rise in the whole level of science in the Soviet Union. I'm afraid my scientist colleagues do not share my opinion and that is why I am writing to you as leader of the Party, since it seems to me that you will better appreciate the correctness of my views. Of course there is a certain amount of propaganda for science, but it is very little and very weak, even by comparison with the West and I propose that it should become the centre of attention.

The problem is I think clear – we must educate the masses to have an interest in science and to appreciate its significance for progress. I don't think this should be difficult, since there is a large natural interest in science among the masses, which is no less than their interest in the theatre. This can be illustrated by many examples. How eagerly people listen to popular lectures, read popular articles, visit scientific exhibitions, etc. But this interest is far from being satisfied and we are not only far behind European countries in this respect but, even worse, the significance of this kind of propaganda is not sufficiently appreciated. A great deal of attention is given to scientific propaganda in the capitalist countries. It is particularly well developed in England . . . and to a certain extent this may explain the exceptionally high standard of science there.

In England special societies were already established 100 years ago for the popularisation of science, such as the Royal Institution and the British Association for the Advancement of Science. Its museums, such as the British Museum and the Science Museum in South Kensington, are the best in the world, and more space is devoted to science and scientific life in its Press than in any other country. This policy is due to a number of circumstances but the most important reason is that English science is and always has been supported by individual benefactors. In this way very large sums of money have been collected and evidently this is only possible if there is a broad interest in science in the country. With us of course the motive for scientific propaganda is different . . . but it is important for us to study critically how it is done in capitalist countries.

So what can be done here and what is being done? Let me list the main points:

(1) *Scientific museums*. Museums are the most powerful and graphic means of educating scientific interest and understanding in the masses. It is sufficient to point out that the single colossal Science Museum in New York has up to 14 million visitors a year, while just the technical department of the South Kensington Science Museum has two million visitors and so on . . . Here there are no museums of

comparable size. Our Academy of Sciences proposes to set up a number of such museums but only in 10 years' time. I am a member of its Museums Committee, but it is a very feeble committee. It does not accept the principle of museums being a means of mass education – I had a big disagreement with them on this point and was in a minority of one. At the general meeting of the Academy on 29 June a number of Academicians proposed that we should have nothing to do with a museum on the history of science and technology. I consider this attitude to museums incorrect and that on the contrary we must start setting them up in the Soviet Union energetically and without delay – we cannot afford to wait ten years.

I therefore proposed that the Academy of Sciences should straightaway organize exhibitions illustrating various scientific problems and in this way gain experience and exhibits, so that as soon as a museum building was complete its contents would be immediately available. I suggested that we should start with aviation since there is exceptional interest in it in the country. It would be a good idea if a building like the Manezh* were made available and the exhibition would show how science has helped aviation and how aviation has helped science. Later we could go on to transport, metallurgy and so on. At first this was approved by the Praesidium of the Academy and a provisional plan for the project was drawn up, but later the Praesidium decided that it could not set aside the 1½ million rubles needed and everything came to a standstill and there seems no prospect of any movement. However I consider that we should immediately devote funds and effort to this project.

(2) *Cinema*. This is the next most important means of scientific propaganda. I am a member of the committee of our cinema organization concerned with scientific films. Of its fifteen members only two attend its meetings. We have few scientific films, mostly for teaching, and usually of rather mediocre quality – they do not reach the screens of the cinema for the public. Although in England almost every cinema shows short films of a scientific, technological or ethnographic character in addition to the main drama features, we have nothing at all of this kind. Moreover in London there is a special cinema which

---

* This is a large building adjacent to the Kremlin. Manezh is the Russian equivalent of the French manège, a riding school. This building, erected in 1817, is, in fact, often used for large exhibitions.

shows only scientific films and again we have nothing of this kind. This powerful means of scientific propaganda is not exploited at all with us and here too we must move away from our "dead point".

(3) *Popular literature and lectures.* We are weak in this respect too and our popularisation of science is rather of the pot-boiler variety. The English example is very instructive here. I have already mentioned the Royal Institution whose object is to organise popular lectures and courses both for adults and children. This is a large organisation and has its own large house in the centre of London. It is considered a great honour to be invited to lecture there and such lectures are better paid than anywhere else . . . The activity of the British Association is somewhat different – it organizes daily scientific meetings of a general kind during the holidays, which attract a wide audience from all layers of society. We have no organizations with similar aims.

(4) *Scientific journalism.* Our newspapers publish scientific material only incidentally and it is often completely garbled. Neither *Pravda* nor *Izvestiya* have a scientifically literate journalist who could put together an interview on a scientific theme independently. When I ask why this is so, I am told it is because there wouldn't be enough work for such a person. But I have met very good journalists of the leading English and American newspapers with whom it is both pleasant and interesting to talk, since they are well informed on scientific topics. However, our own newspapers are very poorly and inadequately informed about scientific life both here and abroad . . .

(5) No broad general scientific interest is inculcated at an early age into our children at school. I was amazed to discover this when I talked to teachers and lectured to children and I am clear that we are badly behind in this respect.

From this short review it is evident that we are very far behind in the matter of instilling scientific interest in the masses. There is a great deal to be done, and it should not be difficult if sufficient attention is devoted to the question. It seems evident that an interest in science by the masses is needed not only to provide the right atmosphere for scientific work but because of its colossal importance in other ways. It would give the possibility of selecting scientists from a much wider public, it would raise the awareness of workers and peasants in their work, it would encourage inventiveness and so on. I am sure that if it were possible to raise the interest of the masses in science, it would arouse enthusiasm among the scientists for their own work. They would become a part of the country and would be proud of their Soviet science. They would take a hand themselves in the organization of science and get over our organizational disease. The Government in its direction of science would have to pull on the reins, as with a good racehorse, rather than having to lay on the whip as at present . . . So we must take up this work energetically to accelerate the process or else we shall not have a healthy science, and this would mean that we could not go forward independently in technology, agriculture, etc. Without a properly developed science we shall be restricted to a purely imitative role in our technical development for far longer than is necessary.

(I) To V.I. Mezhlauk   Moscow, 19 November 1937

There are a number of journals such as *Nature, La Science et la Vie* and others, which publish all that is new in science and the scientific world. These journals are very important for us since they publish all the latest discoveries briefly and rapidly, so that they appear well before the full accounts in the large scientific journals. They also report on all conferences, meetings and discussions. For the last two months these journals have ceased to reach us. I made enquiries and was told that they were held up by the censorship. But last week when I was in Leningrad I found that the censor there was a more reasonable person and all the journals had got through. That the journals should be held up at all is in itself a shame but it's even worse that it is done in such an absurd way. I read through the relevant issues in Leningrad to discover why the censor had held them up, but I didn't find it an easy task.

For instance in one issue it mentions that there was a total absence of Soviet scientific delegates at a congress in Paris organized on the occasion of the World Fair. What is there in that to forbid the journal coming through? We all know that we are like schoolgirls who are kept in secluded boarding schools so that they cannot be deprived of their virtue or abducted. In another issue there was an account of the discussions between [N.I.] Vavilov and Lysenko, based on shorthand records published in the Soviet Union. I enclose a translation of their account and as you can see there is some unfriendly comment, but what is remarkable about that? And why hide it from Soviet scientists? . . .

I should like to say a few words about Lysenko's dispute with Vavilov. In my view every dispute and discussion in science, however far apart the opinions of the two sides may be, provides an extremely useful stimulus. For after all what is science but a generalization of experimental results? The experimental material, like Nature itself, is immutable for all time, provided it is correct, but the theoretical generalizations change gradually as new experimental material becomes available. Thus any theoretical edifice will eventually change, however sturdily it is built. Usually scientific disputes are about theories, but what is important is that a dispute should be based on experimental material. Only then is the dispute productive.

Our discussions, however, have started to use methods which are not only absurd, but harmful. This is shown not only in the disputes of the geneticists but also in physics – as in the scientifically illiterate article by A.A. Maksimov*, and in history as, for instance, in a criticism published in *Pravda*, of Tarlé's book on Napoleon. Schematically, the argument goes as follows. If you are not a Darwinist in biology, a Materialist in physics, or a Marxist in history, then you are an "enemy of the people". This argument sticks in the throats of 99% of our scientists. Of course such methods of dispute are not only harmful for science but compromise such strong theoretical edifices as Darwinism, Materialism and Marxism. What should be said to the scientists in dispute is that they should base their arguments on the force of their scientific knowledge rather than the forces of Comrade Ezhov [the then head of the NKVD, the secret police] . . . Greetings.

As recounted earlier (p. 67), the outstanding theoretical physicist L.D. Landau (see fig. 35) was arrested in April 1938 and Kapitza wrote immediately to Stalin (see next letter) and nearly a year later to Molotov pleading for his release. Landau was released soon after the second letter, subject to the guarantee given by Kapitza in a brief formal letter to L.P. Beria, then head of the NKVD (the secret police).

---

* This was a polemical article in No 7 *Under the Banner of Marxism* on "The philosophical views of Academician V.F. Mitkevich".

(H, I) To J.V. Stalin   Moscow, 28 April 1938

This morning L.D. Landau, a scientist in this Institute, was arrested. Though he is only 29 years old, he and Fok are the most prominent theoretical physicists in the Soviet Union. His papers on magnetism and quantum theory are frequently cited in the scientific literature, both here and abroad. Only last year Landau published a remarkable paper in which he identified for the first time a new source of stellar energy. This paper gives a possible explanation of why the energy of the sun and other stars has not yet been exhausted. Bohr and other leading scientists predict a great future for these ideas of Landau.

There is no doubt that the loss to our Institute, to Soviet and to world science of Landau's scientific activity will not go unnoticed and will cause deep concern. Of course knowledge and talent, however remarkable, do not confer any right to break the law and if Landau is guilty he must pay the price. But in view of his exceptional gifts, I beg you to give appropriate orders that his case should be very carefully considered. It seems to me that Landau's character which, frankly speaking, is bad, should also be taken into account. He is quarrelsome and enjoys looking for mistakes in others. If he finds a mistake he delights in irreverent teasing, especially if the mistake has been made by a pompous greybeard, such as one of our Academicians. This has made him many enemies. It has not been easy to get on with him in the Institute, though lately he has taken some notice of our complaints and improved a little. I forgave him his pranks because of his extraordinary talents. But with all his shortcomings of character I find it difficult to believe that Landau could ever do anything dishonest. Landau is young and he can still do a great deal of science. No one can judge all this better than another scientist, and that is why I am writing to you.

(H, I) To V.M. Molotov   Moscow, 6 April 1939

In my recent studies on liquid helium close to the absolute zero, I have succeeded in discovering a number of new phenomena which promise to shed light on one of the most puzzling areas of contemporary physics. I am planning to publish part of this work in the course of the next few months, but to do this I need theoretical help. In the Soviet

Union it is Landau who has the most perfect command of the theoretical field I need, but unfortunately he has been in custody for a whole year.

All this time I have been hoping that he would be released because, frankly speaking, I am unable to believe that he is a state criminal. My disbelief is because such a brilliant and talented young scientist as Landau who, though only 30 years old, has won a European reputation and is moreover very ambitious and has been completely occupied by his scientific work, could hardly have had the motive or found the time and energy for any other kind of activity. It is true that he has a very sharp tongue, the misuse of which together with his intelligence has won him many enemies who are only too glad to do him a bad turn. But for all his bad character, which I myself have had to cope with, I have never noticed any sign of dishonest behaviour.

Of course I realize I may be meddling in something which is none of my business, since it lies within the competence of the NKVD, but I must point out the following abnormal facts:

(1) Landau has been in prison for a year but the investigation is still incomplete. This is an abnormally lengthy period of investigation.

(2) Although I am Director of the Institute where he had been working, I have been told nothing of the crimes of which he is accused.

(3) The main point is that for unknown reasons, science both in the Soviet Union and worldwide, has been deprived of Landau's brains for a whole year.

(4) Landau is in poor health and it will be a great shame for the Soviet people if he is allowed to perish for nothing.

I therefore make the following requests:

(1) Is it not possible to draw the attention of the NKVD to the special desirability of accelerating Landau's case?

(2) If this is not possible, perhaps Landau's brains could be used for scientific research while he is in the Butyrki prison? I have heard that this procedure has been followed in the case of engineers.

(H, I)   To L.P. Beria
         People's Commissar for Internal Affairs
         [Head of the NKVD]   Moscow, 26 April 1939

I hereby request the release from prison under my personal guarantee of the arrested Professor of Physics, Lev Davidovich Landau. I guarantee to the NKVD that Landau will not engage in any kind of counter-revolutionary activities against the Soviet government in my Institute and I will also take all necessary measures within my power to ensure that he should not engage in any such activity outside the Institute either. In case I should notice any remark of Landau which could be harmful to the Soviet state I shall immediately inform the organs of the NKVD.

P. Kapitza

(I) To J.V. Stalin   Moscow, 14 June 1940

A very talented young scientist from our Institute, A.B. Migdal, was selected by an expert committee as the leading candidate in physics for a doctoral Stalin Studentship. But at the last minute the overall committee withdrew his candidature for the following reason. Migdal had good references from the relevant Party and social organizations, but it seems that six or seven years ago he was arrested by mistake and was in prison for two months. He is not being blamed for this, but what is being held against him is that the organization proposing his candidature did not know about this arrest. Migdal told me, "After all, the arrest was not my mistake so why should I have to tell about it?". But the committee (Schmidt, Kaftanov and others) say, "This was a deliberate and malicious withholding of information and even if Migdal is a genius his name cannot appear at the head of the list of Stalin Studentships" . . .

I feel that Migdal has not been fairly treated and I cannot be indifferent. The explanation given to me by Comrade Schmidt did not satisfy me and so I am writing to ask you if you can find it possible to instruct the relevant Party organizations to go into the matter, and most important of all to act in such a way as to avoid damaging Migdal as a human being by embittering him, for in any case everyone would know that if he is excluded it is not because of lack of talent.

Stalin's secretary Poskrebyshev rang Kapitza on 20 June to say that the chairman of the committee had received an appropriate instruction from Stalin and the chairman soon discovered that "there had been a misunderstanding". Academician Migdal recalls that Kapitza told him the outcome the same day, adding, "Well, now you really have got a Stalin Studentship".

(I) To V.M. Molotov
Personal   Moscow, 10 November 1940

I have just had a telegram from Cambridge [from Dirac] to say that Professor P. Langevin [see fig. 79] is in prison in Paris.* Langevin is a great scientist and a great friend of the Soviet Union. I love him dearly and he is a very decent person. I should like our friends to know that we value their friendship and I hope something can be done for him . . .

(I) To J.V. Stalin   Moscow,
    Chairman of the State Defence Committee   19 April 1943

On 28 February I reported to the Council of People's Commissars that the liquid oxygen installation has gone into experimental exploita-tion . . . The question of introducing such equipment into industry has been under discussion for two months, but I consider everything is proceeding in an unhealthy manner. At present, in wartime, we can afford to take up such big projects as a new method of producing li-quid and gaseous oxygen on a large scale only if they are really essen-tial. If it has been decided that a project is really essential, then it should be taken up seriously and at full tempo, however difficult this

---

* Langevin was arrested by the Germans on 30 October 1940 and was in the Santé prison for 38 days. He was then sent to Troyes and eventually escaped to Switzerland on 8 May 1944. In January 1941, through the Commissariat for Foreign Affairs, Kapitza was able to send Langevin a warm invitation on behalf of the Academy of Science to come and work in the Soviet Union, but Langevin was unable to come. See Kapitza's post-war letter to him of 23 January 1945 (p. 363).

may be. It can't be tackled half-heartedly. So what is happening in practice?

For the last five or six months I have abandoned my Institute [then evacuated to Kazan] and my scientific work and sat here in Moscow helping the Autogenic Board which was entrusted with the construction of this machine. A healthy technical organization could have done this work on their own without any help from me. But often an innovative project fails with us because our industry is as yet incapable of mastering new ideas and principles. Like a spoilt child it has to keep chewing what it puts in its mouth and is reluctant to swallow it. It is this chewing process that has occupied us for the last five or six months. We have had to educate them, teach them precision, we have had to make all the difficult joints ourselves and even the working drawings for the installation had to be made in our Institute. But if now we asked them to repeat the whole installation they would still not be sufficiently prepared to do it on their own. We are making five tons of liquid oxygen a day, which is more than industry can make use of. Often when we are testing how reliably the machine runs over a period, we have to pour tons of liquid oxygen into the street. It's partly that they didn't believe the machine would actually work and partly their general inability to do anything on time.

I am completely upset since I see all these ugly things but have no power to make the Autogenic Board work properly, and for the time being I can exert only a moral influence. Yet there are some good workers among them and it's only because of them that we have managed to get a successful result. Moreover, there are now almost no sceptics but on the contrary there are many enthusiasts. The chief shortcomings lie in the organization and its managers. We can't go on like this. The absurdity of the situation is not the fact that I, a scientist, have to bother about questions of supply, elementary organization and so on – after all in wartime it's not possible to turn down any kind of work. But it is absurd to take up the oxygen problem only by half.

Now that I, and also the members of Kaftanov's commission, consider the problem of producing liquid oxygen on a large scale solved and ready to be introduced into industry, I have to take up the problem of producing gaseous oxygen on a large scale by our methods. In normal conditions, this last step should not be difficult and there is every reason to suppose it will be successfully achieved. This would then open up new perspectives for improving metallurgy, gasification of coal, production of aluminium, carbides and in general for over-coming present difficulties caused by insufficient cheapness of oxygen. All this can of course be done and I will happily undertake it – but it can only be done if we don't do it by halves but at full tempo. But for the last three months I have been unable to obtain the construc-tional materials, transport, lathe repairs, etc., essential for bringing my Institute back from evacuation [in Kazan]. The Institute [in Moscow] has been occupied by troops and a lot of damage has been done.

So I put the question to you: is it worth taking on all these oxygen projects in wartime, bearing in mind that it will mean a big effort and some detriment to one or another branch of industry? Moreover there is some risk involved, as in any innovation. In order that you should be able to assess this risk I asked for an objective commission to be set up . . . If it is decided that the project is worthwhile, then we must be supported boldly and resolutely and fully rather than by halves. A big scale is essential for the scientific work and an even bigger scale for in-stalling the project into industry. A factory has to be built, staff must be transferred from other manufactures, and materials, lathes, etc., will be needed. Good responsible people must be recruited for the organization of the project. All this must be done under direct and systematic supervision and control and the project should be regarded as one of the most important.

If you have the least doubt it would be better to postpone the whole thing until peace returns. There are always many interesting and im-portant problems in science which can be usefully carried on, even in the most difficult conditions and with very modest resources – but un-fortunately the oxygen problem is not one of them. If it is taken up only by half it will still demand effort and resources, but it will not pro-duce any useful result during the war. And that would be worse than not taking it up at all.

(I) To V.M. Molotov   Moscow, 14 October 1943

Much respected Vyacheslav Mikhailovich,

Today I learned that the Danish physicist Niels Bohr has escaped to Sweden. He is a very great scientist, the founder of modern atomic theory, a Nobel laureate and an honorary member of many academies including our own. Bohr is a good friend of the Soviet Union and has visited three times to deliver lectures, etc. I know him very well and regard him as a great scientist and a good person. I think it would be appropriate and correct if we could offer him and his family hospitality in the Soviet Union while the war continues. Even if he is unable to make use of such an invitation, it should still be offered. If you approve of this idea then the invitation could come either from the Academy of Sciences . . . or less formally, from me personally.

By the way I heard about Bohr's escape quite by chance so you can see how little interest or attention our Information Bureau pays to keeping us scientists informed – this makes it difficult for us to take any initiatives of a general kind.

P.S.   How soon shall we be allowed to have radio receivers again? Perhaps when we have captured Minsk? Or must we wait for Berlin? P.K.

Kapitza was authorized to send a personal invitation two weeks later and this is the letter that follows.

(I) To Niels Bohr   Moscow, 28 October 1943

We learnt here that you left Denmark and are now in Sweden. Indeed we do not know all the circumstances of your departure but, considering the mess which the whole of Europe is now in, all we Russian scientists feel very anxious about your fate. Of course, you are the best judge of what path you must take through all this tempest but I want to let you know that you will be welcome in the Soviet Union, where everything will be done to give you and your family a shelter and where we now have all the necessary conditions for carrying on scientific work. You only have to let me know your wishes and what practical means are open to you and I have all reasonable hope that we could help you any time you find it convenient for yourself and family.

As you already know, we had a pretty hard time at the beginning of the war but now the worst is well over. I think that it is in no way an exaggeration to say that all the people of our country are so closely united in their effort to free themselves from the barbaric invasion that throughout history it would be almost impossible to find a similar case. Now our complete victory is evidently only a question of time. We scientists have done everything in our power to put our knowledge at the disposal of this war cause. Our living conditions are now much easier. We are all back in Moscow and have time to spare for scientific work. At our Institute we hold a scientific gathering every week where you will find a number of your friends. The Academy of Sciences has also started its activities in Moscow and has just concluded its session at which a number of new members were elected. If you come to Moscow you will find yourself joined with us in our scientific work. Even the vague hope that you might possibly come to live with us is most heartily applauded by all our physicists: Joffé, Mandelshtam, Landau, Vavilov, Tamm, Alikhanov, Semenov and many others, who all ask me to send you their kindest regards and best wishes.

Mrs. Kapitza and the boys are all well and the latter have grown very big. Peter [i.e. Sergei] has already joined a technical school. We are most anxious to know how Mrs Bohr and your boys are. We have very little information about English physicists apart from occasional exchange of telegrams. They, like us, are hard at work fighting for our common cause against Nazism. Accept my best wishes for the future. Most kind regards from myself and Mrs. Kapitza to you and your family. Let me assure you once more that we consider you not only a great scientist but also a friend of our country and we shall count it a great privilege to do our utmost for you and your family. And from my personal point of view I always couple your name with that of Rutherford and the great affection we both feel for him is a strong bond between us. It will be the greatest pleasure to me to help you in any respect.

P.S.   You may answer this letter through the channel by which you receive it.

The letter from Kapitza to Bohr caused considerable commotion in British government circles, since Bohr had just then been initiated into the secrets of the American atomic bomb and the authorities read into Kapitza's letter much more devious motives than the straightforward human concern that Kapitza had intended. It is said that the wording of Bohr's reply [see following letter] had to be agreed with the British secret service.

(A, I) From Niels Bohr   London, 29 April 1944

I do not know how to thank you for your letter of October 28th [1943] which, through the Counsellor of the Soviet Embassy in London, Mr Zinchenko, I received a few days ago on my return from a visit to America. I am deeply touched by your faithful friendship and most grateful for the extreme generosity and hospitality expressed in the invitation to my family and me to come to Moscow. You know the deep interest with which I have always followed the cultural endeavours within the Soviet Union, and I need not say what pleasure it would be to me for a time to participate with you and my other Russian friends in the work on our common scientific interests. At the moment, however, my plans are quite unsettled . . .

Notwithstanding my urgent desire in some modest way to try to help in the war efforts of the United Nations, I felt it my duty to stay in Denmark as long as I had any possibility to support the spiritual resistance against the invaders and to assist in the protection of the many refugee scientists who after 1933 had escaped to Denmark and found working conditions there. When, however, last September I learned that they all and besides them a large number of Danes like my brother and myself were to be arrested and taken to Germany, my family and I had the great luck of escaping in the last moment to Sweden among the many others who, due to the unity of the whole Danish population against the Nazis, succeeded in counterfoiling the most elaborate measures of the Gestapo. For many reasons indeed I am hoping that I shall soon be able to accept your most kind invitation and come to Russia for a longer or shorter visit, and as soon as I know a little more about my plans I shall write to you again. To-day it is above all upon my heart to express my deepfelt thanks to you and my warmest wishes to you and your family and our common friends in Moscow.

(A) To the General Secretary
      of the Central Committee   Moscow, 8 November 1943

Comrade Stalin,

Here is a small matter but one which is of some importance in principle. *Pravda* asked me for a comment on [the victory at] Kiev. I wrote the following:

Kiev

For us, Kiev – the beginning of our State.
For Ukrainians, Kiev – the heart of their native land.
For the Red Army, Kiev – a supreme victory.
For our Allies, Kiev – an example of bravery and determination in the conduct of war.
For the invaders, Kiev – the loss of the Dnieper and an omen of defeat in the war.

In the opinion of the editors of *Pravda* the phrase underlined [in italics] is not consistent with your words and, unless it is changed, the comment cannot be published.

We feel keenly that the Allies are dragging their feet, this I cannot doubt. I told the editors that they should not force public opinion into a single mould. That is why Pravda is so boring. You need to read only a few words of the leading article to predict all the rest of it. Are the *Pravda* editors behaving correctly? After all, in my case the statement comes from a definite person [i.e. Kapitza], who is expressing his own definite point of view.

Yours,
P. Kapitza.

(A) From J. Stalin   Moscow, 11 November 1943

Comrade Kapitza!

I have received your letter.
    Certainly, it is you who are correct and not *Pravda.* You must have noticed that *Pravda* has already corrected its mistake.

With respect

J. Stalin

A fascimile of the above letter, together with the only other letter from Stalin to Kapitza appears in fig. 78. In the Russian text there is a play on words which is lost in translation. The Russian word for ''correct'' is ''pravy'' which is very similar to *Pravda,* which means ''truth''.

(I) To J.V. Stalin   Moscow, 24 February 1944

It is already 18 days since I asked Comrade G.M. Malenkov [then Secretary General of the Central Committee of the Communist Party] to receive me (copies of my letters enclosed). Please ask him to receive me, so that I can discuss necessary business – it's a pity to waste time. I feel very foolish – not like a scientist who is trying to influence big industry in our country, but rather as if I was begging for privileged rations.*

---

* This was of course a time of great shortages and food rations varied according to categories of importance.

Товарищ Капица!

Письмо Ваше получил.

Конечно правы Вы, а не "Правда".

Вы должно быть заметили, что "Правда" уже
исправила свою ошибку.

С уважением *[signature]*

11 ноября 1943 г.

Тов. Капица!

Все Ваши письма получил. В письмах много поучи-
тельного, - думаю как-нибудь встретиться с Вами и по-
беседовать о них.

Что касается книги Л. Гумилевского „Русские ин-
женеры", то она очень интересна и будет издана в ско-
ром времени.

[signature]

4.1У.46г.

78. Two letters from Stalin (the translation of the top one appears on p. 358 and of
the second one on p. 378).

(I)   To G.M. Malenkov
      Secretary of the Central Committee
      of the Communist Party   Moscow, 3 March 1944

Much respected Comrade Malenkov,

. . . I take this opportunity to make a few remarks about inventions*
in the Soviet Union. I once had occasion to meet an inventor of a very
unusual kind. This was Leo Szilard, a Hungarian Jewish physicist
whom I met in London where he had arrived like many other Jews
after escaping from the fascism which was beginning to grow in
Hungary. He was well versed in science and had studied with Einstein, but his particular strength was an exceptionally inventive imagination. He invented pumps, new methods of printing, improvements of talking films and so on. But the most interesting side
of his inventive activity was that he never tried to bring his ideas into
practical reality. He would take out a patent and sell it for one or two
hundred pounds to some firm who was glad to buy it for this modest
sum, if only to protect itself against the possible use of the invention
by competitors. A dozen or so such inventions a year were quite sufficient to keep Szilard in clover without any great labour on his part.
I never heard of a single idea of his being exploited in practice, but
apparently he was entirely indifferent to this.**
    Such people can of course exert an influence on the development of
science and technology only if there is someone who can take up their
ideas and realize them in practice. An interesting question is how
would such a Szilard get on in the Soviet Union? It must be said that
it would be difficult to imagine a worse system of encouraging inventiveness than we have in the Soviet Union. The only real good thing
here is the system of Stalin Prizes; but even in that respect there are
elements of degeneration – there is an increasing tendency for the
prizes to be used in the narrowly departmental interests of the People's
Commissariats. All my experience, both abroad and in the Soviet
Union, has shown me that no nation has a more outstanding inventive
genius than ours, and that this has been so from time immemorial . . .
Although inventive genius is one of our basic strengths, we do less to

---

* The earlier part of the letter was concerned with the merits of a particular invention
which Kapitza had been asked to assess.
** Kapitza was of course unaware of Szilard's important role in promoting the atomic
bomb project in the USA.

organize it, to cultivate it and to turn it into a powerful force than we do for anything else, for instance art, the cinema, etc. . . .

(I) To G.M. Malenkov   Moscow, 23 March 1944

I must ask you very insistently to instruct Comrade Ivanov and other officials of your apparatus regarding the following points:

(1) There is no need to prod me to get on with my work and during all my life I have never permitted this. In the present instance it is particularly absurd, since it is I myself who originated the oxygen project. So far I have not sworn at them, but I shall soon not be able to restrain myself.

(2) Impress on them that I am a real scientist and that I am treated with respect by cultured people, not only here but all over the world. They must not summon me and treat me as an inferior and they must value my time.

(3) It seems to me that officials of the Central Committee should treat scientists with due respect, i.e., *sincerely* and not condescendingly and patronizingly as they usually do. Such treatment makes my gorge rise.

(I) To J.V. Stalin   Moscow, 13 October 1944

I don't know what to do. It's three weeks today since I wrote to Comrade Malenkov asking to see him about Oxygen Board business but without any result, although he said he would discuss matters with me once a month. To keep ringing his secretary might help, but that would mean a loss of the respect due to a scientist – respect which it is so important to encourage. Yet to leave things as they are would be bad for the project. This and a number of other circumstances raise doubts in my mind about the whole oxygen project.

If someone retreats from reality and tries to realize an idea for which the time is not yet ripe, he is exerting himself to no purpose, even if the project is very innovative. No one stood in the way of Polzunov, Yablochkov, Lodygin or Popov* and they even had some support.

---

* Russian inventive pioneers in the areas of heat engines, arc lamps, incandescent lamps and radio, respectively.

Their misfortune was that instead of taking up something which was capable of realization, they were ahead of their time and consequently were essentially failures – somewhere between Don Quixote and Leskov's Levsha*. Their main service has been to edify posterity.

The intensification of industrial processes by the use of oxygen is an enormous problem and is relevant to every branch of industry. It can only be solved if everyone involved – from the lowest to the highest – is aware of its significance. I think the main thing needed is that we should all understand that it is only by following the path of new Soviet technology that we can win peace and that is the problem of the day. For we can only consolidate our victory by technological and cultural excellence. But many do not really yet believe in our creative skills and prefer the path of imitating other countries. We need propaganda and of course we are trying to do this in a small way (through the journal *Oxygen*, meetings, etc.).

Comrade Malenkov is attentive, tries to help, is quick at understanding and appreciates the essence of the oxygen problem, but is he really fascinated by it? . . . If the oxygen project really captivated him as a big Government problem, would it be necessary every time to wait for weeks to be received by him. And, as I have told him more than once, if the oxygen project really captivated him, wouldn't he be interested in visiting the factory and looking at the machines? But if such important people are not captivated by a new problem, what will be the attitude of lesser people? A new technology is a victory over Nature and as in any battle one of the main requirements is passion. Thus we only started beating the Germans when everyone of us got really angry with them. The development of a big technological project is not a personal matter but one for society and with us that means for the Government . . .

I have an exceptional respect for you and particularly as a skilful fighter for innovation, so tell me – would it not be better to postpone this colossal project. Are we not perhaps running ahead of our time and isolating ourselves from society? As far as I am concerned it would be a great load off my shoulders to drop it and indeed the laboratory

---

* In Leskov's famous story, the Tula craftsman Levsha tries to improve on a tiny dancing flea made of metal, brought by Tsar Alexander I from England. He succeeds in putting horseshoes on the flea's feet, but alas the flea can no longer dance!

would, as it has in the past, give me sufficient happiness and fully satisfy my demands on life.

The following letter was probably broadcast by radio rather than sent by post. The translation is from the original French.

(A, I)   Moscow, 23 January 1945

My dear Monsieur Langevin,

I was happy to learn that you have survived the terrible time of the occupation of France and that you have returned to your beloved Paris. Your decision to remain in occupied France in order to continue the struggle against the barbarian Germans filled our hearts with anxiety. This required great courage and I know what you have had to endure from the Germans. Evidently these monsters have respect neither for science nor for the justice and equality for which you have led the struggle. The common victory of our democratic countries is now close and soon the horrors of the war will belong to the past. You will return to your scientific work and to your students who will be happy to have their beloved teacher with them again. The younger generation will benefit from the example not only of your outstanding scientific work but of your courage.

I have the most pleasant recollections of our meetings in the past and we shall soon meet again now that the common success of our armies is about to break down the barriers which separate us. While waiting for that happy day I take the occasion provided by your birthday to send you my best wishes for strength and good health and for the rapid renaissance of French science to which you have contributed so much . . .

A bientôt mon cher ami,

(I) To J.V. Stalin   Moscow, 14 March 1945

Two months ago (20 January) I wrote to you that certain stages of the oxygen project were successfully accomplished and further stages must

79. P. Langevin.

now be undertaken. To develop this big project, new forms of organization are necessary and I asked to have a discussion with you, but have
had no reply. In such a situation I don't know what to do, for there is
no one to whom I can complain about you! But since I have taken on
this project, I have no right to remain silent either.

Intensification of processes with oxygen offers the possibility of
transforming the whole profile of industry, since there is little doubt
that with the same expenditure of work it should be possible to double
the production of metals and chemical products . . . But perhaps we
are not yet ready to tackle such great projects and should undertake
them a bit at a time, say in ten-year stages . . .

In order to master such projects it is essential that everyone involved, from the lowest to the highest, should vividly feel the necessity to seek new paths in technology and should understand their significance. But we have not yet got this enthusiasm. Although we have built a great deal and mastered many techniques in the 27 years since the Revolution, how little of major significance of our own have we introduced into technology! Personally I can think of only one major achievement – synthetic rubber. This was an achievement of world significance and we were the pioneers, but, alas, the Americans and the Germans have already overtaken us. And how little we have appreciated the significance of this enormous achievement. Academician Lebedev, the pioneer and inventor, should have been a national hero, but instead he died of typhus in 1934 following a railway journey in a "hard" class coach* – an event of which we should be heartily ashamed. It must be said, that if Lebedev had died in a capitalist country, it would have been in his own private railway coach and only as a result of a train collision. This is not just a matter of chance, but shows that we don't yet appreciate the need of people who can create new technology . . . Perhaps it is just that there is so much genius in our people that we treat it so boorishly.

In the course of these 27 years the capitalist countries have, according to my estimate, produced about 20 fundamental new developments in technology, of a significance comparable to that of synthetic rubber. For instance, synthetic fuels, plastics, internal combustion turbines, television, superhard alloys (tungsten carbide), rocket planes and so on. And we have produced only one. This can't be allowed to continue. Basically it is we, the scientists, who are responsible for not having been able to demonstrate the power of new technology and for not fighting for a healthy path in its development. I think it is time to take up the ideological guidance of the development of new technology . . . We must use ideological leadership to create new directions in the development of industry, exploiting the advantages offered by the socialist system . . . The oxygen project is increasingly becoming a State and political problem and demands someone of appropriate stature to lead it – it is beyond my

* At that time first and second class were called "soft" and "hard", to avoid the appearance of class distinction. The "hard" coaches were just wooden benches without upholstery and usually very crowded, increasing the risk of infection.

power. So I ask you to give instructions that the organizational forms of the oxygen project should be examined and that I should receive a reply.

(I) To J.V. Stalin   Moscow, 13 April 1945

Yesterday the Bureau of the Council of People's Commissars formally recorded delivery of the oxygen turbine installation, thus confirming that we have constructed equipment with an output several times greater than that of previous methods. Moreover the shortage of oxygen in Moscow has ceased three months ago and we have more oxygen than we can use. But during recent weeks certain comrades have created such an atmosphere that I have felt the proceedings of the Bureau to be rather like those of a Court of Law with myself as the accused. The Bureau has not only not recorded that our work represents a new achievement for Soviet technology, but none of its members, except Comrade Mikoyan, have even acknowledged that we have done something useful. The members of the Bureau seem to have been much more worried by the question of what should be done with the surplus oxygen. After the meeting I felt like declaring ''Allow me to assure you, Citizen Judges, that in future I will behave better by not looking for new paths in technology and I promise not to make any inventions''.

This apathetic attitude to the development of really new ideas in technology is characteristic of our age. This is very sad, since in our society the motive of personal enrichment is not available as a stimulus to an inventor. Therefore in our country appreciation of creative work by the Government and society is the strongest motive power and stimulus for an inventor. Without such appreciation our inventive talents will not develop and therefore I consider the formal resolution of the Bureau which contains no real appreciation [of our achievement] to be unhealthy in principle and I wish to bring this to your attention as Chairman of the Council of People's Commissars. After all, a singer would not only find it tedious to sing to a deaf-mute audience, but could hardly develop his talent in such a situation . . . I have already written to you twice about these matters. Although I have not had a word of reply from anyone I must once more raise the question of the oxygen project. For in such an atmosphere it is hard to struggle with enthusiasm for its development. So I am writing to ask that, if only out of respect for my scientific work, you should give instructions for the questions I have raised in these letters to be answered.

(I) To G.M. Malenkov   Moscow, 3 May 1945

Deeply respected Georgii Maksmilyanovich,

I am deeply moved by the exceptional attention which has been accorded to the success we have achieved with the oxygen project and by the award of decorations to myself and my collaborators.* I should like also to express my deep appreciation of your part in the realization of this grandiose project and should like to think that we shall surmount all difficulties and bring it to a successful conclusion.

Permit me to thank you and on your behalf to congratulate all those who have been decorated.

Yours respectfully,
P. Kapitza

P.S.   The final chord of the Balashikha symphony is excellent. So let us play the next one, the Tula symphony, in tempo *presto vivo*. P.K.

The oxygen factory was at Balashikha near Moscow and the larger scale plant was to be in Tula, famous for its samovars and its steel. The Balashikha factory, which continued operating up to 1975, was an important contribution to the war effort and a vindication of Kapitza's turbine method. In a letter to Stalin on 20 January 1945, Kapitza mentioned that even at the start of its activity in late 1944, it produced 40 tons of oxygen in 24 hours, about one-sixth of the total oxygen produced in the Soviet Union by 400 other plants.

---

* By decree of the Praesidium of the Supreme Council of the USSR, Kapitza was awarded the title of Hero of Socialist Labour for his oxygen work on 30 April 1945 and the Institute for Physical Problems was awarded the Order of the Red Banner of Labour and various orders and medals were awarded to a large group of Kapitza's collaborators.

The following series of letters to Stalin illuminates the bizarre events of 1945 and 1946, starting ironically just at the time his work had been recognized by prestigious awards to him and his collaborators (see letter to Malenkov of 3 May 1945) and culminating in Kapitza's dismissal on 17 August 1946. Once again it illustrates Kapitza's boldness – although in fact it was not a reckless boldness, but rather a calculated one, since he knew that he was in great danger once Beria had become his enemy. In fact his gamble of playing off Beria against Stalin probably saved his life. Long afterwards Kapitza learnt from General A.V. Khrulev of a conversation between Beria and Stalin which Khrulev had overheard. Beria was demanding Kapitza's arrest, but Stalin replied, "I'll dismiss him if you like, but *you* mustn't touch him" – a nice example of Stalin's "cat and mouse" tactics. Prior to this, Stalin had sent an apparently amiable letter to Kapitza (see p. 378), and then only a few weeks later (14 May 1946) signed a decree of the Council of Ministers which appointed all Kapitza's main opponents to be members of the Commission investigating the Oxygen Board.

The threat of arrest or a fatal "accident" continued to hang over Kapitza all the time he was at Nikolina Gora until Beria's arrest and execution in 1953. Kapitza was reminded of his danger in 1948 when his friend, the famous Jewish actor Mikhoels, was "accidentally" killed by a lorry in Minsk. Many years later Kapitza told Dirac (1980) of how he had been particularly at risk after Stalin's death in 1953. Two men had appeared at Nikolina Gora and asked to see his lab. Kapitza showed them round but soon became convinced that they were not physicists, and had come for some political reason which he could not guess. At 12 noon the men suddenly said they had seen enough and left. Kapitza heard later that it was at 12 noon on that day that Beria had been arrested (he was executed six months later). He believed that the two men had been sent to protect him in case Beria had taken some last-minute action against him.

(H, I) To J.V. Stalin   Moscow, 3 October 1945

The decree of the Sovnarkom regarding the Oxygen Board dated 29 September 1945 and signed by you had already been under discussion for about six months . . . During this time it has been discussed by seven commissions and three meetings of the Sovnarkom Bureau, but a suitable industrial site has not yet been found and this question has been postponed for another two months. Such an attitude to the oxygen problem clearly demonstrates that we are not yet ready for it. Our culture still has to mature to the point that at least those comrades responsible for putting the decisions into practice should believe in the problem and understand that progress can be made only by using the achievements of our own science rather than by copying the techniques of other countries.

In the process of working out the decision to transfer the Glavavtogen to the Oxygen Board, there has been considerable friction with Sukov, who up to now has acted as a brake on the development of the turbine method of producing oxygen. Sukov has written a letter to you in your capacity of Secretary of the Central Committee and the contents of this letter have become fairly widely known – for instance, Comrade Beria quoted it at a meeting of the Sovnarkom Bureau. This letter contains a number of libellous accusations of a personal nature directed against me. I am very surprised that some of the comrades see nothing unusual in this and Comrade Beria insists that Sukov should be my deputy in the Oxygen Board. I, however, consider that Sukov should be made to answer for these libels and I have written to say so to the Central Committee in the person of Comrade Malenkov.

The above clearly shows that Comrade Beria cares little for the reputation of our scientists, rather as if saying "your business is to invent and investigate, so why do you need a reputation". Having recently had experience of Comrade Beria at the Special Committee,* I now feel particularly clearly how impermissible is his attitude to scientists. When he brought me into the project he simply ordered his secretary to summon me to his office. By contrast, when Witte, the Tsarist Finance Minister, invited Mendeleev to work for the Department of Weights and Measures he came himself to visit Dmitrii Ivanovich. Again, when I was in Comrade Beria's office on 28 September, he suddenly decided to end our conversation and simply shook hands, saying "Well, goodbye". Such matters, though trivial, are significant external signs of the degree of respect for the individual, for the scientist and scholar. It is after all by external signs that we convey ideas to each other.

This immediately raises the question of whether the position of a citizen in our country should be determined solely by his political

---

* This was the high level committee set up in July 1945 to oversee the development of the atomic bomb.

weight. After all there was a time when the Patriarch stood at the side of the Emperor and the Church was then the repository of culture. The Church has had its day and Patriarchs have gone out of circulation, but our country cannot do without leaders in ideas. Even in the realm of social science, however important were the ideas of Marx, they still have to develop and grow . . . Sooner or later our scientists and scholars will have to be promoted to "patriarchal" rank. This will be necessary since otherwise they cannot be made to serve their country with enthusiasm – for we have nothing to buy them with. Capitalist America can do it, but we can't. Without such a patriarchal status for scholars, culture cannot grow independently in our country – this was pointed out by Bacon in his *New Atlantis*. So it is high time that comrades like Comrade Beria should begin to learn respect for scientists and scholars.

All this compels me to feel that so far the time has not yet come for a closer and more fruitful collaboration between our politicians and our scientists. For the present the oxygen problem is just Utopia. I am sure that for the time being I shall be of greater service to my country and to the people if I direct all my efforts to purely scientific work and I propose to devote myself entirely to this. For after all that is the work I enjoy and for which I have earned people's respect. Therefore I ask you to agree to my release from all my duties with the Sovnarkom, apart from my work in the Academy of Sciences.

In a word, it is evidently too early for me to be one of the "Patriarchs" so let me remain in my monastic cell for the present. Comrade [K.S.] Gamov will be able to fill my place quite successfully in the Oxygen Board and Comrade Beria will feel more comfortable in the Special Committee without me. Of course, as always, I shall continue to try to help my country through my scientific knowledge.

(A, I) To Niels Bohr   Moscow, 22 October 1945

It is a great relief to feel that the ordeal of the war is over and we may resume our peaceful life. We all are very happy to know that you and your family went safely through all your adventures and are now united in Copenhagen. I was always happy to have news from you and your family but every time it came with great delay . . . We are all back in Moscow. It is already two years since the Institute resumed normal scientific work. As before the war, we have liquid helium two times a week and have found some curious things at low temperatures. I hope you have seen the theoretical work of Landau on the superfluidity of helium which probably you remember we have discovered just before the war . . . Landau also proved that two kinds of elastic waves must propagate simultaneously in superfluid helium; therefore in helium-II two kinds of sound velocities must exist, one at 150 m/s (that already known), and another (a new one) at 17 – 20 m/s, Peshkov has discovered experimentally the second velocity of sound in helium-II. Besides this peaceful work we helped the country in its war efforts and I am proud to tell you that the Institute received the Order of the Red Banner. It is the only Institute in the Academy of Sciences which received this honour.

At the moment I am much worried about the question of the international collaboration of science which is absolutely necessary for the healthy progress of culture in the world. The recent discoveries in nuclear physics and the famous atomic bomb I think prove once more that science is no more the hobby of university professors but is one of the factors which may influence world politics. Nowadays it is dangerous that scientific discoveries, if kept secret, will not just broadly serve humanity but may be used for selfish interests of particular political or national groups. Sometimes I wonder what must be the right attitude of scientists in these cases. I should very much like at the first opportunity to discuss these problems with you personally and I think it would be wise as soon as possible to bring them up for discussion at some international governing body of scientists. Maybe it will be worth while to think over what measures should be included in the statutes of the "United Nations" which will guarantee a free and fruitful progress of science. I should be glad to hear from you what is the general attitude on these questions of leading scientists abroad. Any suggestions about means to discuss these questions from you I shall welcome warmly. I can indeed inform you what can be done in this line in Russia . . .

By coincidence this letter crossed one from Bohr, dated 21 October, which expressed very similar sentiments

(H, I, f) To J.V. Stalin   Moscow, 25 November 1945

It is now nearly four months since I have been actively participating
in the work of the Special Committee and Technical Council on the
atomic bomb. I have decided to set out my views on the organisation
of this work in detail in this letter and also to ask again to be released
from my participation.

In my opinion there is much that is abnormal in the organization
of the work on the atomic bomb. In any case, what is being done at
present does not provide the shortest or cheapest path to a successful
outcome. The problem facing us is the following. In the space of only
three or four years and with the expenditure of two billion dollars
America has produced the atomic bomb, which is now the most
powerful of all weapons of war and destruction. The world's known
reserves of thorium and uranium would be sufficient to destroy
everything on the land surface of the globe about six times over. But
it would be foolish and indeed absurd to imagine that the main possible use of atomic energy is exploitation of its destructive power. Its
potential role in our economy is no less important than that of oil, coal
and other sources of energy and moreover this energy is concentrated
in a mass ten million times smaller than that of a conventional
fuel . . . We do not know the secret of the atomic bomb, which is a
very carefully guarded state secret known only to the Americans. The
information available at present is not sufficient to make an atomic
bomb and I am sure that some of the information has been provided
only to divert us from the correct road.

Almost all of the two billion dollars spent by the Americans must
have gone on construction and machine building. We could hardly afford to spend such a sum within two or three years during our post-
war reconstruction . . . The only advantage we have over the
Americans is that we know the problem has a solution, while the
Americans had to take a risk . . . Although it will be difficult, we must
try to produce an atomic bomb quickly and cheaply, but not by the
path we are following at present, which is completely unmethodical
and without any plan. The main deficiencies of our present approach
are that it fails to make use of our organizational possibilities and that
it is unoriginal. We are trying to repeat everything done by the
Americans rather than trying to find our own path. We forget that to
follow the American path is not within our means and would take too
long . . .

It would be easier if we knew the path to be followed, but since it is not known, our first need is scientific research to discover it. And to bring the project to practical realization we need a powerful industrial base and organization. It might be thought that until the path is known these cannot be created, but this is not so. In fact it is possible to predict sufficiently accurately what kind of factories and industry will be needed and to start preparing them now. For instance it can be said with certainty that large-scale metallurgical plants for producing thorium and uranium will be required.

I consider that the following plan of action should be adopted. As of now, to work out a two-year plan for preparation of the necessary industry and at the same time to carry out the essential experimental and theoretical scientific research. It is already clear that our main constructional effort should be devoted to re-establishing key factories such as those for compressors, chemical engineering, tube making, . . . production of pure uranium, thorium, aluminium, niobium, beryllium, helium, argon, etc. etc. During these two years we must take a number of measures to improve our scientific base, such as improving the level of our scientific institutes and institutes of higher education and the well-being of our scientists. At present these measures are being only poorly developed – lifelessly and with poor organization – but we cannot manage without them. They are needed not only for the atomic bomb, but for other important developments initiated during the war, such as jet engines, radar and so on . . .

It should be noted that there is great enthusiasm for the atomic bomb project among our scientists and engineers – from the very best down to the crooks and charlatans and including some in prison. Although this enthusiasm has been aroused for different reasons it could be usefully exploited. But if we let them work any old how, we are only frittering away their efforts and we will achieve nothing. To organize all their scientific work is the most important and difficult task. Since we have few means at our disposal it is important that we should make proper use of our scientists and in a well thought out way. Only then is there a chance of finding new paths which will lead to quicker and cheaper solutions than the American ones. The problem is like that of a Commander-in-Chief who has been offered several proposals for taking a fortress. He would hardly tell each general "Use your own plan to take it" in the hope that one of them might succeed. Invariably only a single plan is chosen and a single general to direct it. That is how we should proceed in science, but unfortunately this is not regarded as obvious and is not usually accepted.

Why not let each scientist follow his own plan and work independently? This would seem to be without risk and it might perhaps work. "After all there have been occasions when all the experts have said it won't work and yet it has" . . . But in reality a well-qualified scientist is always able to give a correct assessment of a serious proposal offered by another well-qualified person. Of course it's essential to choose the right expert rather than an irresponsible ignoramus (of whom we have plenty). We have no strict selection of research themes according to a plan, and a whole circus, made up not only of crooks and adventurers, but also of honest people, is beginning to dance around the atomic bomb project. Of course something will eventually come of all this dancing, but evidently not the short and cheap route by which we could overtake America. Apparently the Americans themselves went along this circus type of road and it cost them a great deal of money. But unfortunately with our clumsy patchwork of a Technical Council it is difficult to fight this policy. If, however, we strive for quick success, it will inevitably involve risk and concentration of the main forces along a well-chosen path in a very limited range of directions. On these questions I do not agree with the other members of the Council. Often they are unwilling to argue with me and carry out their measures behind my back. But the only useful path is a single agreed solution and a more restricted council of war such as would be used by a Commander-in-Chief . . .

The next question, the selection of leaders to direct the work, is also a big problem. I take the view that the basis of the choice should not be what a person promises to do, but what he has already achieved. Just as in war, each new order should be based on the successful fulfilment of previous ones. However, maybe because of the Russian meditative character, we prefer far-reaching promises which are so attractive that they make your mouth water. "Ah, but what if it succeeds?" says the responsible official, who is usually quite ignorant of what is really involved, but is gullible and lets the idea go ahead, even though it is evident to any competent person that it has no future. This means that we have much overcrowded working space, many overloaded instruments and machine tools and many overloaded people – all for nothing. The result is failure, and the responsible official then begins to go to the other extreme and to lose all faith in scientists and in science.

The correct organization of this work is possible only subject to one condition which is not at present satisfied. Without this condition we shall not be able to overcome the problems of the atomic bomb quickly and indeed we may not independently succeed in overcoming them at all. This condition is that greater trust must be established between the scientists and the Government. This is an old story, a survival from the revolution. The anomaly has been appreciably smoothed out by the war, and if it still survives it is because there has been insufficient encouragement of respect for the scientist and for science.

It is true that the participation of our scientists in problems of our national economy and defence has always been considerable and important, but the scientist's help was from the outside, by consultations and by solving such problems as were presented to him. It must be noted with much regret that this was because our industry and armaments have developed only along the line of improving existing techniques. For instance Yakovlev, Tupolev and Lavochkin are our most important designers, but they have not gone beyond perfecting already existing types of aircraft. To develop new types, such as the jet engine, demands a different type of designer – more creative and bolder. Such people find little freedom to develop their talents in the Soviet Union. Thus technical advances based on ideas which are new in principle, such as the atomic bomb, the V-2 rocket, radar, gas turbines and so on are either feeble or completely immobile in the Soviet Union.

My turbine oxygen producing installation, an initiative which is new in principle, only got going when I, quite abnormally for a scientist, became the head of the Oxygen Board. Only because of the trust implied by this appointment and the influence it brought, was I able to bring the oxygen installation into being so quickly. This was quite an abnormal and indeed absurd situation and the power it brought weighed heavily on me, but I put up with it because there was a war on and I had to do all I could to achieve success.

Experience showed that I was able to make people listen to me only as Kapitza, Director of a Board under the Sovnarkom, but not as Kapitza, scientist of world reputation. Our cultural education is still inadequate to put Kapitza, the scientist, above Kapitza, the head of a Board . . . This is exactly the situation today in solving the problems of the atomic bomb. The scientists' opinions are greeted with scepticism and behind their backs [the government officials] have it their own way . . . The Special Committee must teach the comrades to believe the scientists and the scientists in their turn will be compelled to feel their responsibility but this is not yet so. This can only be achieved if the responsibilities of the scientists and the comrades in the Special Committee are placed on an equal footing. And this will only be possible when science and the scientist are accepted as the main authority, rather than a subordinate one, as at present.

Comrades Beria, Malenkov and Voznesenski behave in the Special Committee as if they were supermen, particularly Comrade Beria. It is true he has the conductor's baton in his hand – this is as it should be, but after him, the first violin should be played by a scientist, for it is the violin that sets the tone of the whole orchestra. The conductor must not only wave his baton but he must also understand the score and that is where Comrade Beria is weak. I think Comrade Beria could cope better with the task if he devoted more time and effort to it. He is very energetic, good at grasping a problem quickly, good at recognizing the difference between the main issue and subsidiary ones and so he doesn't waste time. Also he certainly has a taste for scientific problems, which he appreciates easily and he formulates his solutions precisely. But he has one failing – an excessive self-confidence, the reason for which is apparently his ignorance of the score.

I tell him quite directly ''You don't understand physics – let us physicists judge these questions'', to which he replies that I don't understand anything about people . . . I offered to teach him some physics if he would come to my Institute. After all you don't have to be an artist yourself to understand how to judge pictures. Our gifted merchant patrons of art, such as the Tretyakov brothers, Shchukin and so on, understood pictures extremely well and recognized great artists ahead of others – they were not artists themselves but they had studied art. If he wasn't so lazy, Beria with his ability and his ''knowledge of people'' could, if he worked, undoubtedly get to understand the creative processes of scientists and engineers and become a first-class conductor of the atomic bomb orchestra. For instance he should learn from the original sources and not just from popular accounts, how the transoceanic cable was laid, how the first turbine was developed, etc. He would then be able to recognize the general laws behind these processes and use this experience to recognize what was important and necessary in developing the atomic bomb. But for this he would need to work and not just sit in his chairman's seat and scribble on the draft proposals – that is not enough to direct the project.

But I can't get anywhere with Beria – as I have already said, his attitude to scientists goes completely against the grain. For instance during the last two weeks he made nine different appointments for me to see him but we haven't yet met, since each time he cancelled the appointment. Apparently this was to tease me, for I can't imagine that he is so unable to organize his time that he can't be sure when he'll be free in the course of two whole weeks . . . All the leading comrades, such as Beria, should let their officials feel that in this business the scientists have the leading rather than a subordinate role.

You only have to listen to the views on science of some of the comrades at the meetings of the Technical Council. Often you listen only out of politeness and conceal a smile at their naïveté. They imagine that once they know that two and two makes four they have plumbed the depths of mathematics and can make authoritative judgements. This is the primary reason for their lack of respect for science, which so hampers progress and should be rooted out. With such conditions of work I don't see any use in my presence at the meetings of the Special Committee and Technical Council. Comrades Alikhanov, Joffé and Kurchatov, who are at least as competent as I am, and indeed more so, can perfectly well replace me in all questions connected with the atomic bomb. Thus, as you can see for yourself, my further participation in the Special Committee and Technical Council is pointless and merely serves to upset me, so interfering with my scientific work. Since I am involved in this project, I naturally feel my responsibility for it, but to turn the project into the direction I think proper, is beyond my powers. Indeed it would be impossible, since Comrade Beria and most of the other comrades disagree with my views, and I cannot be a blind executor of others' policies.

My relations with Comrade Beria are getting worse and worse and he will doubtless be glad to see me go. Friendly agreement (without the sort of atmosphere that there is between a general and lower ranks) is essential for such a creative project and is possible only on an equal footing. This is not there and in any case I can't work in such an atmosphere. After all, from the very first, I asked not to be involved in this business because I knew beforehand how it would turn out. So I ask you once more and very insistently to release me from participation in the work of the Committee and the Council. I count on your agreement since I know that forcing a scholar against his wishes is not your habit.

P.S.   The turbine plant for producing gaseous oxygen has now been achieved in the Institute and tried out several times . . . Thus all my debts to the country and the Government have been repaid in full and I shall continue to insist more and more that I should be released from the Oxygen Board and allowed to return fully to my scientific work. . . .

P.P.S.   I should like Comrade Beria to see this letter, since it is not a denunciation but a useful criticism. I would tell it all to him in person but it's extremely tiresome to arrange to see him.

The letter below is the second of the only two written replies Kapitza ever received from Stalin; the first appears on p. 358 and facsimiles of both are shown in fig. 78. In spite of the amiable remark no meeting ever took place, although Stalin did occasionally telephone Kapitza.

(I) From Stalin   Moscow, 4 April 1946

Comrade Kapitza,

I have received all your letters. There is much that is instructive in them and I should like to meet you at some time to have a chat.

As regards L. Gumilevski's book* *Russian Engineers*, it is indeed very interesting and will be published in the near future.

J. Stalin

---

* This refers to a long letter dated 2 January 1946 from Kapitza to Stalin (not included in the present book), enclosing the manuscript of Gumilevski's book and recommending it particularly for its emphasis on the leading role of Russians in the history of engineering.

(H, I) To J.V. Stalin   Moscow, 13 April 1946

Two weeks ago I wrote to Comrade Beria that we have developed a turbine method of producing gaseous oxygen, which in my opinion opens up the possibility of getting it in the large quantities required for use in hearth furnaces . . . Having developed and studied this new type of equipment, we have in essence completed the last stage of the oxygen problem. I asked for a commission to be appointed in order to evaluate objectively what has been achieved. This is needed not only for me personally but is also important for the other participants in the work so that they should not feel their labours have been for nothing . . .

P.S.   Thank you for your kind letter, which I was very happy to have.

(H, I, f) To J.V. Stalin   Moscow, 29 April 1946

The time limit you set for the Saburov Commission to report on the production of gaseous oxygen by the turbine method has now expired. So a month has been lost in the development of this project. If the military staff were to dilly-dally with the agreed battle plans in this fashion in time of war it would be clear to everyone that it amounted to criminal delay. Why can't our comrades understand that in the battle for "new technology" it is also intolerable to fight half-heartedly?

It is clear that if such a languid attitude persists the "new technology" will be mastered by others who are more expeditious. The responsible comrades must be made to realize that the most important principles of any struggle, wherever it takes place – in the arena, in the laboratory or at the front – are speed and attack coupled with courage and determination. If we move at a leisurely pace all our striving to acquire "new technology" will not go beyond paper work and we shall be doomed to imitate developments made elsewhere. Once again I ask for your help to prevent further postponement of a decision by the Council of Ministers.

P.S.   Forgive my persistence but when I feel the responsibility is mine, there is nothing else I can do. I telephone Comrade Saburov every day.

Kapitza was greatly upset by the hostile decree of 14 May and wrote to Stalin twice on the same day. As was often his practice in writing on important matters he would write once "for the record", so that a formal resolution could be made, and back this up by a second and more personal letter.

(I) To J.V. Stalin   Moscow, 19 May 1946

The commission appointed by you on 13 April has virtually completed its investigation of my oxygen technique. The report of the experts was favourable and the commission accepted it. It would be difficult to infer any partiality towards me in the conclusions since I took no part in selecting the members of the commission. The experts were not chosen by me nor did they work in my laboratory. I therefore thought that this matter had been exhaustively dealt with.

However, your new decree of 14 May, which took me completely by surprise, broadens the scope of the commission and appoints Professors Gersh, Gelperin and Usyukin as new members – so the only oxygen specialists on the commission are people openly opposed to the lines on which I am working. These three are offended because I did not want to draw them into our work since in my opinion they have not only achieved nothing of any significance but they are, moreover, unprincipled and harmful people who like to fish in troubled waters. My attempts to challenge them to an open discussion in the Technical Council of the Oxygen Board were unsuccessful for they either remained silent or did not attend the Council meetings, and indeed their scientific ability is very limited. Their basic mode of action is to write to members of the Government behind the back of a colleague, emphasising only the many doubts and uncertainties that are inevitable in any novel method.

To invite these people to judge my activities cannot be considered objective, it is simply insulting. Such behaviour is possible only if the intention is to put obstacles in my way, ruin my work and undermine my authority. I have always been in sympathy with control by society and every governmental commission, including this one (there have been three in all), was appointed at my request. Hitherto I have been asked for my opinion on the questions to be considered and on the composition of the commission. In the present instance this was not done and I was not even invited to attend the meeting of the new commission. Even a criminal is given the right to challenge jurors and to be present in court. Obviously, I am under attack. What have I done to deserve this? . . .

P.S.   . . . As a result of your decree there are already government inspectors in both the Institute and the Board and the general atmosphere is depressing. I am writing to you about this since it damages and slows down our work. I should also like it to be made clear to Comrade Saburov that a new scientific achievement, however it is assessed, is not a crime and a scientist is not a criminal and should not be treated like one.

(I) To J.V. Stalin
Personal   Moscow, 19 May 1946

. . . The objective facts about the oxygen project show that we have evidently overtaken everyone else and I am proud of this. But now, when it seemed that the long road had been travelled and we could rejoice, it has turned out otherwise. Gersh, Gelperin and Usyukin, who have continually hindered my work and intrigued behind my back while flattering me to my face, have suddenly been appointed my judges. How did they manage it? Perhaps they felt that now is the time to stake everything and they followed Iago's road. This has produced a Shakespearian tragedy – but what will be the denouement? That depends on you.

During these years there have been many struggles, many difficulties to overcome and much work. At critical moments, just as now, I wrote to you. I have often felt somewhat of a Don Quixote and more than once I have felt like dropping everything and returning to my scientific work. But usually I felt your supporting hand and fought on. Of course you couldn't, like Rutherford, enter into the technical details of my daring, but I felt that you, like him believed in me and that was the main thing I needed. Sometimes I even felt that you understood the difficulties of my struggle – for who, if not you, knows what a struggle is. Sometimes, however, I had doubts about the strength of your support, for you have never wanted to talk to me. But now your decree [of 14 May] and the whole turn of events have made me doubt and have greatly hurt me. Without your support it would be best for me to leave at once and in essence it wouldn't even be shameful to go now, since the scientific problem is solved, I am very tired and I find it very difficult to fight – for I have to fight with tricky people.

It is rightly said that there is no good business which remains unpunished and your decree has punished me properly. However badly others may regard what I have achieved, I nevertheless feel happy and proud that I have produced gaseous oxygen by the turbine method. History will not reproach me that I didn't carry the project through to the end.

(I) To J.V. Stalin   Moscow, 2 June 1946

At the meeting of experts for the assessment of the turbine method of oxygen production it transpired that Professor Usyukin had sent a letter to the Central Committee criticizing my work on the turbine method. It appears that, as a result of this letter, the composition of the commission and its terms of reference have been changed. I do not know the contents of this letter.

As proof of Professor Usyukin's dishonesty I enclose a certified copy of his address to the Council of the Institute of Chemical Engineering, in which he gave a laudatory review of my oxygen equipment and put it forward for the award of a Stalin Prize. He did this of his own volition and without my knowledge.

During the past year the equipment at Balashikha has worked continuously and successfully, giving results no worse than during the acceptance tests. Half of the oxygen used in Moscow comes from this plant. During this year, for the first time, Moscow's thirst for oxygen has been satisfied, even though it now consumes considerably larger quantities than before the war, so there are no objective data that could have caused Professor Usyukin to change his opinion so suddenly.

I want to draw your attention again to the fact that such dishonest people are unable to give an objective assessment. The new members appointed to the commission gave an unfavourable opinion about the plant without having looked at it even once. All the meetings were held without my participation. None of the chairmen of the expert commission – not Pervukhin nor Saburov nor Malyshev – ever consulted me. The whole affair was conducted like the investigation of a crime and not of a scientific and technical problem. All this could ruin a novel and important initiative.

P.S.   It is exactly two months today since I asked for an assessment of our work but no decision has yet been taken.

(I) To J.V. Stalin   Moscow, 16 July 1946

I have just received the papers from Saburov that I asked for exactly a month ago. I see from them that the experts worked dishonestly and misled the commission. I have written in detail about this to Saburov and I enclose a copy. This again emphasizes the lack of integrity of the experts appointed, as I wrote to you at the very beginning.

If it were not so disgraceful that rogues like Usyukin, Gersh, Gelperin and others still have an easy life here and that they spoil and often ruin forward-looking scientific developments and technical applications and if I lacked confidence that they would be brought to book – as our duty demands (Bardin quite rightly said to me, that they should be given a punch on the nose) – then I would have thrown everything up long ago and pursued my own real interests.

The oxygen project, like any other new and important initiative will only find its feet if we really work hard at it. I believe I am on the right track and I am ready to work at full stretch on the scientific aspects and to accept personal responsibility, but more than that is needed – I must be trusted as a scientist and respected as a person. Just now, at the commission meeting, I was treated with such contempt, both personally and as a scientist, by your ministers, especially Malyshev and Pervukhin, that I had only one wish – to get away and give up working with them. Trying to work under these conditions is pointless.

I am therefore asking you urgently that the Government should decide as soon as possible, if only out of respect to me as a scientist, what is to be the fate of the oxygen project. I must either be given honest and strong support or else simply taken off the oxygen work altogether. A compromise decision should not, and indeed must not be taken.

The decision came on 17 August 1946 in the form of a Sovnarkom decree signed by Stalin, dismissing Kapitza not only from the oxygen project but also from his Directorship of the Institute for Physical Problems (see p. 66). Kapitza had prepared himself for the former but the latter came as a great shock.

(I)   To S.I. Vavilov   Nikolina Gora,
        President of the Academy of Sciences   end of April 1947

Deeply respected Sergei Ivanovich,

I have received your letter of 15 April containing a request from the editors of the Jubilee volume that I should write an account of my work on helium, but it places me in a difficult position. In the decision of the Praesidium dated 20 September 1946, protocol 23 states: "The most urgent contemporary problems in physics were not reflected in the programme of the Institute for Physical Problems". I was removed from the Directorship of the Institute and deprived of the possibility of continuing these scientific investigations about which you now want me to write – as you put it, "about your investigations which are of outstanding significance". The invitation to write an article about the work which I was denied the chance of continuing seems to me, from the human point of view, as cruel an irony as to invite a starving man to describe a feast.

Moreover, I do not consider it proper to describe work already completed by myself and my students, or what I am doing now, after removal from the Directorship, until such time as a critical review of my scientific activities is conducted by the physico-mathematical division of the Academy of Sciences. I have asked for this several times and it seems to me that such a request from a member of the Academy is entirely lawful, but it has not yet been granted. After all, scientific achievements have an absolute value against the background of the development of science throughout the world and the Academy of Sciences, as the leading scientific organization in the Soviet Union, ought to have its own objective assessment, independent of where and how these achievements were made.

(I) To J.V. Stalin   Nikolina Gora, 6 August 1948

It is now two years since I was deprived of the possibility of doing serious scientific work. During this period the course of technical development throughout the world has made it increasingly obvious that my point of view is now generally accepted on the benefits to be derived from the use of oxygen in the main branches of industry . . . This intensified use of oxygen is considered by some to be no less valuable than atomic energy and even more urgent. This is confirmed by a great revolution in scientific work in the USA and large scale investment in the construction of oxygen plants. A number of scientists acknowledge in their publications that the low-pressure turbine method of extracting oxygen which I was the first to propose really opened up the possibility of using oxygen on a large scale in industry . . .

The information in the press during recent months continues to tell of new installations in the USA and now a huge oxygen plant of my design is reported in France. Installations of this type in the USA are used for the synthesis of liquid fuels, each having an output of more than 250 000 tons per year, of which 70% is high octane. History teaches us that when new techniques are being introduced time will inevitably establish the scientific truth of a new method, so I am waiting patiently for the moment when it will become clear beyond dispute that the decision two years ago to close down my line of work completely and instead copy the German high-pressure equipment, was not only wrong in itself but irrevocably wrecked our own very important and original entry into the field of advanced technology, about which we really ought to have been proud. When that time comes my ''disgrace'' will be revoked for it will inevitably be recognized that my scientific judgment was correct and that I fought honestly for the propagation in our country of one of the most important developments of our times.

I fully understand that while I remain a ''disgraced scientist'', shunned by colleagues, someone whom people are afraid to help, I cannot think of seeking wider scope for my scientific activities and must accept limits if I am to work successfully by myself on a modest scale. Even with the small facilities now available to me I have been able to do some theoretical and experimental work. In the meantime

I have completed four papers and shall send reprints of those* already published under separate cover. I have even made a small discovery – a new form of wave motion in liquids. But the work goes slowly since I have to do everything myself, even making the necessary apparatus with my own hands, helped only by my family.

To improve the conditions for my work I need to have a permanent assistant seconded to me, who could be S.I. Filimonov, to have my laboratory officially recognized, to be given a staff of two or three, to have regular supplies and equipment sent out and to have urgent repairs done.

I have already asked the Academy of Sciences for these things several times and they would be agreeable to this modest request but they tell me that the Council of Ministers does not approve and that is why I have approached you. Knowing your respect for scientific creativity I believe you will be kind enough to agree to instruct the Academy of Sciences to improve the conditions of my work . . .

(I) To S.I. Vavilov   Nikolina Gora, 18 October 1949

I received your letter of 30 September in which the Praesidium of the Academy of Sciences asks me to state "the reason for my non-appearance at the session of 17 September and in the event of being unable to attend a meeting of the General Assembly in future, to send an apology in good time."** I have to inform you that the reason for my "non-appearance" was the state of my health, especially of my nerves.

---

* On one of these reprints Kapitza adapted a quotation from Tolstoy and wrote that he was "Not a scientist who does scientific work, but one who is unable *not* to do scientific work". The boldness of this remark is reminiscent of the inscription he wrote many years previously on the reprint he presented to Rutherford (see p. 16 and fig. 7).
** The Academy was anxious to avoid a demonstrative absence of Kapitza from the special meeting in December 1949 in honour of Stalin's 70th birthday. This is probably why Kapitza replied to Vavilov's request at such length. In fact, Kapitza attended neither the Academy celebration nor a similar occasion at the Physico-Technical Faculty of the Moscow State University, where he still held a Chair. He was punished by being dismissed from his Chair (see pp. 389 and 392).

I think that you should understand, simply as a human being, that my state of mind could hardly be normal after all that has happened to me since 1946. I was relieved of all my responsibilities and, what was most important for a scientist, deprived of the possibility of continuing in those fields such as low temperature, high magnetic fields, liquid helium, etc., where my work was in full swing and my achievements and discoveries had already won wide recognition and this prohibition is still in force after more than three years. It is equivalent to depriving a musician of his instrument, an artist of his paints or a writer of paper and pencil.

You must know that my method of producing oxygen, which was rejected here, is now recognized throughout the world as an important technical achievement but here, in spite of this I am still in disgrace. They are either afraid or not permitted to mention my earlier achievements. For example, the article on "Physics" in the 1948 edition of the Great Soviet Encyclopaedia, does not mention a single one of my papers, not even my discovery of superfluidity in liquid helium, a discovery that certainly received general recognition.

Four months ago I was evicted from the apartment I had lived in for 13 years and from time to time they try to evict me from my dacha. My request to the authorities, made on your advice, to restore the possibility of my carrying out the scientific work that I think interesting and important, has remained unanswered. The Academy is even curtailing the very modest means that they provide for my personal research and thus further restricting the minimal conditions for scientific work that I have created by myself in my home laboratory.

Obviously such conditions would drive any scientist into a state of nervous depression. Only the fact that I live in seclusion, and keep busy with physical work, not allowing myself to think about my former scientific interests, and concentrating on theoretical researches, enables me to preserve my mental balance.

Any contribution from me to the Academy meetings under the existing conditions would not only be valueless but would place a heavy burden on my spirit. I hope that the time will come when scientific truth will triumph and the Praesidium of the Academy will re-examine the resolution of 20 September 1946 and its negative assessment of all my scientific work, and I can then renew my closer contact with academic life.

(I) To V.A. Engelhardt   Nikolina Gora, 2 December 1949

Deeply respected Vladimir Aleksandrovich,

I am very grateful for the books and articles which you so kindly lent me two years ago and which I now return. I then thought that in the position I found myself, the only possibility of continuing scientific research work was to take up biology. I thought that there was no branch of physics where it would be possible to look for really new and significant phenomena. But I was wrong. Almost immediately after I was deprived of my Institute and its facilities for low temperatures and high magnetic fields, I came across an interesting question in hydrodynamics – the flow of a thin layer of a viscous liquid. Ever since the time of Poiseuille this was considered as a classical case of laminar flow, but I realized that there are a number of indications that this is not so. In fact, if surface forces are taken into account, theory shows that the motion will be wavelike. It is rare to discover a new form of wave motion and I decided to look into it. With the modest means at my disposal in my dacha, and of course with the help of my son [Sergei], I succeeded in observing and studying this type of wave flow and in confirming my theory. I enclose a reprint as a keepsake [this is paper (28)].

Having finished this paper, I came across yet another bit of "virgin territory", which I also hope to be able to conquer with the modest means available to me.* At present I am working alone, without assistants and, as in my student days, I have to make my apparatus with my own hands. It turns out that I have not lost the knack – the cutting tool of the lathe still obeys me – so there is hope of success, and I still have interesting problems for another year of "physics at home". But nevertheless I may perhaps still eventually come to biology. I feel however I have no right to retain your books any longer, but if I do turn to you again in the future I hope you will treat my request once again as amicably and kindly as before. Once again many thanks and greetings.

---

* This refers to the high-power microwave project outlined on p. 390.

The fawning style of the references to Stalin in the following letter is typical of that used during the time of the "personality cult".

(I)   From S.A. Khristianovich   Moscow,
      Pro-Rector of Moscow State University   28 December 1949

On the occasion of the 70th birthday of the Great Stalin, our whole country demonstrated its loyalty to our Soviet state and to the construction of communism in our country and its passionate love for Comrade Stalin, leader of the Soviet people and of the whole of progressive mankind. There were celebratory meetings and sessions devoted to the 70th birthday of Comrade STALIN at the Physico-Technical faculty of Moscow State University and at the Academy of Sciences, where you work, but you were not present at any of them. This has greatly puzzled our scientific community and we are unable to find any satisfactory explanation [of your absence]. You must agree that it is not possible to entrust the education of our scientific youth to someone who sets himself against the whole of our people in such a demonstrative manner . . .

(I) To S.A. Khristianovich   Moscow, 29 December 1949

My absence from the meetings [you mention] was for the single reason that the state of my nervous system continues to be very poor and is far from recovery. It is therefore extremely trying for me to be among a crowd of people at meetings, at the theatre, etc., as I wrote to the President of the Academy [see p. 386]. The interpretation you have put on my absence, as some sort of demonstration, is totally incorrect and absurd and is entirely without foundation.

The breakthrough referred to in the next few letters was in the development of a new source of high-power microwaves. This opened up the possibility in principle of producing powerful beams with military potential. These ideas were somewhat similar to those of the present American

Strategic Defence Initiative programme, although with the technology of 40 years ago, even more remote from practical realization than the SDI of to-day. Kapitza's emphasis on the potential military applications of his work was probably intended not only to obtain material backing for his work at Nikolina Gora, but also to protect himself against Beria, since Malenkov was known to be Beria's rival in the struggle for first place in Stalin's favour.

(I) To G.M. Malenkov   Nikolina Gora, 25 June 1950

I am approaching you not just as one of the leaders of the Party but also because I have always greatly appreciated your interest in my work. I think that the significance of the question I am writing about justifies my giving you a detailed account.

During the war I was already thinking a lot about methods of defence against bombing raids behind the lines more effective than anti-aircraft fire or just crawling into boltholes. Now that atomic bombs, jet aircraft and missiles have got into the arsenals, the question has assumed vastly greater importance. During the last four years I have devoted all my basic skills to the solution of this problem and I think I have now solved that part of the problem to which a scientist can contribute. The idea for the best possible method of protection is not new. It consists in creating a well-directed high-energy beam of such intensity that it would destroy practically instantaneously any object it struck. After two years work I have found a novel solution to this problem and, moreover, I have found that there are no fundamental obstacles in the way of realizing beams of the required intensity.

The next stage consists in finding the precise experimental conditions for producing one of the types of radiation I have discovered. Research on a new phenomenon does not usually require very complex or powerful equipment and my present small laboratory was quite adequate for the solution of the relevant problem. However, original research is always dogged by failures and mistakes and is best done without too many observers. For a whole year I searched unsuccessfully for a way to produce one of the types of radiation that I had identified by theory. It was only at the end of December last year that I got on the right track and since then the work has gone well and may now be considered complete . . .

The next stage should consist in gradually increasing the power of the beam in order to reveal what technical difficulties may arise, but it would be very difficult to carry out this stage of the work with my present facilities. There are good grounds for thinking that a very promising way has been found which should be pursued to its end in the interests of Soviet science. I therefore decided to write to you for authoritative guidance as to how a person in my situation should now proceed. But first I want to tell you about some interesting problems that can be attacked by the use of the phenomena I have discovered. The process of radiation emission from a plane surface that I have discovered also has high efficiency so it not only permits the transformation of electrical power into an energetic beam but also the reverse process. It can be used to transform electromagnetic radiation into conventional electric power, thus opening up the possibility of long distance transmission of large amounts of energy without conductors . . .

It seems to me that electronics is in the same situation today as electrical engineering was in the middle of last century. At that time the practical use of electrical distribution along conductors began with the development of electrical communications and it was only when people learned how to construct high-power generators that electrical power became the basic means of energy transfer over long distances and this led to the electrification of the country. Today we make wide use of electronics only for radio communication because we do not know how to operate with large power. It can be reliably predicted that the future of electrical engineering in the second half of this century will be closely linked with high-power electronics and with the solution of problems like those I have outlined above. The work I have done opens up this path and that is why it is of fundamental significance.

Let me now return to the question of the future of the work I have described. Discovering and understanding new phenomena is very interesting and attractive for a scientist but if he convinces only himself about the significance of his work this does not amount to a scientific achievement. First, my achievements have to be recognized by other scientists if they are to be of value in scientific work and, second, the significance of the discovery has to be confirmed by practical applications. In my situation it is virtually impossible to fulfil these two conditions. It would be easy to fulfil the first if I could publish my theoretical and experimental work but the present international situation requires us to keep our work secret,* often to our own disadvantage. The practical applications, moreover, would exceed the limits of the facilities at my disposal.

There is no doubt that the main difficulty in the way of the development of my work is the ban placed on me, which makes it impossible to arrange normal collaborative work. I experienced this very clearly during my professorship at Moscow State University. When I got the Chair in the Physico-Technical Faculty three years ago, my chief difficulty was in finding the basic staff. Even the most junior workers would not take the risk of association with a banned person.

For instance, I offered the job of being my deputy to three young physicists, two of whom had been my own students, but they all refused. During my two years at the University I not only failed to find sufficiently qualified staff to work with me but it wasn't possible to get even only slightly novel lecture demonstration experiments set up satisfactorily and there were occasions when, in despair, I made the apparatus for a demonstration with my own hands at home and took it with me to the lecture.

I cannot blame people for not coming to work with me. As they told me themselves, they did not feel that my position on the faculty was very firm and in fact, after two years work I received a letter [see p. 389] from the Pro-Rector (Academician Khristianovich) informing me that owing to my absence from the meeting in honour of Stalin's 70th birthday I was to be dismissed and my written explanation was considered inadequate. My appointment was officially terminated that same month on the grounds of "absence of pedagogical workload" . . .

I am told there is no chance of my being allowed to return to the Institute I created. The most valuable aspect of my Institute was the staff, not only the scientists, but also the technicians, mechanics, and other supporting staff. I chose them all carefully and trained them. If I am able to find working space and facilities somewhere I shall have to start from scratch in finding staff. Of course, I can train young people again but this takes two or three years, and to do it alone is not easy and uses up a lot of energy.

It is largely my own fault that I have got into this position since,

---

* Secrecy was partially lifted in 1962 [see publication (33)].

obviously, a man must learn to adjust his behaviour to conform with the actual situation in which he finds himself. I am now ready to forget any feeling of resentment and I want to do all I can to change the existing position and to achieve good conditions for the development of this important work, but I don't know how best to set about it. I have approached you in the hope that, when you have assessed the existing situation, you will be able to show me the way that will lead to a full development of my scientific work.

(I) To J.V. Stalin   Nikolina Gora, 22 November 1950

About a year ago I wrote to tell you that I was working on a new method of producing high-energy beams of great power and that I considered this problem very important. In the spring I finally obtained the effect I was seeking and I reported this to the President of the Academy of Sciences. Vavilov readily agreed to make it possible for me to produce beams of power greater than ever before. In accordance with his instructions accommodation is now being completed and the necessary equipment provided for my work. This is a modest project but it will suffice since the process is easily demonstrated. However, there are limits to the help Vavilov can give me. It will be very difficult for me to carry out experiments alone as I have to work with high voltages at high power and for this I need an assistant whom I can trust. Vavilov has asked the Director of my former Institute several times to second my former assistant to work with me but without result. He tells me he has no power to order this since the Institute is not controlled by the Academy. . . .

The only way left to me is to appeal to you for help and to ask you urgently to give instructions for my former assistant to be made available to me. I take this bold step because the phenomenon I have discovered is already of general scientific value and so my request for help has a firm foundation. It would be no exaggeration to say that I have hit upon a scientific pathway of great significance so it is most important to determine as quickly as possible what are the limits to the perspectives it reveals. This requires more powerful equipment than was needed to discover the phenomenon. If I am given real practical help now I should begin to obtain clearly demonstrable results in a few months. But the work has been held up for six months while I have been trying to obtain equipment. I need little, but I need it quickly.

The following letter was written as a result of a curious incident. Kapitza's
dacha was not on the telephone but Malenkov telephoned a Central Committee dacha not far away and asked them to take an urgent message to Kapitza,
asking him to get in touch with him. This Kapitza did by telephoning from
a sanatorium not far from his dacha and Malenkov asked him to send a
detailed report on his work to Stalin at once. Then Kapitza said that he
doubted if his letters ever reached Stalin or if he ever read them, Malenkov
said: "Comrade Stalin reads not only the letters you write to him but also
those you write to me". It is possible that Kapitza owed his life to the fact
that Stalin enjoyed getting letters from him.

(I) To J.V. Stalin   Nikolina Gora, 30 December 1950

Comrade Malenkov told me I should send you a detailed account of
my present work in the field of electronics and I am sending a report
of what I have done during the last four years and what I propose to
do in the immediate future.

On 5 May 1950, I sent the President of the Academy of Sciences
an account (copy enclosed) of the scientific basis of the problem of obtaining very powerful beams for defence purposes and of the results
I had obtained. Following this approach to Vavilov, the Academy of
Sciences began to give me more active help in my work. Another
reason for sending this account of my work to Vavilov was to draw the
attention of the Academy to this matter, which I believe to be timely
and capable of realization. It would be a pity if someone else solved
the problem before us.

I should like to emphasize once again that this line of research is exceptionally important for us. Of course, in scientific research one can
never guarantee what is going to happen and even finding the answers
to the questions already before us can be difficult. One thing for which
a scientist must take responsibility is that the line of research he has
chosen is the correct one. The more I delve into this problem the more
essential it appears to me and it is my duty to the country to draw attention to it. In the enclosed memo I have therefore given a very broad
picture of possible developments in electronics.

In my work at present I am trying to produce high-power beams as
quickly as possible so as to demonstrate the significance of the line I
have chosen, since experiment is the only final and convincing proof.
I now have almost everything I need for beginning these experiments.

The main hindrances are the supply of energy and the difficulty of working without competent assistants. If these obstacles can be overcome I can start the experiments in February. Once more I ask for your help – very modest help, merely to speed up work on the problems in hand. A scientist needs to feel that he is trusted to some extent – without that he finds it very hard to carry on.

(I)   To A.N. Nesmyeanov   Nikolina Gora,
      President of the Academy of Sciences   20 October 1951

Deeply respected Aleksandr Nikolayevich,

. . . As soon as the long awaited electricity supply arrived, I have been totally immersed in my work, which has turned out extraordinarily interesting and absorbing. I have not worked so intensively for a long time – it is usually only in one's youth that one works with such absorption and stubbornness – and the work is going well. Moreover, the problem is very important for us [i.e. for the country] and, as far as can be foreseen in such cases, it can be solved. If you can help accelerate its solution I shall be very grateful.

The main things I need are another pair of reliable hands to help* and more perfect and up-to-date equipment. As the Russian proverb** has it "poverty makes you cunning" and it is true that even with the very modest resources available, it's possible to think of various "tricks" to overcome technical difficulties, but unfortunately this always involves loss of time . . .

(I)   To G.M. Malenkov   Nikolina Gora,
      President of the Council of Ministers   22 July 1953
      [formerly Sovnarkom]

Two months ago, on the 15 May, I attended a meeting at the Academy of Sciences with the President and others to discuss the

---

* He has in mind more help from S.I. Filimonov who was then assisting only on a half-time basis.
** A rough English equivalent is "Necessity is the mother of invention".

organization of my work in electronics. In view of recent events* I think I should write to you about it. Since the President was hesitant about supporting my work I raised the question of why a scientist such as I, should have been prevented for the last seven years from giving of his best. I asked him to point to a single important error in my scientific work which might justify such an attitude to me.

The facts tell a different story. When I was punished in 1946 the main charge against me was that there were errors in the new method of oxygen production I had developed, that is, the low pressure cycle. Along with some other scientists, I could not agree with this and the passage of time clearly showed that the decision of the Council of Ministers was a mistake. Today my work on oxygen is widely recognized not only abroad but also here. The large installation in Tula is constructed on the basis of my low-pressure cycle and uses my turbo expansion machines. As was to be expected, all the economic indicators for this equipment have proved much higher than those for the former German high pressure machine.

It appears that I was one of the first scientists to speak openly about the practicability of using atomic energy to create an atomic bomb of enormous destructive power and I pointed out the need to work on this problem. (I have in mind my address to the antifascist meeting . . . on 12 October 1941). So I ask why have I been cut off for seven years from my students and colleagues and from the unique machines I created for studies at low temperature and in high magnetic fields? It is generally recognized that this work elevated Soviet science to a leading position in one of the most important areas of modern physics.

I asked Nesmeyanov as President what could I have done that was so bad as to justify the current systematic suppression of my scientific work. He replied that he thought the reason did not arise out of my scientific activities and he, as President of the Academy, had not been informed what it was. I asked him to approach the Government with an official request to clarify the reason because, not only as a scientist but also as a citizen, I have the right to know where I have gone wrong and his position as President obliges him to answer such a question

---

* The reference is to the arrest of Beria soon after Stalin's death.

from a member of the Academy. After I had put the question so bluntly the President asked me whether I had had a clash with any members
of the Government. I told him there had been no clash, except for one
with Voznesenki. Nesmeyanov remarked that this would not now be
significant but asked whether I had not had a clash with Beria. I said
there had certainly been a number of serious disagreements on matters
concerning scientific work but surely that could be no reason for deny-
ing a scientist the possibility of continuing his work. I was told not to
deceive myself and Academician Topchiev, who was present, said that
Beria was now in full charge of the division of new technology, to
which my present work on electronics as well as my former work
would belong.

Following this conversation I still could not believe that
Nesmeyanov and Topchiev were right and that in our country work
on scientific problems could be prevented because a particular
member of the Government had taken a dislike to a scientist. The
above-mentioned conversation takes on a completely new significance
in the light of the events of the past few days. Nesmeyanov and Topchiev obviously possessed information leading them to conclude that
Beria personally had hindered and blocked my scientific work all those
years. I believe I now have grounds to petition the Council of
Ministers to review their attitude to my scientific work and I make the
following request.

The fact that industry has successfully changed to the low-pressure
cycle that I suggested for producing bulk quantities of oxygen clearly
indicates that the assessment of my work in the decision of the Council
of Ministers on 17 August 1946 and in the statement by the Academy
on 20 September 1946 was wrong. Recognition of this mistake would
not only be salutary for the advancement of Soviet science and would
give some assurance to our scientists that in the Soviet Union scientific
truth finally triumphs, but it would also be important for the successful development of my present work in the field of electronics. I
do not intend to return to scientific work on oxygen but the decision
I ask for would also enable me to publish my book on theoretical
aspects of the oxygen problem . . .

I am most concerned about the fate of the work on high-power electronics on which, as you know, I am now engaged. In spite of all the
difficulties and obstacles I encountered during those years, I managed
to get so far with the solution to the problem that scientific opinion,

in the form of the academic commission you know about, recommended unanimously a year ago that my work be supported and developed. However, nothing has yet been done and I am not receiving the help I need. From my discussion with Nesmeyanov and Topchiev the reason appears to be Beria's intervention and there is now no reason for further delay.

The principal task for me now is to test a series of new ideas experimentally, especially concerning the cumulative acceleration of charged particles and I have been demanding help for this project from the Academy. I am asking for very little in the way of expenditure – the main thing is to get the work done very quickly. In original research the greatest danger is loss of time, and we have already lost a lot of time. Nesmeyanov told me that without a decision by the Government on these questions he could not take the responsibility on his own shoulders. So without help from the Council of Ministers this project will never get going properly. I am therefore asking you to give the following instructions:

(1) To construct the necessary special laboratory rapidly. It is not large, about 2000 cubic metres, and the design for it is ready. A site of some 2 acres must be secured and the construction done rapidly. All the information is known to Academician I.P. Bardin who is the acting President of the Academy.

(2) Arrange for the rapid construction of special equipment (solenoid, etc.). The design and ordering details are ready and are with Academician A.V. Topchiev.

(3) My work must be legalised and I must be given a staff of about 10 or 12. To prevent me being distracted from my scientific work I specially request that all the administrative services of the laboratory should be organized by the Institute for Physical Problems.

I hope that the questions I am raising are in tune with the times and . . . that you will show consideration for my request even though this must be a very busy time for you. I recall the help you gave me in my scientific work with gratitude.

His requests were granted very rapidly, and following an order of the Academy dated 28 August 1953, a new "Physical Laboratory" was authorised. This was, at first, at Nikolina Gora, but after 1955 in a special building of the Institute for Physical Problems.

Soon after Khrushchev came to power, Kapitza established a correspondence with him and on one occasion when they met even succeeded in being "one up". This was when at a reception in the Kremlin, Khrushchev rebuked Kapitza for wearing a Ukrainian shirt (see fig. 80) rather than formal attire. Kapitza replied "But surely diplomatic protocol recognizes national dress as appropriate for formal occasions?". This was immediately confirmed by I.M. Maisky (former Ambassador to Great Britain) who happened to be standing by and Khrushchev had to withdraw his criticism.

80. Wearing his Ukrainian shirt.

(I)   To N.S. Khrushchev   Nikolina Gora,
       First Secretary of the Central Committee   12 April 1954

Deeply respected Nikita Sergeyevich,

It is a good thing that a number of the most important aspects of our
way of life have recently been subjected to a genuinely open and
critical scrutiny. As a result people have become more confident that
the development of our country is proceeding along the right lines.
However, the organization of science has, for some reason, remained
in the shadows and it has to be said bluntly that the situation in this
respect is unsatisfactory.

I have therefore read with great interest the leading article which
has just appeared in *Kommunist* under the title "Science and Life".
This is a good article since it is the first which presents a direct and
accurate diagnosis of a number of the ills that afflict our science,
though it does not investigate the necessary remedies. However, there
is a more substantial criticism that can be made, I think, not only of
this article but also of a simplistic understanding of the links between
science and practical affairs that is very widely held here.

It is taken for granted that the principal task for science is the solution of the urgent difficulties facing our economy – and science, after
all, certainly ought to be able to do this – but that is not its most important aspect. Really advanced science, by studying the laws governing the natural world around us, will search for and create fundamentally new avenues for the material and spiritual development of society. Advanced science is not limited to meeting practical demands but
makes independent contributions to culture and so changes our way
of life. I refer to such fundamental innovations as for instance radio
was in its time, and atomic energy and antibiotics are now. These innovations were derived from new scientific discoveries and theories
made in laboratories, without any stimulus from the world of ordinary
practical affairs. Of course, the solution of these problems is closely
linked with the demands of life but the link is not a simple one for it
can be understood and correctly assessed only gradually, firstly by
"scientists" and much later by "practical" people.

For many years the practical people among us were scornful about
research on the atomic nucleus because they could envisage no

immediately useful applications. If we had kept to the narrowly practical recipe for reinvigorating science that the leading article in *Kommunist* proposes we would never have found a way of harnessing atomic energy. It will need a bold initiative and the solution by scientists of equally novel problems to bring science in our country up to the level of real leadership.

Following new paths in science is troublesome since you don't immediately find the way and in the search for it you may be led up blind alleys. You must never let failure get you down and this requires courage, enterprise and perseverance. These are just the conditions that we have not yet succeeded in creating and we must frankly acknowledge that here we lag behind the capitalist countries. In ordinary scientific research aimed at solving urgent problems arising in practical work the research can be organized along well established lines, so bureaucratic methods don't hinder the work too badly and a successful result immediately brings a dividend. In that type of research we do reasonably well, so well even that the successes in our scientific establishments often conceal the absence of real creativity.

I am deeply worried by this state of affairs. We undoubtedly have sufficient potential creativity and it all comes down to a question of organizing it. To develop first-rate science it is necessary to begin by setting some fundamental theoretical problems as targets. Then the conditions for scientific work must be made more agreeable than those we have here at present. One has to remember that the list of fundamental problems in science is not a long one and that there are very few people able and willing to solve them, so it is most important to choose people carefully and to look after them intelligently.

The Council of Ministers has given the Academy the task of selecting the leading problems in all branches of science. It seemed at first that they would choose some two or three such problems in each field but it has turned out differently. Already 80 problems have been selected of the most varied character but most of them cannot be considered fundamental problems of the type I referred to, while a number of really fundamental problems have been overlooked. For example, one of the selected problems, which is certainly important but hardly fundamental is "Fighting the waste in electric lamp factories" while the basic problem of the efficient combustion of coal in power stations is not included. This is a fundamental problem of electrochemistry linked with the possibility of oxygenated gasification of coal and is of pressing importance. I personally think that it will be solved within the next few decades and will then revolutionize energy supply since it will change the aspect of contemporary power stations, increase their efficiency, make them simpler and eliminate steam turbines, boilers and so on.

The timid and hostile attitude that our scientists adopt to new basic problems is not a matter of chance. They have been intimidated by often being "attacked", painfully and for nothing, until they have lost all confidence. This is because their work is assessed by bureaucrats instead of by the concensus of scientific opinion. The unhealthy secrecy surrounding the results of scientific work has totally eliminated any general scientific opinion. It is obvious that first-rate scientists cannot develop in these conditions.

The fundamental stimulus for any form of creativity is dissatisfaction with the existing order. An inventor is not satisfied with existing processes and thinks up new ones, a scientist doesn't like current theories and searches for a more general one and so on. People who are actively dissatified are restless and temperamentally disinclined to be obedient sheep, that is, the kind of scientists beloved by our bureaucracy because it is less troublesome to work with them. It is, after all, more comfortable to ride a gelding but it is the restive thoroughbred that wins the race.

With respect,

Yours sincerely,
P. Kapitza

(I) To N.S. Khrushchev   Nikolina Gora, 16 January 1955

I have been recently asked more than once to write an article about atomic energy for one of our journals. I couldn't fulfil this request because I had no opportunity to see relevant foreign publications. When I asked for foreign socio-political journals I was usually told this was unnecessary since they themselves could tell me exactly what to write, i.e., they proposed to follow the well-known recipe – "The ideas are ours – the signature yours". Recently, however, the journal *New Times* sent me the necessary foreign journals and I was able to write the article. In it I give my point of view on the contemporary position concerning both the peaceful and military uses of nuclear energy. Much that I have written is along the lines of a recent conversation with you so I think you may be interested to look through the article. I am afraid that the editor of the journal may wish to delete some of my comments. To prevent this, experienced comrades have advised me to ask your opinion on the suitability of the material for publication before I send it to the journal.*

A month has gone by since you told me that I could count on help with my work and on a friendly attitude to myself but nothing has changed yet in the conditions under which I work. The long-standing request by the Academy of Sciences to the Central Committee to give me back my Institute has remained unanswered. All this disturbs me, mainly because it suggests that the problem I have taken up – "High-power electronics" has not really been given serious consideration. I was evidently mistaken in thinking that you wished to resolve matters quickly when you found that my sentence of five years was excessive. The way things are going it looks as if it will be still longer.

In fact, he was reinstated as Director of the Institute just 12 days later.

(H, I)   To N.S. Khrushchev   Nikolina Gora,
         Personal   22 September 1955

I enclose copies of letters that I wrote to Comrade Stalin 10 years ago. They reflect the abnormal conditions for scientific work which obtained at that time and which have not yet been completely eliminated today, so these letters may be of some interest to you.

I draw attention to the letter of 25 November 1945 [see p. 372] in which I reiterate my request to be released from work on the atomic bomb, as a result of which I was in fact released on 21 December. It is perfectly clear from this letter that the sole reason which forced me to opt out of this work was the intolerable attitude of Beria to science and scientists. It seems to me that the criticisms I made at that time about our initial steps into the field of atomic development were subse-

---

* This article was not in fact published.

quently taken into account and proved useful. There is therefore no foundation for the reproach that I am a pacifist and refused to work on the atomic bomb for that reason – although, personally, I do not see why it should be reckoned a crime if someone firmly believes that he should take no part in making destructive and murderous weapons. During the war I took an active part in our defence work and I believe it is natural and right for a man to defend his country against external aggression. As regards my struggle with Beria, I consider it was not only justified but also of some use.

Among the letters are some that cast light on the oxygen problem and on the path Beria chose to ruin it. This may be of interest since the Central Committee [of the Communist Party] is now reviewing the former decision of the Council of Ministers on the oxygen problem . . .

(I) To N.S. Khrushchev   Nikolina Gora, 15 December 1955

As you may remember, I visited you exactly a year ago and the main subject of our conversation was the fate of Soviet science. The situation is still not a happy one. Now that I have returned to active academic life and am more closely in touch with what is going on in the Academy of Sciences and in its Institutes, I have observed that the quality of scientific work has not only not improved over the past year, but has somewhat deteriorated. So I have decided to draw your attention again to the fate of our science . . .

In these days, if our footballers score a goal against a foreign team, our boxers fight well or our ballerinas twirl and leap better than any others, this may be very pleasant and flattering but it is not a convincing proof of cultural leadership. Only when we achieve a leading position in science will we be recognised internationally as a country which has created the most advanced social structure.

Our leaders usually consider that the comparatively large sum earmarked for science in our state budget is sufficient evidence of their concern for science. Certainly, a large amount of money is required for the successful development of science, but it must also be used effectively and, regrettably, that is not the case here. What is happening now is similar to what happens in agriculture if you begin expensive enrichment of the soil without bothering about the plants to be grown.

If the ground is too liberally manured the weeds will grow even better than the useful plants. This is roughly what is beginning to happen to us in science. Thanks to high wages and privileges for scientific workers, weeds are shooting up and engulfing the genuine scientists. The privileges now established for scientists here can give positive results only if there is a very efficient process for rooting out weeds and this we do not have. The situation which has arisen here during recent years is as follows: the weeds, profiting from the weakness of our bureaucratic system, have grown so strong that they have begun to retard the development of healthy science and the situation has become threatening.

The only well-tried method of eradicating these weeds is through a healthy public opinion and this we lack. What must we do to encourage the growth of a healthy public opinion on scientific questions? First and foremost is the need for a natural desire among scientists to have free discussion. To foster such a desire a man should never feel afraid of expressing his opinion even if it is rejected. Our leaders need not fear that in the search for scientific truth some erroneous scientific hypotheses may arise. Scientific truth is unique and a correct hypothesis always finds a way to survive. The very fact that it survives is proof of its correctness. This is the dialectical law governing the growth of our knowledge of Nature. It is not only useless but extremely harmful to define scientific truth by decree as has sometimes been done by the scientific department of the Central Committee, especially when [Yu.A] Zhdanov was its chairman.

A scientific idea has to be born and take root in competition with other ideas, for only in this way can its correctness be tested. If competition is suppressed scientific knowledge turns into dogma and the development of science ceases. However great and significant a dogma may be, it is static. This has been most strikingly illustrated by the development of materialistic philosophy. To a large extent the whole content and achievements of the Marxist classics have been transformed into a series of dogmas and consequently philosophy has ceased to develop here. We should not be afraid to admit that in our country science is dominated by legalistic scribes like Nudnik in Korneichuk's play *Wings* rather than by lively scientist-philosophers, eagerly trying to understand Nature in a way that is relevant to the demands of our time.

What has happened to philosophy here is something like what

would happen to chess if every player who lost a match was never allowed to play again. In the end there would be only one player left . . . and the game would cease to exist. With us only materialist philosophers are left and they have forgotten how to argue or to think. This would inevitably happen to even the most eminent scientist if he never had to defend his ideas. Had it not been for Bogdanov, Mach and others, and the possibility at that time of free and unfettered discussion, Lenin could never have written his classic work *Materialism and empirio-criticism*. If we are not to give up the chance of playing a game we must respect and take care of the players who lose. Lenin gave an excellent example of respect for his opponents though he attacked their ideas mercilessly.

These are all simple truths. Why ever did our scientists lose the urge to discuss new ideas? Why have scientific discussions at general meetings of the Academy of Sciences degenerated into popular lectures? Academy meetings nowadays differ little from the collective farmers' meeting in Korneichuk's play. "Nudniks" in the Academy read lectures with no relevance to real life, generally on some historical topic, celebrating the memory of great scholars or great events, with no discussion and no conclusions. You can read these lectures more comfortably and profitably by sitting at home. Nowadays a meeting of Academicians is not a leading scientific forum for the solution of important scientific problems relevant to the development and growth of our culture, but is more like a religious rite conducted in accordance with a prearranged liturgy. This not only brings discredit on the leading socialist state and its science but is also a menacing symptom of the decline of a healthy public opinion and consequently of a healthy leading science. The second condition for the development of science is that its leaders must take note of public opinion and organize our scientific life on that basis. Of course, public opinion cannot be spontaneous – it must be organised and guided along the right lines, but never by decree. We must never forget Lenin's testament – that the bonds between party and society must not be based on decrees.

What is happening here in biology can hardly be ignored and the consequences of our errors in the organisation of science stand out more clearly there than in any other field. Undoubtedly the neglect of sound public opinion and the attempt to define scientific truth by decree has allowed powerful weeds to flourish (Boshyan, Lepeshinskaya and others). Healthy biological science has almost died out and even to scientists like myself working in other fields it is obvious that our biology has never before been at such a low ebb as it is now. This is the more disgraceful since formerly we held a leading position internationally in several basic areas of biology, associated with the names Sechenov, Pavlov, Timiryazev, Tsvyet, Mechnikov, Vinogradsky and many others.

A number of our leading scientists have recently sent a letter* to the Central Committee describing the sad state of biology here and their initiative must be welcomed as an indication of the renaissance of public opinion. There is much in their letter that is good but one aspect is bad – that the writers are again asking the Central Committee to issue a decree about biology, but merely in a different direction. It would have been better to publish the letter and then organize an honest discussion about it . . .

It seems to me that the most important step now for the development of both biology and other branches of science is to organise a demonstration of healthy public opinion by raising a number of interesting controversial questions in the fields of genetics, cybernetics, cosmology, science and life, nuclear energy, theories of space and time, and so on. Discussion of these questions at meetings of the Academy of Sciences and in print would have to be open and sharp, with the participation of foreign scientists and philisophers representing a wide range of opinions. Our philosophers must learn to give up the habit of fighting only opponents whose hands are tied behind their backs and they should try to win in a free fight. Anyone who fears the outcome of such a free discussion has no faith in the strength of Leninism. Victory in such a contest would have much greater importance for us than winning any international sporting event.

The award of Stalin Prizes is one other very effective way of exerting a healthy influence on the development of science through public opinion and I don't know why we have ceased to make such awards.** I am convinced that the former custom of annual assessment of scientific work by the Stalin Committee, composed of leading scientists,

---

* There were about 300 signatories, including Kapitza.
** No Stalin prizes had been awarded since 1952 but, renamed as State Prizes, they were again awarded annually from 1966. In parallel with these, Lenin Prizes, originally awarded from 1925 to 1937, were revived in 1957 and are regarded as rather more prestigious than State Prizes.

and the award of prizes was very worth while. This committee must be reconstituted in one form or another, emphasising its democratic character, for example, by election of its members and for a limited term of office . . .

(H, I)   To N.S. Krushchev
            Personal   Moscow, 23 August 1956

I believe I am justified in raising with you the question of the conditions that are necessary to encourage successful scientific work. No creative worker, whether he be scientist, writer or artist, can work energetically and boldly unless he feels respected and trusted and believes people are interested in his work . . . Could one really imagine that a musician would develop and perfect his talent if he had to play only to a hall full of deaf mutes?

The fact that I do not enjoy a ''good relationship'' is evident from a number of events. Calcutta University has just awarded me the Sarvadhikari Gold Medal, which is apparently the most important scientific award in India. I should have gone to Calcutta to receive the award at a ceremonial meeting on 1 September. Moreover, I received invitations from several Indian scientific institutions in Bombay and Delhi. I have not been allowed to travel to India. The same kind of distrust is apparent in the following incident. Not long ago a number of English scientists arrived in Moscow, among whom were several of my old friends from Cambridge University, where I worked for thirteen years. It was only natural that I should invite them to my home but subsequently I was specially summoned by the President of the Academy, Academician Nesmeyanov, who warned me, in the presence of Academician Topchiev, that I must not associate with foreign scientists unless a third party was present. Excuse my bluntness but this conversation left an unpleasant taste. I had the impression that I was talking not with scientific colleagues but with officers of the security police.

The following fact is even more insulting. In the Praesidium of the Academy there is only one physicist (Kurchatov) but three chemists and three mathematicians. We consider this situation abnormal since just now physics is playing a leading role and absorbing by far the largest part of the budget. Nesmeyanov agreed with this and so it was decided to increase the number of physicists in the Praesidium. The division of physico-mathematical sciences put me forward as a candidate but when Nesmeyanov brought up this question in the Central Committee (it is said with Suslov) he was told to "refrain" from electing me and no election was held.

Let me add the following instance. In 1949 I was dismissed from heading a Chair at Moscow University because I had not attended the meetings celebrating Stalin's 70th birthday. The dismissal procedure was so peculiar that I am enclosing copies of the letter [see p. 389] from Academician Khristianovich explaining the reason for the dismissal and also the dismissal order signed by the Rector of Moscow State University (at that time Nesmeyanov). Recently, Academician Petrovsky, the present Rector, evidently wanted to smoothe over this affair and when we happened to meet he proposed to make me a member of the scientific council of the University at the first opportunity, but nothing came of this. The Ministry of Higher Education refused to confirm my candidature.

What oppresses me most, however, is the oxygen affair. By the decree of 17 August 1946, the Council of Ministers condemned my work on oxygen and criticised me both as a scientist and as head of the Oxygen Board. I was taken off everything and have been excluded from participation in the oxygen work ever since . . . But only two or three years later I was proved right when my low-pressure method of producing oxygen began to be exploited in England, France and America. These countries needed my patents and many requests to buy them were received. Since these approaches from foreign industry were the best proof of the originality and progressive nature of my oxygen work the then President of the Academy, S.I. Vavilov, wrote in the name of the Academy to the Council of Ministers asking that the unfavourable decree about my work should be reconsidered. Moreover, he advised me to write to Comrade Mikoyan about the need to sell my patents. But neither the Academy nor I received any reply . . . Last year the Academy again (for the third time) asked the Central Committee and the Council of Ministers to re-examine the decree about my oxygen work, but for over a year there has been no progress . . .

Even during the years when I was cut off from major scientific activity I continued to feel that the wider scientific community valued my achievements highly. It is not just that my work has got into the

textbooks, both here and abroad, but there is not a major country that
has not recognized my scientific work by electing me to honorary
membership of their academy or to an honorary doctorate or by awar-
ding me a medal. This is objective proof that my work is highly
valued. Of course, in normal circumstances these foreign honours
serve chiefly to satisfy personal vanity, but in the situation I was in at
that time they served to confirm my confidence in the correctness of
my views and they helped to sustain my morale . . .

    I am writing in such detail to draw your attention to this question
since it affects not only me but others among our outstanding creative
workers, who are not working at full stretch because they are not properly appreciated. I think this is one of the principal reasons why we
are increasingly losing the leadership both in science and in art. To
stimulate the development of any kind of creative activity, an ap-
preciative atmosphere is more important than any material benefits.

(I)   To Academician P.N. Fedoseyev   Moscow,
       Vice-President of the Academy of Sciences   7 January 1964

Deeply respected Pyotr Nikolayevich,

I enclose a photocopy of an article by F.N. Kaplunenko entitled "Is
Kapitza's picture of the future of science in line with the Marxist position?" (*Journal of Kishinev State University*, No. 3, 1963). Of course,
every scholar has the right to criticize the views expressed by another
but Kaplunenko has taken two small points from my article* and
created the impression that my whole article was devoted to these
alone, and that is not permissible. Most of all, however, I am disturbed by the spirit of his article – "prohibiting and forbidding". F.N.
Kaplunenko expresses doubts as to whether Academician Kapitza has
any right to put forward his views publicly. In his opinion the publication of such an article as mine should be accompanied by an editorial
comment explaining that this article is published only to stimulate

---

* The article by Kaplunenko was devoted to two publications by Kapitza: "The
Future of Science", *Nauka i zhizn*, 1962, No. 3, p. 18 and "Theory, Experiment,
Practice", *Ekonomicheskaya Gazeta*, 26 March 1962.

discussion, since "it does not express the established point of view of scientific circles in our country on the problems in question".

I should like to remark that Kaplunenko's opinion is not original. The matter was examined in sufficient detail by Kozma Prutkov* in his well-known "Plan for introducing unanimity in Russia". Comrade Kaplunenko ought to know that the only articles worthy of publication are those that stimulate discussion. If an article does not stimulate discussion its proper place is in the waste paper basket. This is indeed the dialectic of progress in science.

Intolerance and the partition of scholars into orthodox and heterodox, with the resulting unpleasant consequences for the heretics, frightens scholars away from discussion and in my view this represents the chief hindrance to the development of friendly cooperation between philosophers and experimental scientists. It is only because Kaplunenko's article illustrates this negative point of view that it deserves to be drawn to your attention.

(I)   To M.V. Keldysh
      President of the Academy of Sciences   Moscow, 5 June 1972

Deeply respected Mstislav Vsevolodovich,

The present conditions for visits abroad are intolerable. You probably know that my visit to Niels Bohr's Institute in Denmark could not take place because permission to travel was granted only a day after the scheduled departure date. . . . The telegram from our ambassador in Copenhagen, N. Egorychev, makes it clear that the cancellation, only a few hours before the previously agreed time of my arrival, offended the Danes, since they had carefully prepared a programme for my visit which, of course, involved them in expense and they have not as yet renewed their invitation.

Naturally, I cannot but feel deeply offended too, since I can put no other interpretation on what has happened than that it indicates a disdainful attitude to me. I am an old man, a scientist, fulfilling a commission given me by the Academy of Sciences and I had a right to

---

* Pseudonym for a group of satirical writers in the mid-19th century.

expect some vestiges of respect. It was much the same with my participation in the meeting now in progress in Brussels on questions of European security. I am a member of our committee on European security and I sympathise strongly with its activities and am glad to help. A few days before my aborted trip to Denmark on 22 May I learned for the first time . . . that I had been included in our delegation to Brussels. If you respect a person and value his participation, you naturally ask him before making up the list of delegates, whether or not he is able and willing to go and you tell him what part he is expected to play. Even now it is not clear to me whether I am expected to speak to the conference in Brussels or merely to be there for the sake of decorum.

My experience of trips to such conferences has taught me that they usually involve me in a heavy load of work. Since I can speak some foreign languages and am well known, many people wish to speak to me, especially journalists, and they put questions that are often difficult to answer. If I am to take part effectively, I need adequate time for preparation and close contact with the leader of the delegation and these did not exist. For instance, the leader of the delegation, Comrade A.P. Shitikov, has not once discussed matters with me. I feel I am being treated like a football, which is also not consulted before being kicked. Naturally, in these circumstances, I was in no position to go to Brussels, although I regret not being able to help in this promising initiative.

As you probably remember, only six months ago the Academy of Sciences asked me to represent it in London at the Royal Society celebrations of the centenary of the birth of Rutherford, who was himself a foreign member of our Academy. I prepared a lecture but a few days before my departure my trip was cancelled. According to what you told me at the time, this was bound up with diplomatic misunderstandings due to the expulsion from England of employees of our trade delegation. To this day I cannot see the logical link between these two events. Why should a diplomatic incident deny us, as scientists, the opportunity of expressing our respect for a great English scientist?

From the examples I have given it is obvious that the conditions regulating contacts with foreign scientists are not only abnormal but their realization often entails an offensive attitude to the scientist involved which borders on the outrageous. Since contacts with foreign scientists are organised by the Academy, of which I am a member, I feel I must inform you, as President, about the abnormal conditions under which such contacts take place. I also beg you to let me know what I must do to protect myself in future from similar personal insults which, moreover, impose a great strain on the nerves.

I had planned some visits abroad for September of this year – to the Pugwash Conference in Oxford and also, as you will have heard from the French Academy, to Paris as representative of our Academy of Sciences at the centenary celebrations of the birth of Paul Langevin. In December I am intending to visit Delhi at the invitation of the Indian Government. All these trips are included in the Academy's programme of international links.

In view of all that has happened, and of the existing conditions for trips abroad, I do not see how I can fulfil all these engagements and I am letting you know this now. I regret having to trouble you but it is well known that international scientific links are essential for the successful development of science in any country including our own. Under the conditions in force here the normal development of such links is impossible and this situation must be changed. A number of other members of the Academy are of the same opinion.

(I) To Yu.V. Andropov
     Chairman of the KGB   Moscow, 22 April 1980

Deeply respected Yuri Vladimirovich,

I enclose with this letter a copy of my book *Science – an international activity*, published in Italy by agreement with VAAP.* It contains translations of basic articles from my book *Experiment, theory, practice* that I sent you some two years ago. My reason for sending you the Italian book, which is one of the copies sent to the author by the publishers through the post, is that, as you can easily confirm, the first 30 pages containing the preface entitled "Peter Kapitza, humanist, scientist and practical revolutionary" are missing from this copy. They were torn out at the post office by our censors.

The third edition of the book *Experiment, theory, practice* will be

---

* This is the Soviet organization for the protection of authors' rights.

published here next year and it has also been published abroad in nine languages, but only the Italian communist publishing house inserted the preface, written by a communist philosopher, Comrade L.L. Radice and torn out by our censors. I learned of the existence of this preface only after the book appeared in print. The question arises: why did our censors think it necessary to prevent me reading the preface to my book, written moreover by a communist scientist?

This is not an isolated instance of our censorship. In foreign journals to which I subscribe through the Academy of Sciences, whole articles are often cut out. Several times I tried to subscribe to the newspaper *Le Monde* and the journal *Newsweek* but they very soon ceased to arrive. It is incomprehensible why it should be necessary to prevent me receiving information from abroad. What is this all about – is it a concern for my morals?

Indeed this censorship is harmful for us. Here is an example. In 1967, Leiden University awarded me the Kamerlingh Onnes Gold Medal. This is usually presented on the birthday of the famous scientist whose name it carries, during the year following the award. When I arrived in Leiden I noticed some embarrassment among the professors. It turned out that this was due to the fact that in the interval between the announcement of the award and the presentation of the medal there had occurred the events in Czechoslovakia. The presentation is a very formal occasion on which students as well as professors are present and they informed me with some distress that the students were intending to organize a demonstration. I first asked whether I would be pelted with rotten eggs or tomatoes but they reassured me that there would merely be a chorus of catcalls. I then asked to be given the opportunity of talking with the students beforehand and they readily agreed to this. They arranged that some twenty representatives of the students would meet me at breakfast next day together with the Dean and one or two Professors. The Dutch speak English fluently and we had a lively discussion lasting about three hours. At the presentation there were no catcalls and when I was in Delft a few days later the students there asked for a discussion with me. They had evidently heard of the events in Leiden. The same thing was repeated when I went to Amsterdam and after the discussion I was touched to receive from the students a small album with reproductions of pictures by Bosch. You may recall that this 16th century painter was a precursor of the surrealists, and that he was famous for his painting *A ship of fools*.

Judge for yourself – to conduct that kind of discussion successfully I have to be well informed about what people in other countries say and think about us. What our censorship is doing may be likened to tying weights to the legs of our Olympic runners or tying the hands of our wrestlers. Besides, in these days measures taken by the censors such as cutting out articles, can no longer be effective since information is now widely distributed by radio, not only, of course, from a *Voice of America* or a *Deutsche Welle* that are not distinguished for trustworthiness. But knowing foreign languages, one can listen directly to broadcasts intended for their own country which, moreover, are not jammed. Of course, this is a bother because in order to hear what you are interested in, you have to put up with a lot of irrelevant stuff. In printed papers you read only what is necessary, so I prefer them . . .

I decided to write to you about the article torn out of the translated edition of my book because this was done by officials who are responsible to you.

P.S.   There was an amusing postscript to the award of the Kamerlingh Onnes Medal. When I returned to Moscow it was discovered that I had spent the foreign currency given to me but I had not retained the necessary confirmatory documents and here no one believes your word. They demanded from me ten times the amount of the advance I had spent, so I decided to make use of my medal. It is made of gold, which is the best kind of currency, and a very small piece of it would suffice to pay for my "unlawful" spending. I proposed to the Finance Department of the Academy of Sciences that they saw off the required amount of gold from the medal but they rejected this operation as too unusual and became reconciled to the overspending.

(I) From Yu.V. Andropov   Moscow, 28 May 1980

Deeply respected Piotr Leonidovich,

Referring to your letter about the circumstances of your receipt of your author's copy of your book from Italy, I have to inform you that the organs of the KGB do not engage in the activity you speak of in your letter. I enclose a new copy of the Italian edition of your book containing the complete text.

 With best wishes,

Yu.V. Andropov

(H, I) To Yu.V. Andropov   Moscow, 11 November 1980

Like many others, I am greatly disturbed by the situation and the fate of our distinguished scientific colleagues, the physicists A.D. Sakharov and Yu.F. Orlov. The present position can be very simply described. Sakharov and Orlov have done a great deal of good through their scientific work but their dissident activities have been considered damaging. They are now so placed that they cannot undertake any kind of activities at all. Thus they can do neither good nor harm. One may ask whether this is good for the country and in this letter I shall try to examine the question objectively.

 If you were to ask the scientific community they would reply decisively that when such distinguished scientists as Sakharov and Orlov are deprived of the possibility of normal scientific work a loss is incurred for humanity. If you were to ask the general public who are usually little acquainted with the work of scientists, they would give the opposite view of the situation.

 Since the time of Socrates the history of human culture provides frequent examples of great hostility to dissidents. To decide the question before us objectively we must, of course, take into account the actual social situation in the country at the particular time. In our own circumstances of building a new social structure, I think it is more correct to rely on Lenin as being more impartial, since he was not only a great intellectual and scholar but also a great man of action. His attitude to scholars in similar situations is well known and can be seen most clearly in his attitude to I.P. Pavlov.

 After the revolution Pavlov's dissident views were well known, not only here but abroad, and his negative attitude to socialism was clearly and openly expressed. He criticised and even scolded the leadership uninhibitedly using the harshest expressions, crossed himself every time he passed a church, wore the tsarist decorations that he had never bothered about before the revolution, and so on. Lenin simply ignored all these symptoms of dissent. To Lenin, Pavlov was a great scientist and he did everything possible to secure good conditions for Pavlov's scientific work. For instance, as is well known, Pavlov used dogs for his fundamental work on conditioned reflexes. During the 1920s food supplies in Petrograd were catastrophically bad but on Lenin's instructions, the food supply for Pavlov's dogs was maintained at a normal level. I remember Academician A.N. Krylov telling me that once, when he met Pavlov in the street, he asked him "Ivan Petrovich, may I beg a favour of you?". "Certainly". "Take me as one of your dogs!" Pavlov replied "You're an intelligent fellow but you do talk such nonsense".

I know many other examples of Lenin showing particular consideration for scientists. This is well known from his letters to K.A. Timiryazev, A.A. Bogdanov, Karl Steinmetz and others. His treatment of D.K. Chernov, the distinguished metallurgist, whose classical work on steel alloys became the foundation of modern scientific metallurgy, also made a great impression on me. At the time of the revolution Chernov was already about 80 and was professor at the Mikhailovsky Artillery Academy, with the rank of general and the title of Court Chamberlain. During the Civil War he lived in his villa in the Crimea which was occupied by Wrangel. When Wrangel had to retreat from the Crimea he invited Chernov to emigrate with him, but Chernov refused and remained alone in his villa, which was surrounded by the Red Army. They asked Lenin: "What shall we do with Chernov?". Lenin ordered Chernov to be protected in every possible way and a special detachment of Red sailors was posted at the villa . . .

I should like also to give an instructive example which occurred in our own time, and involved the heterodox opinions of a dissident creative worker. In 1945 Broz Tito came to visit me at the Institute through the initiative of Paul Savić, a physicist who had been scientific attaché at the Yugoslav Embassy during the war and who worked with us in our Institute in his spare time. Savić is now President of the Academy of Sciences in Belgrade. While showing Comrade Tito round the Institute I asked what he thought of I. Meštrović, whose works I admired. Meštrović was a pupil of Rodin and one of the greatest sculptors of our time. Tito began to speak very negatively about Meštrović, saying that he regarded him as an enemy who had spoken out sharply against the established regime in Yugoslavia. Moreover, he was religious and a friend of the Pope, and so on. I began to argue with Tito, saying that a person who had reached such creative heights in art, as had Meštrović, should not be judged by ordinary standards, and I used Lenin's treatment of Pavlov as an example. Tito loved an argument so the conversation became heated and we both spoke sharply. But suddenly Tito broke off the discussion and said that I had convinced him and that he would change his treatment of Meštrović on his return to Yugoslavia.

Then, as you know, we had a break with Tito and normal relations were not resumed until Khrushchev came to power. During that period I too had some unpleasant experiences, but later my position improved and I was able to travel to other socialist countries . . .

In 1966 I decided to visit Yugoslavia, paying my own way through Intourist. I was warmly welcomed at the frontier and told that Marshal Tito wished to see me. We arrived late at night, but at 10 the next morning Tito sent a car for us. The first thing he said to me when we met was that he was very grateful for the argument we had in Moscow, which made him change his attitude to Meštrović; consequently, Meštrović had changed his attitude to the Government of Yugoslavia and indeed to such an extent that he began to work actively for Yugoslavia. A special museum was built in Split for his sculptures, and he is now buried beside it . . .

When leaving Yugoslavia I was told that I had been awarded their highest honour – the Order of *Zastava za lentom*. Meštrović had already died by then and evidently he never knew why his treatment had changed. Tito commissioned the leading sculptor Augustinčić to make a bust of me. Back in Moscow I was informed that the bust was finished and would be forwarded together with the Order but the matter dragged on because our authorities did not approve of the Order. I then consulted the Vice-President of the Academy, B.P. Konstantinov, and he told me that after his personal approach to you, approval had been given. And, indeed, some time later, the Yugoslav Ambassador in Moscow handed me the sculpture and the Order.

These examples show that you have to treat dissidents with care and consideration, as Lenin did. Dissident thinking is closely linked with valuable creative activity, and creative activity in any branch of culture ensures the progress of mankind. Some basic dissatisfaction with the existing order must be there to stimulate the wish to create – that is, one has to be a dissident and this applies to any branch of human activity. Of course, there are many who are dissatisfied, but to be really creative one must also possess talent. Life shows that great talent is very rare and therefore it must be valued and protected, which is difficult to achieve even under good leadership. Great creativity calls for considerable temperament and this leads to extreme forms of discontent, and consequently talented individuals usually have what is called "a difficult character" . . .

The dialectic of the development of human culture consists in the conflict between conservatism and dissidence, and this has been so at all times and in all areas of human culture. If one examines the behaviour of a man like Sakharov it is clear that, basically, his creative activity also comes from dissatisfaction with what exists. When this concerns physics, for which he has great talent, his activity is exceptionally valuable. But when he takes up social problems this does not lead to such useful results and he excites a strongly negative reaction among people with a bureaucratic turn of mind, who usually lack any creative imagination. The consequence is that, instead of following Lenin and simply ignoring the results of his independent thinking in this field, they attempt to restrain him by administrative measures and give no thought to the fact that they are simultaneously destroying his useful creative activity – they throw away the baby with the bathwater. Great creative work belongs to the realm of ideas and does not respond to administrative enforcement. The proper way to act in such cases was well demonstrated by Lenin in his treatment of Pavlov . . . who eventually softened his non-conformist views . . .

Today, for some reason, we are forgetting Lenin's precepts with regard to scientists. The examples of Sakharov and Orlov show that this leads to unfortunate consequences. This is much more serious than might appear at first glance since it leads, in the long run, to our falling behind the capitalist countries in the field of big science and this is due in large measure to insufficient recognition of the need to protect the creative activity of a great scientist. Today, by comparison with Lenin's time, there is considerably less concern for the proper treatment of scientists; instead, their treatment often takes the form of imposing bureaucratic uniformity. To win a race a good horse is needed, but prize-winning horses are rare and usually restive and they also need skilful riders and careful attention. An ordinary horse is easier and more comfortable to ride but, of course, it won't win a race.

We have achieved nothing by increasing the administrative action against Sakharov and Orlov – rather it has merely strengthened their dissent. The pressure on them has now reached such proportions that it has even excited negative reactions abroad. By punishing Orlov with twelve years deprivation of freedom [seven in camps and five in exile] for dissent we completely remove him from the possibility of scientific work and it is hard to justify the need for such a ferocious sentence. That is why it evokes general bewilderment and is often interpreted as a sign of our weakness. In foreign countries, for instance, there is now an ever-widening boycott of scientific links with us. At the Centre for European Nuclear Research (CERN) in Geneva where some of our scientists are working, their colleagues wear sweaters carrying the name Orlov. All this, of course, is a passing phenomenon but it puts a brake on the development of science . . .

I cannot imagine what further action is going to be taken against our dissident scholars. If it is intended to increase forceful methods still further, this gives no promise of any satisfactory outcome. Wouldn't it be better simply to put the engine into reverse?*

---

* It was, however, only after the advent of "glasnost" that they were released.

Name Index

Apart from the usual function of tracking down references, this name index is intended to supply a few biographical details which would otherwise have to appear as footnotes. Names mentioned only in passing and especially those well known in the West are often omitted, but names of Russian cultural and political figures, who feature significantly in the text, are usually included, since less about them is likely to be familiar to Western readers. The term Academician means full member of the Academy of Sciences, roughly equivalent to Fellow of the Royal Society (F.R.S.). Where a name occurs very frequently in the text only a few leading page references or sometimes only chapter references are given. Notation such as 126–141 means that reference is made frequently over the range indicated, but not necessarily on every page. Page numbers in italics refer to letters to or from the person concerned.

Adrian, E.D., later Lord Adrian (1889–1977). F.R.S. Cambridge physiologist. President of the Royal Society 1950–1955. Nobel Prize 1932. 50, 214, 252, *267.*

Aleksandrov, A.P. (b. 1903). Academician. Director of the Institute for Physical Problems 1946–1954; from 1960 Director of the Institute for Atomic Energy. President of the Academy of Sciences 1976–1986. 72.

Alikhanov, A.I. (1904–1970). Academician. Experimental nuclear and elementary particles physicist; also worked on nuclear reactor technology. 355, 377.

Allen, J.F. (b. 1908). F.R.S. Did pioneer experiments on liquid helium II in the Mond Laboratory. Later Professor at St. Andrews University. 43, 61.

Allibone, T.E. (b. 1903). F.R.S. Worked with Rutherford at the Cavendish. Later Chief Scientist at the Central Electricity Generating Board. 74, 75.

Andronikashvili, E.L. (1910–1989). Georgian low-temperature physicist who demonstrated the two-fluid nature of liquid helium II at the Institute for Physical Problems. Later Director of the Physical Institute of the Georgian Academy of Sciences in Tbilisi. 64, 66.

Andropov, Yu.V. (1914–1984). Head of the KGB 1967–1982. General Secretary of the Central Committee after 1982. *413, 415, 416.*

Aston, F.W. (1877–1945). F.R.S. Cambridge physicist, pioneer on isotopes. Nobel Prize 1922. 21, 122, 155, 302.

Austin, Lord (1866–1941). Pioneer motor car manufacturer and philanthropist. 289, 302.

Bakh, A.N. (1857–1946). Academician. Founder of the Soviet school of biochemistry. 211, 220.

421